MATH FOR LIBERAL ARTS STUDENTS

DARLENE DIAZ, SANTIAGO CANYON COLLEGE

This textbook was originally adapted from David Lippman's *Math in Society* in addition to The Monterey Institute for Technology and Education's content for *Geometry and Measurement*, Larry Ottman and Ellen Lawsky's content for *Normal Distribution*, and my own written content to fit the course outline of record. This is the textbook's second edition, where Chapter 5: Measurement and Geometry in version [0.99] is now two chapters, Chapter 5: Measurement and Chapter 6: Geometry. Exercises were added at the end of Chapter 5 and Chapter 6. Content was added in Chapter 11 for better use of the z-score table (also added at the end of section 11.2) in addition to more exercises at the end of the chapter.

The projects written at the end of the textbook are written by me, Darlene Diaz, and can be used in any form as the instructor desires. Supplementary materials listed on page iii can be found in MyOpenMath (or LumenOHM); this includes, Class Notes, the Practice Workbook, online homework exercises and quizzes, and discussion/forums. The Class Notes and Practice Workbook are authored by me, Darlene Diaz.

- Myopenmath.com
- ohm.lumenlearning.com
- http://www.opentextbookstore.com/mathinsociety/
- http://nrocmath.org/products/higher-ed/developmental-mathematics
- http://www.ck12.org

Copyright © 2017 Darlene Diaz
Version [1.99]

D1157433

MATH FOR LIBERAL ART STUDENTS

Table of Contents

Projects

Preface

The traditional high school and college mathematics sequence leading from algebra through calculus could leave one with the impression that mathematics is all about algebraic manipulations. This book is an exploration of the wide world of mathematics, of which algebra is only one small piece. The topics are chosen because they provide glimpses into other ways of thinking mathematically, and because they have interesting applications to everyday life. Together, they highlight algorithmic, graphical, algebraic, statistical, and analytic approaches to solving problems.

This textbook is meant to expose students to different topics in mathematics rather than the usual topics a student usually sees in their mathematical sequence of courses. This textbook is meant to motivate students' interests in mathematics from a liberal arts perspective and obtain an appreciation for mathematics (or increase it). Each topic is presented in a sufficient amount of depth so that the student has an increased understanding of the history and artistry of mathematics, and the application of these topics to the real-world and everyday life. The applications are extended concepts given through the projects for selected topics in addition to the explorations in selected chapters.

Supplements

The Washington Open Course Library (OCL) project helped fund the creation of a full-course package for this book in addition to materials authored by me, Darlene Diaz, which contains the following features:

- Online homework for all the chapters (algorithmically generated, free response)
- Online quizzes for all the chapters (algorithmically generated, free response)
- Written assignments and discussion forum assignments for most chapters
- Class Notes for traditional classes with lectures
- Practice Workbook for each chapter for in-class (group) assignments
- Projects based from concepts of selected chapters with real-world applications

The course shell was built for the IMathAS online homework platform, and is available for Washington State faculty at www.wamap.org and mirrored for others at www.myopenmath.com and ohm.lumenlearning.com.

Acknowledgements

I want to acknowledge my peers at Santiago Canyon College for assisting in version [1.99] of this textbook. With suggestions, I made edits and reorganized the first version of this textbook, version [0.99], to better serve the students with content, exercises, and homework.

Chapter 1: Problem Solving

In previous math courses, you've no doubt run into the infamous "word problems." Unfortunately, these problems rarely resemble the type of problems we actually encounter in everyday life. In math books, you usually are told exactly which formula or procedure to use, and are given exactly the information you need to answer the question. In real life, problem solving requires identifying an appropriate formula or procedure, and determining what information you need (and won't need) to answer the question.

In this chapter, we will review several basic but powerful algebraic ideas: percents, rates, and proportions. We will then focus on the problem-solving process, and explore applying these ideas to solve problems where we don't have perfect information.

Table of Contents

1.1 Percents

In the 2004 vice-presidential debates, Edwards claimed that US forces have suffered "90% of the coalition casualties" in Iraq. Cheney disputed this, saying that in fact Iraqi security forces and coalition allies "have taken almost 50 percent" of the casualties[1]. Who is correct? How can we make sense of these numbers?

Percent literally means "per 100," or "parts per hundred." When we write 40%, this is equivalent to the fraction $\frac{40}{100}$ or the decimal 0.40. Notice that 80 out of 200 and 10 out of 25 are also 40%, since $\frac{80}{200} = \frac{10}{25} = \frac{40}{100}$.

---

[1] http://www.factcheck.org/cheney_edwards_mangle_facts.html

Example 1

243 people out of 400 state that they like dogs. What percent is this?

$$\frac{243}{400} = 0.6075 = \frac{60.75}{100}. \text{ This is } 60.75\%.$$

Notice that the percent can be found from the equivalent decimal by moving the decimal point two places to the right.

Example 2

Write each as a percent: a) $\frac{1}{4}$ b) 0.02 c) 2.35

a) $\frac{1}{4} = 0.25 = 25\%$ b) $0.02 = 2\%$ c) $2.35 = 235\%$

Percents
If we have a *part* that is some *percent* of a *whole*, then

$$\text{percent} = \frac{\text{part}}{\text{whole}}, \text{ or equivalently, } \text{part} = \text{percent} \cdot \text{whole}$$

To do the calculations, we write the percent as a decimal. Recall, to rewrite the percent as a decimal, we move the decimal over two places to the left, or, equivalently, divide the percent number by 100.

Example 3

The sales tax in a town is 9.4%. How much tax will you pay on a $140 purchase?

Here, $140 is the whole, and we want to find 9.4% of $140. We start by writing the percent as a decimal by moving the decimal point two places to the left (which is equivalent to dividing by 100). We can then compute:
$$\text{tax} = 0.094(140) = \$13.16$$

Example 4

In the news, you hear "tuition is expected to increase by 7% next year." If tuition this year is $1,200 per quarter, what will it be next year?

The tuition next year will be the current tuition plus an additional 7%, so it will be 107% of this year's tuition:

$$\$1200(1.07) = \$1,284.$$

Alternatively, we could have first calculated 7% of $1200: $1200(0.07) = $84. Notice this is *not* the expected tuition for next year (we could only wish). Instead, this is the expected *increase*, so to calculate the expected tuition, we'll need to add this change to the previous year's tuition:

$$\$1200 + \$84 = \$1,284.$$

Try it Now 1
A TV originally priced at $799 is on sale for 30% off. There is then a 9.2% sales tax. Find the price after including the discount and sales tax.

Example 5

The value of a car dropped from $7,400 to $6,800 over the last year. What percent decrease is this?

To compute the percent change, we first need to find the dollar value change: $6800-$7400 = -$600. Often, we take the absolute value of this amount, which is called the **absolute change**: $\left|-600\right| = 600$.

Since we are computing the percent decrease relative to the starting value, we compute this percent out of $7,400:

$$\frac{600}{7400} = 0.081 = 8.1\% \text{ decrease. This is called a } \textbf{relative change}.$$

Absolute and Relative Change
Given two quantities,
Absolute change = $\left|\text{ending quantity} - \text{starting quantity}\right|$

Relative change: $\dfrac{\text{absolute change}}{\text{starting quantity}}$

Absolute change has the same units as the original quantity.
Relative change gives a percent change.
The starting quantity is called the **base** of the percent change.

The base of a percent is very important. For example, while Nixon was president, it was argued that marijuana was a "gateway" drug, claiming that 80% of marijuana smokers went on to use harder drugs like cocaine. The problem is, this isn't true. The true claim is that 80% of harder drug users first smoked marijuana. The difference is one of base: 80% of marijuana smokers using hard drugs, vs. 80% of hard drug users having smoked marijuana.

These numbers are not equivalent. As it turns out, only one in 2,400 marijuana users actually go on to use harder drugs[2].

Example 6

There are about 75 QFC supermarkets in the U.S. Albertsons has about 215 stores. Compare the size of the two companies.

When we make comparisons, we must ask first whether an absolute or relative comparison. The absolute difference is $215 - 75 = 140$. From this, we could say "Albertsons has 140 more stores than QFC." However, if you wrote this in an article or paper, that number does not mean much. The relative difference may be more meaningful. There are two different relative changes we could calculate, depending on which store we use as the base:

Using QFC as the base, $\dfrac{140}{75} = 1.867$.

This tells us Albertsons is 186.7% larger than QFC.

Using Albertsons as the base, $\dfrac{140}{215} = 0.651$.

This tells us QFC is 65.1% smaller than Albertsons.

Notice both of these are showing percent *differences*. We could also calculate the size of Albertsons relative to QFC: $\dfrac{215}{75} = 2.867$, which tells us Albertsons is 2.867 times the size

of QFC. Likewise, we could calculate the size of QFC relative to Albertsons: $\dfrac{75}{215} = 0.349$,

which tells us that QFC is 34.9% of the size of Albertsons.

Example 7

Suppose a stock drops in value by 60% one week, then increases in value the next week by 75%. Is the value higher or lower than where it started?

To answer this question, suppose the value started at $100. After one week, the value dropped by 60%:
$100 - $100(0.60) = $100 - $60 = $40.

In the next week, notice that base of the percent has changed to the new value, $40. Computing the 75% increase:
$40 + $40(0.75) = $40 + $30 = $70.

In the end, the stock is still $30 lower, or $\dfrac{\$30}{\$100} = 30\%$ lower, valued than it started.

---

[2] http://tvtropes.org/pmwiki/pmwiki.php/Main/LiesDamnedLiesAndStatistics

Try it Now 2

The U.S. federal debt at the end of 2001 was $5.77 trillion, and grew to $6.20 trillion by the end of 2002. At the end of 2005 it was $7.91 trillion, and grew to $8.45 trillion by the end of 2006[3]. Calculate the absolute and relative increase for 2001-2002 and 2005-2006. Which year saw a larger increase in federal debt?

Example 8

A Seattle Times article on high school graduation rates reported "The number of schools graduating 60 percent or fewer students in four years – sometimes referred to as "dropout factories" – decreased by 17 during that time period. The number of kids attending schools with such low graduation rates was cut in half."

a) Is the "decrease by 17" number a useful comparison?

b) Considering the last sentence, can we conclude that the number of "dropout factories" was originally 34?

Solution:

a) This number is hard to evaluate, since we have no basis for judging whether this is a larger or small change. If the number of "dropout factories" dropped from 20 to 3, that'd be a very significant change, but if the number dropped from 217 to 200, that'd be less of an improvement.

b) The last sentence provides relative change which helps put the first sentence in perspective. We can estimate that the number of "dropout factories" was probably previously around 34. However, it's possible that students simply moved schools rather than the school improving, so that estimate might not be fully accurate.

Example 9

In the 2004 vice-presidential debates, Edwards claimed that US forces have suffered "90% of the coalition casualties" in Iraq. Cheney disputed this, saying that in fact Iraqi security forces and coalition allies "have taken almost 50 percent" of the casualties. Who is correct?

Without more information, it is hard for us to judge who is correct, but we can easily conclude that these two percents are talking about different things, so one does not necessarily contradict the other. Edward's claim was a percent with coalition forces as the base of the percent, while Cheney's claim was a percent with both coalition and Iraqi security forces as the base of the percent. It turns out both statistics are in fact fairly accurate.

---

[3] http://www.whitehouse.gov/sites/default/files/omb/budget/fy2013/assets/hist07z1.xls

In the 2012 presidential elections, one candidate argued that "the president's plan will cut $716 billion from Medicare, leading to fewer services for seniors," while the other candidate rebuts that "our plan does not cut current spending and actually expands benefits for seniors, while implementing cost saving measures." Are these claims in conflict, in agreement, or not comparable because they're talking about different things?

We'll wrap up our review of percents with a couple cautions. First, when talking about a change of quantities that are already measured in percents, we have to be careful in how we describe the change.

Example 10

A politician's support increases from 40% of voters to 50% of voters. Describe the change.

We could describe this using an absolute change: $|50\% - 40\%| = 10\%$. Notice that since the original quantities were percents, this change also has the units of percent. In this case, it is best to describe this as an increase of 10 **percentage points**.

In contrast, we could compute the percent change: $\dfrac{10\%}{40\%} = 0.25 = 25\%$ increase. This is the relative change, and we'd say the politician's support has increased by 25%.

Lastly, a caution against averaging percents.

Example 11

A basketball player scores on 40% of 2-point field goal attempts, and on 30% of 3-point of field goal attempts. Find the player's overall field goal percentage.

It is very tempting to average these values, and claim the overall average is 35%, but this is likely incorrect since most players make many more 2-point attempts than 3-point attempts. We don't actually have enough information to answer the question. Suppose the player attempted 200 2-point field goals and 100 3-point field goals. Then they made 200(0.40) = 80 2-point shots and 100(0.30) = 30 3-point shots. Overall, they made 110 shots out of 300, for a $\dfrac{110}{300} = 0.367 = 36.7\%$ overall field goal percentage.

1.2 Proportions and Rates

If you wanted to power the city of Seattle using wind power, how many windmills would you need to install? Questions like these can be answered using rates and proportions.

> **Rates**
> A rate is the ratio (fraction) of two quantities.
> A **unit rate** is a rate with a denominator of one.

Example 12

Your car can drive 300 miles on a tank of 15 gallons. Express this as a rate.

Expressed as a rate, $\dfrac{300\,\text{miles}}{15\,\text{gallons}}$. We can divide to find a unit rate: $\dfrac{20\,\text{miles}}{1\,\text{gallon}}$, which we

could also write as $20\dfrac{\text{miles}}{\text{gallon}}$, or just 20 miles per gallon.

> **Proportion Equation**
> A proportion equation is an equation showing the equivalence of two rates or ratios.

Example 13

Solve the proportion $\dfrac{5}{3} = \dfrac{x}{6}$ for the unknown value x.

This proportion is asking us to find a fraction with denominator 6 that is equivalent to the

fraction $\dfrac{5}{3}$. We can solve this by multiplying both sides of the equation by 6, giving

$x = \dfrac{5}{3} \cdot 6 = 10$.

Example 14

A map scale indicates that ½ inch on the map corresponds with 3 real miles. How many

miles apart are two cities that are $2\dfrac{1}{4}$ inches apart on the map?

We can set up a proportion by setting equal two $\dfrac{\text{map inches}}{\text{real miles}}$ rates, and introducing a

variable, x, to represent the unknown quantity – the mile distance between the cities.

$$\dfrac{\frac{1}{2}\,\text{map inch}}{3\,\text{miles}} = \dfrac{2\frac{1}{4}\,\text{map inches}}{x\,\text{miles}}$$ Multiply both sides by x

and rewriting the mixed number

$$\frac{\frac{1}{2}}{3} \cdot x = \frac{9}{4}$$ Multiply both sides by 3

$$\frac{1}{2}x = \frac{27}{4}$$ Multiply both sides by 2 (or divide by ½)

$$x = \frac{27}{2} = 13\frac{1}{2} \text{ miles}$$

Many proportion problems can also be solved using **dimensional analysis**, the process of multiplying a quantity by rates to change the units.

Example 15

Your car can drive 300 miles on a tank of 15 gallons. How far can it drive on 40 gallons?

We could certainly answer this question using a proportion: $\dfrac{300 \text{ miles}}{15 \text{ gallons}} = \dfrac{x \text{ miles}}{40 \text{ gallons}}$.

However, we earlier found that 300 miles on 15 gallons gives a rate of 20 miles per gallon. If we multiply the given 40-gallon quantity by this rate, the *gallons* units reduces and we're left with a number of miles:

$$40 \text{ gallons} \cdot \frac{20 \text{ miles}}{\text{gallon}} = \frac{40 \text{ gallons}}{1} \cdot \frac{20 \text{ miles}}{\text{gallon}} = 800 \text{ miles}$$

Notice if instead we were asked "how many gallons are needed to drive 50 miles?" we could answer this question by inverting the 20-mile-per-gallon rate so that the *miles* units reduces and we're left with gallons:

$$50 \text{ miles} \cdot \frac{1 \text{ gallon}}{20 \text{ miles}} = \frac{50 \text{ miles}}{1} \cdot \frac{1 \text{ gallon}}{20 \text{ miles}} = \frac{50 \text{ gallons}}{20} = 2.5 \text{ gallons}$$

Dimensional analysis can also be used to do unit conversions. Here are some unit conversions for reference.

Unit Conversions

Length

1 foot (ft) = 12 inches (in)	1 yard (yd) = 3 feet (ft)
1 mile = 5,280 feet	
1000 millimeters (mm) = 1 meter (m)	100 centimeters (cm) = 1 meter
1000 meters (m) = 1 kilometer (km)	2.54 centimeters (cm) = 1 inch

Weight and Mass

1 pound (lb) = 16 ounces (oz)	1 ton = 2000 pounds
1000 milligrams (mg) = 1 gram (g)	1000 grams = 1 kilogram (kg)

> 1 kilogram = 2.2 pounds (on earth)
>
> **Capacity**
> 1 cup = 8 fluid ounces (fl oz)*   1 pint = 2 cups
> 1 quart = 2 pints = 4 cups   1 gallon = 4 quarts = 16 cups
> 1000 milliliters (ml) = 1 liter (L)
>
> *Fluid ounces are a capacity measurement for liquids. 1 fluid ounce ≈ 1 ounce (weight) for water only.

Example 16

A bicycle is traveling at 15 miles per hour. How many feet will the bicycle travel in 20 seconds?

To answer this question, we need to convert 20 seconds into feet. If we know the speed of the bicycle in feet per second, this question would be simpler. Since we don't, we will need to do additional unit conversions. We will need to know that 5280 ft = 1 mile. We might start by converting the 20 seconds into hours:

$$20\,\text{seconds} \cdot \frac{1\,\text{minute}}{60\,\text{seconds}} \cdot \frac{1\,\text{hour}}{60\,\text{minutes}} = \frac{1}{180}\,\text{hour}$$ Now we can multiply by the 15 miles/hr

$$\frac{1}{180}\,\text{hour} \cdot \frac{15\,\text{miles}}{1\,\text{hour}} = \frac{1}{12}\,\text{mile}$$ Now we can convert to feet

$$\frac{1}{12}\,\text{mile} \cdot \frac{5280\,\text{feet}}{1\,\text{mile}} = 440\,\text{feet}$$

We could have also done this entire calculation in one long set of products:

$$20\,\text{seconds} \cdot \frac{1\,\text{minute}}{60\,\text{seconds}} \cdot \frac{1\,\text{hour}}{60\,\text{minutes}} \cdot \frac{15\,\text{miles}}{1\,\text{hour}} \cdot \frac{5280\,\text{feet}}{1\,\text{mile}} = 440\,\text{feet}$$

Try it Now 4

A 1,000-foot spool of bare 12-gauge copper wire weighs 19.8 pounds. How much will 18 inches of the wire weigh, in ounces?

Notice that with the miles per gallon example, if we double the miles driven, we double the gas used. Likewise, with the map distance example. If the map distance doubles, the real-life distance doubles. This is a key feature of proportional relationships, and we must confirm before assuming two things are related proportionally.

Example 17

Suppose you're tiling the floor of a 10 ft by 10 ft room, and find that 100 tiles are needed. How many tiles are needed to tile the floor of a 20 ft by 20 ft room?

In this case, while the width the room has doubled, the area has quadrupled. Since the number of tiles needed corresponds with the area of the floor, not the width, 400 tiles will be needed. We could find this using a proportion based on the areas of the rooms:

$$\frac{100 \, \text{tiles}}{100 \, \text{ft}^2} = \frac{n \, \text{tiles}}{400 \, \text{ft}^2}$$

Other quantities just don't scale proportionally at all.

Example 18
Suppose a small company spends $1,000 on an advertising campaign, and gains 100 new customers from it. How many new customers should they expect if they spend $10,000?

While it is tempting to say that they will gain 1,000 new customers, it is likely that additional advertising will be less effective than the initial advertising. For example, if the company is a hot tub store, there are likely only a fixed number of people interested in buying a hot tub, so there might not even be 1000 people in the town who would be potential customers.

Sometimes when working with rates, proportions, and percents, the process can be made more challenging by the magnitude of the numbers involved. Sometimes, large numbers are just difficult to comprehend.

Example 19

Compare the 2010 U.S. military budget of $683.7 billion to other quantities.

Here we have a very large number, about $683,700,000,000 written out. Of course, imagining a billion dollars is very difficult, so it can help to compare it to other quantities.

If that amount of money was used to pay the salaries of the 1.4 million Walmart employees in the U.S., each would earn over $488,000.

There are about 300 million people in the U.S. The military budget is about $2,200 per person.

If you were to put $683.7 billion in $100 bills, and count out 1 per second, it would take 216 years to finish counting it.

Example 20

Compare the electricity consumption per capita in China to the rate in Japan.

To address this question, we will first need data. From the CIA[4] website we can find the electricity consumption in 2011 for China was 4,693,000,000,000 KWH (kilowatt-hours), or 4.693 trillion KWH, while the consumption for Japan was 859,700,000,000, or 859.7 billion KWH. To find the rate per capita (per person), we will also need the population of the two countries. From the World Bank[5], we can find the population of China is 1,344,130,000, or 1.344 billion, and the population of Japan is 127,817,277, or 127.8 million.

Computing the consumption per capita for each country:

---

[4] https://www.cia.gov/library/publications/the-world-factbook/rankorder/2042rank.html
[5] http://data.worldbank.org/indicator/SP.POP.TOTL

China: $\dfrac{4,693,000,000,000\,\text{KWH}}{1,344,130,000\,\text{people}} \approx 3491.5$ KWH per person

Japan: $\dfrac{859,700,000,000\,\text{KWH}}{127,817,277\,\text{people}} \approx 6726$ KWH per person

While China uses more than 5 times the electricity of Japan overall, because the population of Japan is so much smaller, it turns out Japan uses almost twice the electricity per person compared to China.

1.3 Geometry

Geometric shapes, as well as area and volumes, can often be important in problem solving.

Example 21
| You are curious how tall a tree is, but don't have any way to climb it. Describe a method for determining the height.
|
| There are several approaches we could take. We'll use one based on triangles, which requires that it's a sunny day. Suppose the tree is casting a shadow, say 15 ft long. I can then have a friend help me measure my own shadow. Suppose I am 6 ft tall, and cast a 1.5 ft shadow. Since the triangle formed by the tree and its shadow has the same angles as the triangle formed by me and my shadow, these triangles are called **similar triangles** and their sides will scale proportionally. In other words, the ratio of height to width will be the same in both triangles. Using this, we can find the height of the tree, which we'll denote by h:
|
| $$\frac{6\,\text{ft tall}}{1.5\,\text{ft shadow}} = \frac{h\,\text{ft tall}}{15\,\text{ft shadow}}$$

Multiplying both sides by 15, we get $h = 60$. The tree is about 60 ft tall.

It may be helpful to recall some formulas for areas and volumes of a few basic shapes.

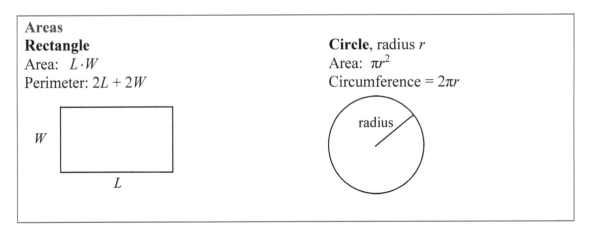

Areas
Rectangle
Area: $L \cdot W$
Perimeter: $2L + 2W$

Circle, radius r
Area: πr^2
Circumference $= 2\pi r$

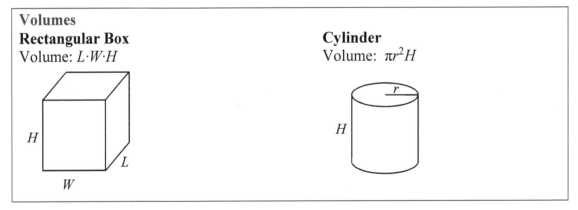

Volumes
Rectangular Box
Volume: $L \cdot W \cdot H$

Cylinder
Volume: $\pi r^2 H$

Example 22

If a 12-inch diameter pizza requires 10 ounces of dough, how much dough is needed for a 16-inch pizza?

To answer this question, we need to consider how the weight of the dough will scale. The weight will be based on the volume of the dough. However, since both pizzas will be about the same thickness, the weight will scale with the area of the top of the pizza. We can find the area of each pizza using the formula for area of a circle, $A = \pi r^2$:

A 12-inch pizza has radius 6 inches, so the area will be $\pi \cdot 6^2 \approx 113$ square inches.
A 16-inch pizza has radius 8 inches, so the area will be $\pi \cdot 8^2 \approx 201$ square inches.

Notice that if both pizzas were 1 inch thick, the volumes would be 113 in^3 and 201 in^3 respectively, which are at the same ratio as the areas. As mentioned earlier, since the thickness is the same for both pizzas, we can safely ignore it.

We can now set up a proportion to find the weight of the dough for a 16-inch pizza:
$$\frac{10\,\text{ounces}}{113\,\text{in}^2} = \frac{x\,\text{ounces}}{201\,\text{in}^2} \qquad \text{Multiply both sides by 201}$$
$$x = 201 \cdot \frac{10}{113} \approx 17.8 \text{ ounces of dough for a 16-inch pizza.}$$

It is interesting to note that while the diameter is $\frac{16}{12} = 1.33$ times larger, the dough required, which scales with area, is $1.33^2 = 1.78$ times larger.

Example 23

A company makes regular and jumbo marshmallows. The regular marshmallow has 25 calories. How many calories will the jumbo marshmallow have?

We would expect the calories to scale with volume. Since the marshmallows have cylindrical shapes, we can use that formula to find the volume. From the grid in the image, we can estimate the radius and height of each marshmallow.

25 calories

Photo courtesy Christopher Danielson

The regular marshmallow appears to have a diameter of about 3.5 units, giving a radius of 1.75 units, and a height of about 3.5 units. The volume is about $\pi(1.75)^2(3.5) = 33.7$ units^3.

The jumbo marshmallow appears to have a diameter of about 5.5 units, giving a radius of 2.75 units, and a height of about 5 units. The volume is about $\pi(2.75)^2(5) = 118.8$ units^3.

We could now set up a proportion, or use rates. The regular marshmallow has 25 calories for 33.7 cubic units of volume. The jumbo marshmallow will have:

$$118.8 \text{ units}^3 \cdot \frac{25 \text{ calories}}{33.7 \text{ units}^3} = 88.1 \text{ calories}$$

It is interesting to note that while the diameter and height are about 1.5 times larger for the jumbo marshmallow, the volume and calories are about $1.5^3 = 3.375$ times larger.

Try it Now 5
A website says that you'll need 48 fifty-pound bags of sand to fill a sandbox that measure 8ft by 8ft by 1ft. How many bags would you need for a sandbox 6ft by 4ft by 1ft?

1.4 Problem Solving and Estimating

Finally, we will bring together the mathematical tools we've reviewed, and use them to approach more complex problems. In many problems, it is tempting to take the given information, plug it into whatever formulas you have handy, and hope that the result is what you were supposed to find. Chances are, this approach has served you well in other math classes.

This approach does not work well with real life problems. Instead, problem solving is best approached by first starting at the end: identifying exactly what you are looking for. From there, you then work backwards, asking "what information and procedures will I need to find this?" Very few interesting questions can be answered in one mathematical step; often times you will need to chain together a solution pathway, a series of steps that will allow you to answer the question.

Problem Solving Process
1. Identify the question you're trying to answer.
2. Work backwards, identifying the information you will need and the relationships you will use to answer that question.
3. Continue working backwards, creating a solution pathway.
4. If you are missing necessary information, look it up or estimate it. If you have unnecessary information, ignore it.
5. Solve the problem, following your solution pathway.

In most problems we work, we will be approximating a solution, because we will not have perfect information. We will begin with a few examples where we will be able to approximate the solution using basic knowledge from our lives.

Example 24
___
How many times does your heart beat in a year?

This question is asking for the rate of heart beats per year. Since a year is a long time to measure heart beats for, if we knew the rate of heart beats per minute, we could scale that quantity up to a year. So, the information we need to answer this question is heart beats per minute. This is something you can easily measure by counting your pulse while watching a clock for a minute.
Suppose you count 80 beats in a minute. To convert this beats per year:

$$\frac{80 \text{ beats}}{1 \text{ minute}} \cdot \frac{60 \text{ minutes}}{1 \text{ hour}} \cdot \frac{24 \text{ hours}}{1 \text{ day}} \cdot \frac{365 \text{ days}}{1 \text{ year}} = 42{,}048{,}000 \text{ beats per year}$$

Example 25
___
How thick is a single sheet of paper? How much does it weigh?

While you might have a sheet of paper handy, trying to measure it would be tricky. Instead we might imagine a stack of paper, and then scale the thickness and weight to a single sheet. If you've ever bought paper for a printer or copier, you probably bought a ream, which contains 500 sheets. We could estimate that a ream of paper is about 2 inches thick and weighs about 5 pounds. Scaling these down,

$$\frac{2 \text{ inches}}{\text{ream}} \cdot \frac{1 \text{ ream}}{500 \text{ pages}} = 0.004 \text{ inches per sheet}$$

$$\frac{5 \text{ pounds}}{\text{ream}} \cdot \frac{1 \text{ ream}}{500 \text{ pages}} = 0.01 \text{ pounds per sheet, or } 0.16 \text{ ounces per sheet.}$$

Example 26

A recipe for zucchini muffins states that it yields 12 muffins, with 250 calories per muffin. You instead decide to make mini-muffins, and the recipe yields 20 muffins. If you eat 4, how many calories will you consume?

There are several possible solution pathways to answer this question. We will explore one.

To answer the question of how many calories 4 mini-muffins will contain, we would want to know the number of calories in each mini-muffin. To find the calories in each mini-muffin, we could first find the total calories for the entire recipe, then divide it by the number of mini-muffins produced. To find the total calories for the recipe, we could multiply the calories per standard muffin by the number per muffin. Notice that this produces a multi-step solution pathway. It is often easier to solve a problem in small steps, rather than trying to find a way to jump directly from the given information to the solution.

We can now execute our plan:

$$12 \text{ muffins} \cdot \frac{250 \text{ calories}}{\text{muffin}} = 3000 \text{ calories for the whole recipe}$$

$$\frac{3000 \text{ calories}}{20 \text{ mini} - \text{muffins}} \text{ gives } 150 \text{ calories per mini-muffin}$$

$$4 \text{ mini muffins} \cdot \frac{150 \text{ calories}}{\text{mini} - \text{muffin}} \text{ totals } 600 \text{ calories consumed.}$$

Example 27

You need to replace the boards on your deck. About how much will the materials cost?

There are two approaches we could take to this problem: 1) estimate the number of boards we will need and find the cost per board, or 2) estimate the area of the deck and find the approximate cost per square foot for deck boards. We will take the latter approach.

For this solution pathway, we will be able to answer the question if we know the cost per square foot for decking boards and the square footage of the deck. To find the cost per square foot for decking boards, we could compute the area of a single board, and divide it into the cost for that board. We can compute the square footage of the deck using geometric formulas. So first we need information: the dimensions of the deck, and the cost and dimensions of a single deck board.

Suppose that measuring the deck, it is rectangular, measuring 16 ft by 24 ft, for a total area of 384 ft^2.

From a visit to the local home store, you find that an 8-foot by 4-inch cedar deck board costs about $7.50. The area of this board, doing the necessary conversion from inches to feet, is:

$$8 \text{ feet} \cdot 4 \text{ inches} \cdot \frac{1 \text{ foot}}{12 \text{ inches}} = 2.667 \text{ ft}^2. \text{ The cost per square foot is then}$$

$$\frac{\$7.50}{2.667 \text{ ft}^2} = \$2.8125 \text{ per ft}^2.$$

This will allow us to estimate the material cost for the whole 384 ft^2 deck

$$\$384 \text{ ft}^2 \cdot \frac{\$2.8125}{\text{ft}^2} = \$1080 \text{ total cost.}$$

Of course, this cost estimate assumes that there is no waste, which is rarely the case. It is common to add at least 10% to the cost estimate to account for waste.

Example 28

Is it worth buying a Hyundai Sonata hybrid instead the regular Hyundai Sonata?

To make this decision, we must first decide what our basis for comparison will be. For the purposes of this example, we'll focus on fuel and purchase costs, but environmental impacts and maintenance costs are other factors a buyer might consider.

It might be interesting to compare the cost of gas to run both cars for a year. To determine this, we will need to know the miles per gallon both cars get, as well as the number of miles we expect to drive in a year. From that information, we can find the number of gallons required from a year. Using the price of gas per gallon, we can find the running cost.

From Hyundai's website, the 2013 Sonata will get 24 miles per gallon (mpg) in the city, and 35 mpg on the highway. The hybrid will get 35 mpg in the city, and 40 mpg on the highway.

An average driver drives about 12,000 miles a year. Suppose that you expect to drive about 75% of that in the city, so 9,000 city miles a year, and 3,000 highway miles a year.

We can then find the number of gallons each car would require for the year.

Sonata:
$$9000 \text{ city miles} \cdot \frac{1 \text{ gallon}}{24 \text{ city miles}} + 3000 \text{ hightway miles} \cdot \frac{1 \text{ gallon}}{35 \text{ highway miles}} = 460.7 \text{ gallons}$$

Hybrid:

$$9000 \text{ city miles} \cdot \frac{1 \text{ gallon}}{35 \text{ city miles}} + 3000 \text{ hightway miles} \cdot \frac{1 \text{ gallon}}{40 \text{ highway miles}} = 332.1 \text{ gallons}$$

If gas in your area averages about \$3.50 per gallon, we can use that to find the running cost:

Sonata: $460.7 \text{ gallons} \cdot \frac{\$3.50}{\text{gallon}} = \$1612.45$

Hybrid: $332.1 \text{ gallons} \cdot \frac{\$3.50}{\text{gallon}} = \$1162.35$

The hybrid will save \$450.10 a year. The gas costs for the hybrid are about $\frac{\$450.10}{\$1612.45} =$ 0.279 = 27.9% lower than the costs for the standard Sonata.

While both the absolute and relative comparisons are useful here, they still make it hard to answer the original question, since "is it worth it" implies there is some tradeoff for the gas savings. Indeed, the hybrid Sonata costs about \$25,850, compared to the base model for the regular Sonata, at \$20,895.

To better answer the "is it worth it" question, we might explore how long it will take the gas savings to make up for the additional initial cost. The hybrid costs \$4,965 more. With gas savings of \$451.10 a year, it will take about 11 years for the gas savings to make up for the higher initial costs.

We can conclude that if you expect to own the car 11 years, the hybrid is indeed worth it. If you plan to own the car for less than 11 years, it may still be worth it, since the resale value of the hybrid may be higher, or for other non-monetary reasons. This is a case where math can help guide your decision, but it can't make it for you.

Try it Now 6
If traveling from Seattle, WA to Spokane WA for a three-day conference, does it make more sense to drive or fly?

Try it Now Answers
1. The sale price is \$799(0.70) = \$559.30. After tax, the price is \$559.30(1.092) = \$610.76
2. 2001-2002: Absolute change: \$0.43 trillion. Relative change: 7.45%
   2005-2006: Absolute change: \$0.54 trillion. Relative change: 6.83%
   2005-2006 saw a larger absolute increase, but a smaller relative increase.
3. Without more information, it is hard to judge these arguments. This is compounded by the complexity of Medicare. As it turns out, the \$716 billion is not a cut in current spending, but a cut in future increases in spending, largely reducing future growth in health care payments. In this case, at least the numerical claims in both statements could be

considered at least partially true. Here is one source of more information if you're interested: http://factcheck.org/2012/08/a-campaign-full-of-mediscare/

4. $18\,\text{inches} \cdot \dfrac{1\,\text{foot}}{12\,\text{inches}} \cdot \dfrac{19.8\,\text{pounds}}{1000\,\text{feet}} \cdot \dfrac{16\,\text{ounces}}{1\,\text{pound}} \approx 0.475$ ounces

5. The original sandbox has volume 64 ft³. The smaller sandbox has volume 24ft³.

$\dfrac{48\,\text{bags}}{64\,\text{ft}^3} = \dfrac{x\,\text{bags}}{24\,\text{ft}^3}$ results in $x = 18$ bags.

6. There is not enough information provided to answer the question, so we will have to make some assumptions, and look up some values.

Assumptions:

a) We own a car. Suppose it gets 24 miles to the gallon. We will only consider gas cost.

b) We will not need to rent a car in Spokane, but will need to get a taxi from the airport to the conference hotel downtown and back.

c) We can get someone to drop us off at the airport, so we don't need to consider airport parking.

d) We will not consider whether we will lose money by having to take time off work to drive.

Values looked up (your values may be different)

a) Flight cost: $184

b) Taxi cost: $25 each way (estimate, according to hotel website)

c) Driving distance: 280 miles each way

d) Gas cost: $3.79 a gallon

Cost for flying: $184 flight cost + $50 in taxi fares = $234.

Cost for driving: 560 miles round trip will require 23.3 gallons of gas, costing $88.31.

Based on these assumptions, driving is cheaper. However, our assumption that we only include gas cost may not be a good one. Tax law allows you deduct $0.55 (in 2012) for each mile driven, a value that accounts for gas as well as a portion of the car cost, insurance, maintenance, etc. Based on this number, the cost of driving would be $319.

Exercises

1. Out of 230 racers who started the marathon, 212 completed the race, 14 gave up, and 4 were disqualified. What percentage did not complete the marathon?

2. Patrick left an $8 tip on a $50 restaurant bill. What percent tip is that?

3. Ireland has a 23% VAT (value-added tax, similar to a sales tax). How much will the VAT be on a purchase of a €250 item?

4. Employees in 2012 paid 4.2% of their gross wages towards social security (FICA tax), while employers paid another 6.2%. How much will someone earning $45,000 a year pay towards social security out of their gross wages?

5. A project on Kickstarter.com was aiming to raise $15,000 for a precision coffee press. They ended up with 714 supporters, raising 557% of their goal. How much did they raise?

6. Another project on Kickstarter for an iPad stylus raised 1,253% of their goal, raising a total of $313,490 from 7,511 supporters. What was their original goal?

7. The population of a town increased from 3,250 in 2008 to 4,300 in 2010. Find the absolute and relative (percent) increase.

8. The number of CDs sold in 2010 was 114 million, down from 147 million the previous year[6]. Find the absolute and relative (percent) decrease.

9. A company wants to decrease their energy use by 15%.
    a. If their electric bill is currently $2,200 a month, what will their bill be if they're successful?
    b. If their next bill is $1,700 a month, were they successful? Why or why not?

10. A store is hoping an advertising campaign will increase their number of customers by 30%. They currently have about 80 customers a day.
    a. How many customers will they have if their campaign is successful?
    b. If they increase to 120 customers a day, were they successful? Why or why not?

11. An article reports "attendance dropped 6% this year, to 300." What was the attendance before the drop?

12. An article reports "sales have grown by 30% this year, to $200 million." What were sales before the growth?

---

[6] http://www.cnn.com/2010/SHOWBIZ/Music/07/19/cd.digital.sales/index.html

13. The Walden University had 47,456 students in 2010, while Kaplan University had 77,966 students. Complete the following statements:
    a. Kaplan's enrollment was ___% larger than Walden's.
    b. Walden's enrollment was ___% smaller than Kaplan's.
    c. Walden's enrollment was ___% of Kaplan's.

14. In the 2012 Olympics, Usain Bolt ran the 100m dash in 9.63 seconds. Jim Hines won the 1968 Olympic gold with a time of 9.95 seconds.
    a. Bolt's time was ___% faster than Hines'.
    b. Hine' time was ___% slower than Bolt's.
    c. Hine' time was ___% of Bolt's.

15. A store has clearance items that have been marked down by 60%. They are having a sale, advertising an additional 30% off clearance items. What percent of the original price do you end up paying?

16. Which is better: having a stock that goes up 30% on Monday than drops 30% on Tuesday, or a stock that drops 30% on Monday and goes up 30% on Tuesday? In each case, what is the net percent gain or loss?

17. Are these two claims equivalent, in conflict, or not comparable because they're talking about different things?
    a. "16.3% of Americans are without health insurance"[7]
    b. "only 55.9% of adults receive employer provided health insurance"[8]

18. Are these two claims equivalent, in conflict, or not comparable because they're talking about different things?
    a. "We mark up the wholesale price by 33% to come up with the retail price"
    b. "The store has a 25% profit margin"

19. Are these two claims equivalent, in conflict, or not comparable because they're talking about different things?
    a. "Every year since 1950, the number of American children gunned down has doubled."
    b. "The number of child gunshot deaths has doubled from 1950 to 1994."

20. Are these two claims equivalent, in conflict, or not comparable because they're talking about different things?[9]
    a. "75 percent of the federal health care law's taxes would be paid by those earning less than $120,000 a year"
    b. "76 percent of those who would pay the penalty [health care law's taxes] for not having insurance in 2016 would earn under $120,000"

---

[7] http://www.cnn.com/2012/06/27/politics/btn-health-care/index.html
[8] http://www.politico.com/news/stories/0712/78134.html
[9] http://factcheck.org/2012/07/twisting-health-care-taxes/

21. Are these two claims equivalent, in conflict, or not comparable because they're talking about different things?
    a. "The school levy is only a 0.1% increase of the property tax rate."
    b. "This new levy is a 12% tax hike, raising our total rate to $9.33 per $1000 of value."

22. Are the values compared in this statement comparable or not comparable? "Guns have murdered more Americans here at home in recent years than have died on the battlefields of Iraq and Afghanistan. In support of the two wars, more than 6,500 American soldiers have lost their lives. During the same period, however, guns have been used to murder about 100,000 people on American soil"[10]

23. A high school currently has a 30% dropout rate. They've been tasked to decrease that rate by 20%. Find the equivalent percentage point drop.

24. A politician's support grew from 42% by 3 percentage points to 45%. What percent (relative) change is this?

25. Marcy has a 70% average in her class going into the final exam. She says "I need to get a 100% on this final so I can raise my score to 85%." Is she correct?

26. Suppose you have one quart of water/juice mix that is 50% juice, and you add 2 quarts of juice. What percent juice is the final mix?

27. Find a unit rate: You bought 10 pounds of potatoes for $4.

28. Find a unit rate: Joel ran 1500 meters in 4 minutes, 45 seconds.

29. Solve: $\dfrac{2}{5} = \dfrac{6}{x}$.

30. Solve: $\dfrac{n}{5} = \dfrac{16}{20}$.

31. A crepe recipe calls for 2 eggs, 1 cup of flour, and 1 cup of milk. How much flour would you need if you use 5 eggs?

32. An 8ft length of 4 inch wide crown molding costs $14. How much will it cost to buy 40ft of crown molding?

33. Four 3-megawatt wind turbines can supply enough electricity to power 3000 homes. How many turbines would be required to power 55,000 homes?

---

[10] http://www.northjersey.com/news/opinions/lautenberg_073112.html?c=y&page=2

34. A highway had a landslide, where 3,000 cubic yards of material fell on the road, requiring 200 dump truck loads to clear. On another highway, a slide left 40,000 cubic yards on the road. How many dump truck loads would be needed to clear this slide?

35. Convert 8 feet to inches.

36. Convert 6 kilograms to grams.

37. A wire costs $2 per meter. How much will 3 kilometers of wire cost?

38. Sugar contains 15 calories per teaspoon. How many calories are in 1 cup of sugar?

39. A car is driving at 100 kilometers per hour. How far does it travel in 2 seconds?

40. A chain weighs 10 pounds per foot. How many ounces will 4 inches weigh?

41. The table below gives data on three movies. Gross earnings is the amount of money the movie brings in. Compare the net earnings (money made after expenses) for the three movies.[11]

Movie	Release Date	Budget	Gross earnings
Saw	10/29/2004	$1,200,000	$103,096,345
Titanic	12/19/1997	$200,000,000	$1,842,879,955
Jurassic Park	6/11/1993	$63,000,000	$923,863,984

42. For the movies in the previous problem, which provided the best return on investment?

43. The population of the U.S. is about 309,975,000, covering a land area of 3,717,000 square miles. The population of India is about 1,184,639,000, covering a land area of 1,269,000 square miles. Compare the population densities of the two countries.

44. The GDP (Gross Domestic Product) of China was $5,739 billion in 2010, and the GDP of Sweden was $435 billion. The population of China is about 1,347 million, while the population of Sweden is about 9.5 million. Compare the GDP per capita of the two countries.

45. In June 2012, Twitter was reporting 400 million tweets per day. Each tweet can consist of up to 140 characters (letter, numbers, etc.). Create a comparison to help understand the amount of tweets in a year by imagining each character was a drop of water and comparing to filling something up.

46. The photo sharing site Flickr had 2.7 billion photos in June 2012. Create a comparison to understand this number by assuming each picture is about 2 megabytes in size, and comparing to the data stored on other media like DVDs, iPods, or flash drives.

---

[11] http://www.the-numbers.com/movies/records/budgets.php

47. Your chocolate milk mix says to use 4 scoops of mix for 2 cups of milk. After pouring in the milk, you start adding the mix, but get distracted and accidentally put in 5 scoops of mix. How can you adjust the mix if:
    a. There is still room in the cup?
    b. The cup is already full?

48. A recipe for sabayon calls for 2 egg yolks, 3 tablespoons of sugar, and ¼ cup of white wine. After cracking the eggs, you start measuring the sugar, but accidentally put in 4 tablespoons of sugar. How can you compensate?

49. The Deepwater Horizon oil spill resulted in 4.9 million barrels of oil spilling into the Gulf of Mexico. Each barrel of oil can be processed into about 19 gallons of gasoline. How many cars could this have fueled for a year? Assume an average car gets 20 miles to the gallon, and drives about 12,000 miles in a year.

50. The store is selling lemons at 2 for $1. Each yield about 2 tablespoons of juice. How much will it cost to buy enough lemons to make a 9-inch lemon pie requiring ½ cup of lemon juice?

51. A piece of paper can be made into a cylinder in two ways: by joining the short sides together, or by joining the long sides together[12]. Which cylinder would hold more? How much more?

52. Which of these glasses contains more liquid? How much more?

In the next 4 questions, estimate the values by making reasonable approximations for unknown values, or by doing some research to find reasonable values.

53. Estimate how many gallons of water you drink in a year.

54. Estimate how many times you blink in a day.

55. How much does the water in a 6-person hot tub weigh?

56. How many gallons of paint would be needed to paint a two-story house 40 ft long and 30 ft wide?

57. During the landing of the Mars Science Laboratory *Curiosity*, it was reported that the signal from the rover would take 14 minutes to reach earth. Radio signals travel at the speed of light, about 186,000 miles per second. How far was Mars from Earth when *Curiosity* landed?

---

[12] http://vimeo.com/42501010

58. It is estimated that a driver takes, on average, 1.5 seconds from seeing an obstacle to reacting by applying the brake or swerving. How far will a car traveling at 60 miles per hour travel (in feet) before the driver reacts to an obstacle?

59. The flash of lightning travels at the speed of light, which is about 186,000 miles per second. The sound of lightning (thunder) travels at the speed of sound, which is about 750 miles per hour.
    a. If you see a flash of lightning, then hear the thunder 4 seconds later, how far away is the lightning?
    b. Now let's generalize that result. Suppose it takes n seconds to hear the thunder after a flash of lightning. How far away is the lightning, in terms of n?

60. Sound travels about 750 miles per hour. If you stand in a parking lot near a building and sound a horn, you will hear an echo.
    a. Suppose it takes about ½ a second to hear the echo. How far away is the building[13]?
    b. Now let's generalize that result. Suppose it takes n seconds to hear the echo. How far away is the building, in terms of n?

61. It takes an air pump 5 minutes to fill a twin sized air mattress (39 by 8.75 by 75 inches). How long will it take to fill a queen-sized mattress (60 by 8.75 by 80 inches)?

62. It takes your garden hose 20 seconds to fill your 2-gallon watering can. How long will it take to fill
    a. An inflatable pool measuring 3 feet wide, 8 feet long, and 1 foot deep.[14]
    b. A circular inflatable pool 13 feet in diameter and 3 feet deep.[15]

63. You want to put a 2" thick layer of topsoil for a new 20'x30' garden. The dirt store sells by the cubic yards. How many cubic yards will you need to order?

64. A box of Jell-O costs $0.50, and makes 2 cups. How much would it cost to fill a swimming pool 4 feet deep, 8 feet wide, and 12 feet long with Jell-O? (1 cubic foot is about 7.5 gallons)

65. You read online that a 15 ft by 20 ft brick patio would cost about $2,275 to have professionally installed. Estimate the cost of having an 18 by 22 ft brick patio installed.

66. I was at the store, and saw two sizes of avocados being sold. The regular size sold for $0.88 each, while the jumbo ones sold for $1.68 each. Which is the better deal?

---

[13] http://vimeo.com/40377128
[14] http://www.youtube.com/watch?v=DlkwefReHZc
[15] http://www.youtube.com/watch?v=p9SABH7Yg9M

67. The grocery store has bulk pecans on sale, which is great since you're planning on making 10 pecan pies for a wedding. Your recipe calls for 1¾ cups pecans per pie. However, in the bulk section there's only a scale available, not a measuring cup. You run over to the baking aisle and find a bag of pecans, and look at the nutrition label to gather some info. How many pounds of pecans should you buy?

Nutrition Facts

Serving Size: 1 cup, halves (99 g)
Servings per Container: about 2

Amount Per Serving

Calories 684   Calories from Fat 596

% Daily Value*

Total Fat 71g   110%

Saturated Fat 6g

68. Soda is often sold in 20 ounce bottles. The nutrition label for one of these bottles is shown to the right. A packet of sugar (the kind they have at restaurants for your coffee or tea) typically contain 4 grams of sugar in the U.S. Drinking a 20 oz soda is equivalent to eating how many packets of sugar?[16]

Nutrition Facts

Serving Size: 8 fl oz (240 mL)
Servings Per Container: about 2.5

Amount Per Serving

Calories 110

	% Daily Value*
Total Fat 0g	0%
Sodium 70mg	3%

For the next set of questions, *first* identify the information you need to answer the question, and *then* turn to the end of the section to find that information. The details may be imprecise; answer the question the best you can with the provided information. Be sure to justify your decision.

69. You're planning on making 6 meatloaves for a party. You go to the store to buy breadcrumbs, and see they are sold by the canister. How many canisters do you need to buy?

70. Your friend wants to cover their car in bottle caps, like in this picture.[17] How many bottle caps are you going to need?

71. You need to buy some chicken for dinner tonight. You found an ad showing that the store across town has it on sale for $2.99 a pound, which is cheaper than your usual neighborhood store, which sells it for $3.79 a pound.   Is it worth the extra drive?

---

72. I have an old gas furnace, and am considering replacing it with a new, high efficiency model. Is upgrading worth it?

73. Janine is considering buying a water filter and a reusable water bottle rather than buying bottled water. Will doing so save her money?

74. Marcus is considering going car-free to save money and be more environmentally friendly. Is this financially a good decision?

For the next set of problems, research or make educated estimates for any unknown quantities needed to answer the question.

75. You want to travel from Tacoma, WA to Chico, CA for a wedding. Compare the costs and time involved with driving, flying, and taking a train. Assume that if you fly or take the train you'll need to rent a car while you're there. Which option is best?

76. You want to paint the walls of a 6ft by 9ft storage room that has one door and one window. You want to put on two coats of paint. How many gallons and/or quarts of paint should you buy to paint the room as cheaply as possible?

77. A restaurant in New York tiled their floor with pennies[18]. Just for the materials, is this more expensive than using a more traditional material like ceramic tiles? If each penny has to be laid by hand, estimate how long it would take to lay the pennies for a 12ft by 10ft room. Considering material and labor costs, are pennies a cost-effective replacement for ceramic tiles?

78. You are considering taking up part of your back yard and turning it into a vegetable garden, to grow broccoli, tomatoes, and zucchini. Will doing so save you money, or cost you more than buying vegetables from the store?

79. Barry is trying to decide whether to keep his 1993 Honda Civic with 140,000 miles, or trade it in for a used 2008 Honda Civic. Consider gas, maintenance, and insurance costs in helping him make a decision.

80. Some people claim it costs more to eat vegetarian, while some claim it costs less. Examine your own grocery habits, and compare your current costs to the costs of switching your diet (from omnivore to vegetarian or vice versa as appropriate). Which diet is more cost effective based on your eating habits?

---

[18] http://www.notcot.com/archives/2009/06/floor-of-pennie.php

Info for the breadcrumbs question

How much breadcrumbs does the recipe call for?
        It calls for 1½ cups of breadcrumbs.
How many meatloafs does the recipe make?
        It makes 1 meatloaf.
How many servings does that recipe make?
        It says it serves 8.
How big is the canister?
        It is cylindrical, 3.5 inches across and
        7 inches tall.
What is the net weight of the contents of 1 canister?
        15 ounces.
How much does a cup of breadcrumbs weigh?
        I'm not sure, but maybe something from the nutritional label will help.
How much does a canister cost? $2.39

```
┌─────────────────────────────────────┐
│  Nutrition Facts                    │
│  Serving Size:  1/3 cup (30g)       │
│  Servings per Container:  about 14  │
│  ───────────────────────────────── │
│  Amount Per Serving                 │
│  ───────────────────────────────── │
│  Calories 110    Calories from Fat  │
│  15                                 │
│  ───────────────────────────────── │
│                         % Daily     │
│  Value*                             │
│  ───────────────────────────────── │
└─────────────────────────────────────┘
```

Info for bottle cap car

What kind of car is that?
  A 1993 Honda Accord.
How big is that car / what are the dimensions? Here is some details from MSN autos:
  Weight: 2800lb   Length: 185.2 in   Width: 67.1 in   Height: 55.2 in
How much of the car was covered with caps?
  Everything but the windows and the underside.
How big is a bottle cap?
  Caps are 1 inch in diameter.

Info for chicken problem

How much chicken will you be buying?
        Four pounds
How far are the two stores?
        My neighborhood store is 2.2 miles away, and takes about 7 minutes. The store
        across town is 8.9 miles away, and takes about 25 minutes.
What kind of mileage does your car get?
        It averages about 24 miles per gallon in the city.
How many gallons does your car hold?
        About 14 gallons
How much is gas?
        About $3.69/gallon right now.

Info for furnace problem
How efficient is the current furnace?
        It is a 60% efficient furnace.
How efficient is the new furnace?
        It is 94% efficient.
What is your gas bill?
        Here is the history for 2 years:

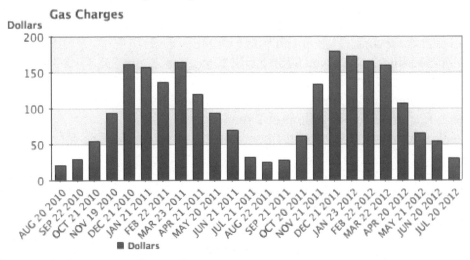

How much do you pay for gas?
        There is $10.34 base charge, plus $0.39097 per Therm for a delivery charge, and
        $0.65195 per Therm for cost of gas.
How much gas do you use?
        Here is the history for 2 years:

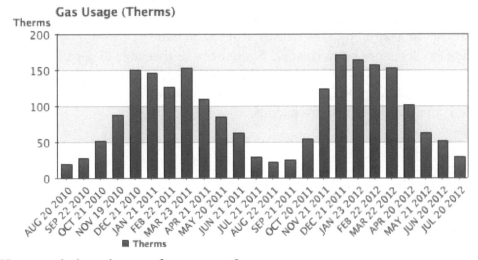

How much does the new furnace cost?
        It will cost $7,450.
How long do you plan to live in the house?
        Probably at least 15 years.

Info for water filter problem

How much water does Janine drink in a day?

> She normally drinks 3 bottles a day, each 16.9 ounces.

How much does a bottle of water cost?

> She buys 24-packs of 16.9 ounce bottles for $3.99.

How much does a reusable water bottle cost?

> About $10.

How long does a reusable water bottle last?

> Basically forever (or until you lose it).

How much does a water filter cost? How much water will they filter?

- A faucet-mounted filter costs about $28. Refill filters cost about $33 for a 3-pack. The box says each filter will filter up to 100 gallons (378 liters)
- A water filter pitcher costs about $22. Refill filters cost about $20 for a 4-pack. The box says each filter lasts for 40 gallons or 2 months
- An under-sink filter costs $130. Refill filters cost about $60 each. The filter lasts for 500 gallons.

Info for car-free problem

Where does Marcus currently drive? He:

- Drives to work 5 days a week, located 4 miles from his house.
- Drives to the store twice a week, located 7 miles from his house.
- Drives to other locations on average 5 days a week, with locations ranging from 1 mile to 20 miles.
- Drives to his parent's house 80 miles away once a month.

How will he get to these locations without a car?

- For work, he can walk when it's sunny and he gets up early enough. Otherwise he can take a bus, which takes about 20 minutes
- For the store, he can take a bus, which takes about 35 minutes.
- Some of the other locations he can bus to. Sometimes he'll be able to get a friend to pick him up. A few locations he is able to walk to. A couple locations are hard to get to by bus, but there is a ZipCar (short term car rental) location within a few blocks.
- He'll need to get a ZipCar to visit his parents.

How much does gas cost?

> About $3.69/gallon.

How much does he pay for insurance and maintenance?

- He pays $95/month for insurance.
- He pays $30 every 3 months for an oil change, and has averaged about $300/year for other maintenance costs.

How much is he paying for the car?

- He's paying $220/month on his car loan right now, and has 3 years left on the loan.
- If he sold the car, he'd be able to make enough to pay off the loan.
- If he keeps the car, he's planning on trading the car in for a newer model in a couple years.

What mileage does his car get?

> About 26 miles per gallon on average.

How much does a bus ride cost?

> $2.50 per trip, or $90 for an unlimited monthly pass.

How much does a ZipCar rental cost?
- The "occasional driving plan": $25 application fee and $60 annual fee, with no monthly commitment. Monday-Thursday the cost is $8/hour, or $72 per day. Friday-Sunday the cost is $8/hour or $78/day. Gas, insurance, and 180 miles are included in the cost. Additional miles are $0.45/mile.
- The "extra value plan": Same as above, but with a $50 monthly commitment, getting you a 10% discount on the usage costs.

Extension: Taxes
Governments collect taxes to pay for the services they provide. In the United States, federal income taxes help fund the military, the environmental protection agency, and thousands of other programs. Property taxes help fund schools. Gasoline taxes help pay for road improvements. While very few people enjoy paying taxes, they are necessary to pay for the services we all depend upon.

Taxes can be computed in a variety of ways, but are typically computed as a percentage of a sale, of one's income, or of one's assets.

Example 1

The sales tax rate in a city is 9.3%. How much sales tax will you pay on a $140 purchase?

The sales tax will be 9.3% of $140. To compute this, we multiply $140 by the percent written as a decimal: $140(0.093) = $13.02.

When taxes are not given as a fixed percentage rate, sometimes it is necessary to calculate the **effective rate.**

Effective rate
The effective tax rate is the equivalent percent rate of the tax paid out of the dollar amount the tax is based on.

Example 2

Joan paid $3,200 in property taxes on her house valued at $215,000 last year. What is the effective tax rate?

We can compute the equivalent percentage: 3200/215000 = 0.01488, or about 1.49% effective rate.

Taxes are often referred to as progressive, regressive, or flat.

Tax categories
A **flat tax**, or proportional tax, charges a constant percentage rate.
A **progressive tax** increases the percent rate as the base amount increases.
A **regressive tax** decreases the percent rate as the base amount increases.

Example 3

The United States federal income tax on earned wages is an example of a progressive tax. People with a higher wage income pay a higher percent tax on their income.

For a single person in 2011, adjusted gross income (income after deductions) under $8,500 was taxed at 10%. Income over $8,500 but under $34,500 was taxed at 15%.

A person earning $10,000 would pay 10% on the portion of their income under $8,500, and 15% on the income over $8,500, so they'd pay:

8500(0.10) = 850	10% of $8500
1500(0.15) = 225	15% of the remaining $1500 of income
Total tax: = $1075	

The effective tax rate paid is 1075/10000 = 10.75%

A person earning $30,000 would also pay 10% on the portion of their income under $8,500, and 15% on the income over $8,500, so they'd pay:

8500(0.10) = 850	10% of $8500
21500(0.15) = 3225	15% of the remaining $21500 of income
Total tax: = $4075	

The effective tax rate paid is 4075/30000 = 13.58%.

Notice that the effective rate has increased with income, showing this is a progressive tax.

Example 4

A gasoline tax is a flat tax when considered in terms of consumption, a tax of, say, $0.30 per gallon is proportional to the amount of gasoline purchased. Someone buying 10 gallons of gas at $4 a gallon would pay $3 in tax, which is $3/$40 = 7.5%. Someone buying 30 gallons of gas at $4 a gallon would pay $9 in tax, which is $9/$120 = 7.5%, the same effective rate.

However, in terms of income, a gasoline tax is often considered a regressive tax. It is likely that someone earning $30,000 a year and someone earning $60,000 a year will drive about the same amount. If both pay $60 in gasoline taxes over a year, the person earning $30,000 has paid 0.2% of their income, while the person earning $60,000 has paid 0.1% of their income in gas taxes.

Try it Now 1
A sales tax is a fixed percentage tax on a person's purchases. Is this a flat, progressive, or regressive tax?

Try it Now Answers
1. While sales tax is a flat percentage rate, it is often considered a regressive tax for the same reasons as the gasoline tax.

Income Taxation

Many people have proposed various revisions to the income tax collection in the United States. Some, for example, have claimed that a flat tax would be fairer. Others call for revisions to how different types of income are taxed, since currently investment income is taxed at a different rate than wage income.

The following two projects will allow you to explore some of these ideas and draw your own conclusions.

Project 1: Flat tax, Modified Flat Tax, and Progressive Tax.

Imagine the country is made up of 100 households. The federal government needs to collect $800,000 in income taxes to be able to function. The population consists of 6 groups:

Group A: 20 households that earn $12,000 each
Group B: 20 households that earn $29,000 each
Group C: 20 households that earn $50,000 each
Group D: 20 households that earn $79,000 each
Group E: 15 households that earn $129,000 each
Group F: 5 households that earn $295,000 each

This scenario is roughly proportional to the actual United States population and tax needs. We are going to determine new income tax rates.

The first proposal we'll consider is a flat tax – one where every income group is taxed at the same percentage tax rate.

1) Determine the total income for the population (all 100 people together)

2) Determine what flat tax rate would be necessary to collect enough money.

The second proposal we'll consider is a modified flat-tax plan, where everyone only pays taxes on any income over $20,000. So, everyone in group A will pay no taxes. Everyone in group B will pay taxes only on $9,000.

3) Determine the total *taxable* income for the whole population

4) Determine what flat tax rate would be necessary to collect enough money in this modified system

5) Complete this table for both the plans

Group	Income per household	Flat Tax Plan Income tax per household	Income after taxes	Modified Flat Tax Plan Income tax per household	Income after taxes
A	$12,000				
B	$29,000				
C	$50,000				
D	$79,000				
E	$129,000				
F	$295,000				

The third proposal we'll consider is a progressive tax, where lower income groups are taxed at a lower percent rate, and higher income groups are taxed at a higher percent rate. For simplicity, we're going to assume that a household is taxed at the same rate on *all* their income.

6) Set progressive tax rates for each income group to bring in enough money. There is no one right answer here – just make sure you bring in enough money!

Group	Income per household	Tax rate (%)	Income tax per household	Total tax collected for all households	Income after taxes per household
A	$12,000				
B	$29,000				
C	$50,000				
D	$79,000				
E	$129,000				
F	$295,000				
				This better total to $800,000	

7) Discretionary income is the income people have left over after paying for necessities like rent, food, transportation, etc. The cost of basic expenses does increase with income, since housing and car costs are higher, however usually not proportionally. For each income group, estimate their essential expenses, and calculate their discretionary income. Then compute the effective tax rate for each plan relative to discretionary income rather than income.

Group	Income per household	Discretionary Income (estimated)	Effective rate, flat	Effective rate, modified	Effective rate, progressive
A	$12,000				
B	$29,000				
C	$50,000				
D	$79,000				
E	$129,000				
F	$295,000				

8) Which plan seems the fairest to you? Which plan seems the least fair to you? Why?

Project 2: Calculating Taxes

Visit www.irs.gov, and download the most recent version of forms 1040, and schedules A, B, C, and D.

Scenario 1: Calculate the taxes for someone who earned $60,000 in standard wage income (W-2 income), has no dependents, and takes the standard deduction.

Scenario 2: Calculate the taxes for someone who earned $20,000 in standard wage income, $40,000 in qualified dividends, has no dependents, and takes the standard deduction. (Qualified dividends are earnings on certain investments such as stocks.)

Scenario 3: Calculate the taxes for someone who earned $60,000 in small business income, has no dependents, and takes the standard deduction.

Based on these three scenarios, what are your impressions of how the income tax system treats these different forms of income (wage, dividends, and business income)?

Scenario 4: To get a more realistic sense for calculating taxes, you'll need to consider itemized deductions. Calculate the income taxes for someone with the income and expenses listed below.

Married with 2 children, filing jointly
Wage income: $50,000 combined
Paid sales tax in Washington State
Property taxes paid: $3200
Home mortgage interest paid: $4800
Charitable gifts: $1200

Chapter 2: Historical Counting Systems

Table of Contents

2.1   Introduction

As we begin our journey through the history of mathematics, one question to be asked is "Where do we start?" Depending on how you view mathematics or numbers, you could choose any of a number of launching points from which to begin. Howard Eves suggests the following list of possibilities.[1]

Where to start the study of the history of mathematics...
- At the first logical geometric "proofs" traditionally credited to Thales of Miletus (600 BCE).
- With the formulation of methods of measurement made by the Egyptians and Mesopotamians/Babylonians.
- Where prehistoric peoples made efforts to organize the concepts of size, shape, and number.
- In pre-human times in the very simple number sense and pattern recognition that can be displayed by certain animals, birds, etc.
- Even before that in the amazing relationships of numbers and shapes found in plants.
- With the spiral nebulae, the natural course of planets, and other universe phenomena.

We can choose no starting point at all and instead agree that mathematics has *always* existed and has simply been waiting in the wings for humans to discover. Each of these positions can be defended to some degree and which one you adopt (if any) largely depends on your philosophical ideas about mathematics and numbers.

Nevertheless, we need a starting point. Without passing judgment on the validity of any of these particular possibilities, we will choose as our starting point the emergence of the idea of number and the process of counting as our launching pad. This is done primarily as a practical matter given the nature of this course. In the following chapter, we will try to focus on two main ideas. The first will be an examination of basic number and counting systems and the symbols that we use for numbers. We will look at our own modern (Western) number system as well those of a couple of selected civilizations to see the differences and diversity that is possible when humans start counting. The second idea we look at is base systems. By comparing our own base-ten (decimal) system with other bases, we quickly become aware that the system that we are so used to, when slightly changed, will challenge our notions about numbers and what symbols for those numbers actually mean.

Recognition of More vs. Less

The idea of number and the process of counting goes back far beyond history began to be recorded. There is some archeological evidence that suggests that humans were counting as far back as 50,000 years ago.[2] However, we do not really know how this process started or developed over time. The best we can do is to make a good guess as to how things progressed. It is probably not hard to believe that even the earliest humans had some sense of *more* and *less*. Even some small animals have been shown to have such a sense. For example, one naturalist tells of how he would secretly remove one egg each day from a plover's nest. The mother was diligent in laying an extra egg every day to make up for the missing egg. Some research has shown that hens can be trained to distinguish between even and odd numbers of pieces of food.[3] With these sorts of findings in mind, it is not hard to conceive that early humans had (at least) a similar sense of more and less. However, our conjectures about how and when these ideas emerged among humans are simply that; educated guesses based on our own assumptions of what might or could have been.

The Need for Simple Counting

As societies and humankind evolved, simply having a sense of more or less, even or odd, etc., would prove to be insufficient to meet the needs of everyday living. As tribes and groups formed, it became important to be able to know how many members were in the group, and perhaps how many were in the enemy's camp. Certainly, it was important for them to know if the flock of sheep or other possessed animals were increasing or decreasing in size. "Just how many of them do we have, anyway?" is a question that we do not have a hard time imagining them asking themselves (or each other).

In order to count items such as animals, it is often conjectured that one of the earliest methods of doing so would be with "tally sticks." These are objects used to track the numbers of items to be counted. With this method, each "stick" (or pebble, or whatever counting device being used) represents one animal or object. This method uses the idea of **one to one correspondence**. In a one to one correspondence, items that are being counted are uniquely linked with some counting tool.

In the picture to the right, you see each stick corresponding to one horse. By examining the collection of sticks in hand one knows how many animals should be present. You can imagine the usefulness of such a system, at least for smaller numbers of items to keep track of. If a herder wanted to "count off" his animals to make sure they were all present, he could mentally (or methodically) assign each stick to one animal and continue to do so until he was satisfied that all were accounted for.

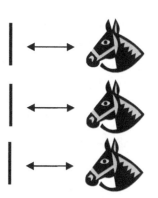

Of course, in our modern system, we have replaced the sticks with more abstract objects. In particular, the top stick is replaced with our symbol "1," the second stick gets replaced by a "2" and the third stick is represented by the symbol "3," but we are getting ahead of ourselves here. These modern symbols took many centuries to emerge.

Another possible way of employing the "tally stick" counting method is by making marks or cutting notches into pieces of wood, or even tying knots in string (as we shall see later). In 1937, Karl Absolom discovered a wolf bone that goes back possibly 30,000 years. It is believed to be a counting device.[4] Another example of this kind of tool is the Ishango Bone, discovered in 1960 at Ishango, and shown below.[5] It is reported to be between six and nine thousand years old and shows what appear to be markings used to do counting of some sort.

The markings on rows (a) and (b) each add up to 60. Row (b) contains the prime numbers between 10 and 20. Row (c) seems to illustrate for the method of doubling and multiplication used by the Egyptians. It is believed that this may also represent a lunar phase counter.

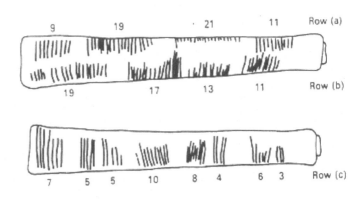

Spoken Words

As methods for counting developed, and as language progressed as well, it is natural to expect that spoken words for numbers would appear. Unfortunately, the developments of these words, especially those corresponding to the numbers from one through ten, are not easy to trace. Past ten, however, we do see some patterns:

Eleven comes from "ein lifon," meaning "one left over."
Twelve comes from "twe lif," meaning "two left over."
Thirteen comes from "Three and ten" as do fourteen through nineteen.
Twenty appears to come from "twe–tig" which means "two tens."
Hundred probably comes from a term meaning "ten times."

Written Numbers

When we speak of "written" numbers, we have to be careful because this could mean a variety of things. It is important to keep in mind that modern paper is only a little more than 100 years old, so "writing" in times past often took on forms that might look quite unfamiliar to us today.

As we saw earlier, some might consider wooden sticks with notches carved in them as writing as these are means of recording information on a medium that can be "read" by others. Of course, the symbols used (simple notches) certainly did not leave a lot of flexibility for communicating a wide variety of ideas or information.

Other mediums on which "writing" may have taken place include carvings in stone or clay tablets, rag paper made by hand (12th century in Europe, but earlier in China), papyrus (invented by the Egyptians and used up until the Greeks), and parchments from animal skins. And these are just a few of the many possibilities.

These are just a few examples of early methods of counting and simple symbols for representing numbers. Extensive books, articles and research have been done on this topic and could provide enough information to fill this entire course if we allowed it to. The range and diversity of creative thought that has been used in the past to describe numbers and to count objects and people is staggering. Unfortunately, we don't have time to examine them all, but it is fun and interesting to look at one system in more detail to see just how ingenious people have been.

2.2   The Number and Counting System of the Inca Civilization

Background

There is generally a lack of books and research material concerning the historical foundations of the Americas. Most of the "important" information available concentrates on the eastern hemisphere, with Europe as the central focus. The reasons for this may be twofold: first, it is thought that there was a lack of specialized mathematics in the American regions; second, many of the secrets of ancient mathematics in the Americas have been closely guarded.[6] The Peruvian system does not seem to be an exception here. Two researchers, Leland Locke and Erland Nordenskiold, have carried out research that has attempted to discover what mathematical knowledge was known by the Incas and how they used the Peruvian quipu, a counting system using cords and knots, in their mathematics. These researchers have come to certain beliefs about the quipu that we will summarize here.

Counting Boards

It should be noted that the Incas did not have a complicated system of computation. Where other peoples in the regions, such as the Mayans, were doing computations related to their rituals and calendars, the Incas seem to have been more concerned with the simpler task of record-keeping. To do this, they used what are called the "quipu" to record quantities of items. (We will describe them in more detail in a moment.) However, they first often needed to do computations whose results would be recorded on quipu. To do these computations, they would sometimes use a counting board constructed with a slab of stone. In the slab were cut rectangular and square compartments so that an octagonal

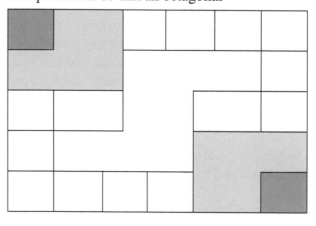

(eight–sided) region was left in the middle. Two opposite corner rectangles were raised. Another two sections were mounted on the original surface of the slab so that there were actually three levels available. In the figure shown, the darkest shaded corner regions represent the highest, third level. The lighter shaded regions surrounding the corners are the second highest levels, while the clear white rectangles are the compartments cut into the stone slab.

Pebbles were used to keep accounts and their positions within the various levels and compartments gave totals. For example, a pebble in a smaller (white) compartment represented one unit. Note that there are 12 such squares around the outer edge of the figure. If a pebble was put into one of the two (white) larger, rectangular compartments, its value was doubled. When a pebble was put in the octagonal region in the middle of the slab, its value was tripled. If a pebble was placed on the second (shaded) level, its value was multiplied by six. And finally, if a pebble was found on one of the two highest corner levels, its value was multiplied by twelve. Different objects could be counted at the same time by representing different objects by different colored pebbles.

Example 1

Suppose you have the following counting board with two different kind of pebbles places as illustrated. Let the solid black pebble represent a dog and the striped pebble represent a cat. How many dogs are being represented?

There are two black pebbles in the outer square regions…these represent 2 dogs.

There are three black pebbles in the larger (white) rectangular compartments. These represent 6 dogs.

There is one black pebble in the middle region…this represents 3 dogs.

There are three black pebbles on the second level…these represent 18 dogs.

Finally, there is one black pebble on the highest corner level…this represents 12 dogs. We then have a total of 2+6+3+18+12 = 41 dogs.

Try it Now 1

How many cats are represented on this board?

The Quipu

This kind of board was good for doing quick computations, but it did not provide a good way to keep a permanent recording of quantities or computations. For this purpose, they used the quipu. The quipu is a collection of cords with knots in them. These cords and knots are carefully arranged so that the position and type of cord or knot gives specific information on how to decipher the cord.

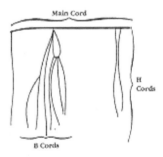

A quipu is made up of a main cord which has other cords (branches) tied to it. See pictures to the right.[7]

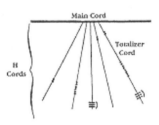

Locke called the branches H cords. They are attached to the main cord. B cords, in turn, were attached to the H cords. Most of these cords would have knots on them. Rarely are knots found on the main cord, however, and tend to be mainly on the H and B cords. A quipu might also have a "totalizer" cord that summarizes all of the information on the cord group in one place.

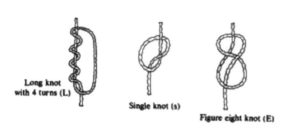

Long knot
with 4 turns (L)

Single knot (s)

Figure eight knot (E)

Locke points out that there are three types of knots, each representing a different value, depending on the kind of knot used and its position on the cord. The Incas, like us, had a decimal (base-ten) system, so each kind of knot had a specific decimal value. The Single knot, pictured in the middle of the diagram[8] was used to denote tens, hundreds, thousands, and ten-thousands. They would be on the upper levels of the H cords. The figure-eight knot on the end was used to denote the integer "one." Every other integer from 2 to 9 was represented with a long knot, shown on the left of the figure. (Sometimes long knots were used to represents tens and hundreds.) Note that the long knot has several turns in it…the number of turns indicates which integer is being represented. The units (ones) were placed closest to the bottom of the cord, then tens right above them, then the hundreds, and so on.

In order to make reading these pictures easier, we will adopt a convention that is consistent. For the long knot with turns in it (representing the numbers 2 through 9), we will use the following notation:

The four horizontal bars represent four turns and the curved arc on the right links the four turns together. This would represent the number 4.

We will represent the single knot with a large dot (•) and we will represent the figure eight knot with a sideways eight (∞).

Example 2

What number is represented on the cord shown?

On the cord, we see a long knot with four turns in it…this represents four in the ones place. Then 5 single knots appear in the tens position immediately above that, which represents 5 tens, or 50. Finally, 4 single knots are tied in the hundreds, representing four 4 hundreds, or 400. Thus, the total shown on this cord is 454.

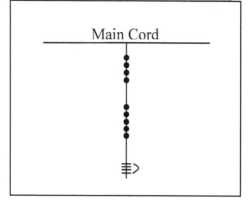

Main Cord

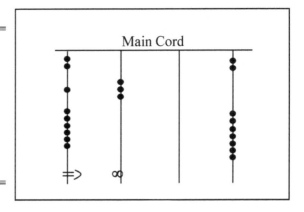

Try it Now 2
What numbers are represented on each of the four cords hanging from the main cord?

The colors of the cords had meaning and could distinguish one object from another. One color could represent llamas, while a different color might represent sheep, for example. When all the colors available were exhausted, they would have to be re–used. Because of this, the ability to read the quipu became a complicated task and specially trained individuals did this job. They were called Quipucamayoc, which means keeper of the quipus. They would build, guard, and decipher quipus.

As you can see from this photograph of an actual quipu, they could get quite complex.

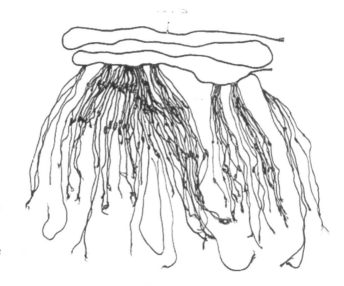

There were various purposes for the quipu. Some believe that they were used to keep an account of their traditions and history, using knots to record history rather than some other formal system of writing. One writer has even suggested that the quipu replaced writing as it formed a role in the Incan postal system.[9] Another proposed use of the quipu is as a translation tool. After the conquest of the Incas by the Spaniards and subsequent "conversion" to Catholicism, an Inca supposedly could use the quipu to confess their sins to a priest. Yet another proposed use of the quipu was to record numbers related to magic and astronomy, although this is not a widely accepted interpretation.

The mysteries of the quipu have not been fully explored yet. Recently, Ascher and Ascher have published a book, *The Code of the Quipu: A Study in Media, Mathematics, and Culture*, which is "an extensive elaboration of the logical-numerical system of the quipu."[10] For more information on the quipu, you may want to check out the following Internet link:

http://www.anthropology.wisc.edu/salomon/Chaysimire/khipus.php

We are so used to seeing the symbols 1, 2, 3, 4, etc. that it may be somewhat surprising to see such a creative and innovative way to compute and record numbers. Unfortunately, as we proceed through our mathematical education in grade and high school, we receive very little information about the wide range of number systems that have existed and which still exist all over the world. That's not to say our own system is not important or efficient. The fact that it has survived for hundreds of years and shows no sign of going away any time soon suggests that we may have finally found a system that works well and may not need further improvement, but only time will tell that whether or not that conjecture is valid or not. We now turn to a brief historical look at how our current system developed over history.

2.3   The Hindu-Arabic Number System

The Evolution of a System

Our own number system, composed of the ten symbols {0,1,2,3,4,5,6,7,8,9} is called the *Hindu-Arabic system*. This is a base-ten (decimal) system since place values increase by powers of ten. Furthermore, this system is positional, which means that the position of a symbol has bearing on the value of that symbol within the number. For example, the position of the symbol 3 in the number 435,681 gives it a value much greater than the value of the symbol 8 in that same number. We'll explore base systems more thoroughly later. The development of these ten symbols and their use in a positional system comes to us primarily from India.[11]

It was not until the 15th century that the symbols that we are familiar with today first took form in Europe. However, the history of these numbers and their development goes back hundreds of years. One important source of information on this topic is the writer al–Biruni, whose picture is shown here.[12] Al–Biruni, who was born in modern day Uzbekistan, had visited India on several occasions and made comments on the Indian number system. When we look at the origins of the numbers that al–Biruni encountered, we have to go back to the third century B.C.E. to explore their origins. It is then that the Brahmi numerals were being used.

The Brahmi numerals were more complicated than those used in our own modern system. They had separate symbols for the numbers 1 through 9, as well as distinct symbols for 10, 100, 1000,…, also for 20, 30, 40,…, and others for 200, 300, 400, …, 900. The Brahmi symbols for 1, 2, and 3 are shown below.[13]

1	2	3
—	=	≡

Brahmi one, two, three

These numerals were used all the way up to the 4th century C.E., with variations through time and geographic location. For example, in the first century C.E., one particular set of Brahmi numerals took on the following form[14]:

1	2	3	4	5	6	7	8	9
—	=	≡	+	h	Ϟ	ʔ	↳	ʔ

From the 4th century on, you can actually trace several different paths that the Brahmi numerals took to get to different points and incarnations. One of those paths led to our current numeral system, and went through what are called the Gupta numerals. The Gupta numerals were prominent during a time ruled by the Gupta dynasty and were spread throughout that empire as they conquered lands during the 4th through 6th centuries. They have the following form[15]:

1	2	3	4	5	6	7	8	9
—	=	≡	ɕ	b	த	η	ſ	ꜱ

How the numbers got to their Gupta form is open to considerable debate. Many possible hypotheses have been offered, most of which boil down to two basic types[16]. The first type of hypothesis states that the numerals came from the initial letters of the names of the numbers. This is not uncommon…the Greek numerals developed in this manner. The second type of hypothesis states that they were derived from some earlier number system. However, there are other hypotheses that are offered, one of which is by the researcher Ifrah. His theory is that there were originally nine numerals, each represented by a corresponding number of vertical lines. One possibility is this:[17]

1	2	3	4	5	6	7	8	9
I	II	III	II/II	III/II	III/III	IIII/III	IIII/IIII	IIII/IIII/III

Because these symbols would have taken a lot of time to write, they eventually evolved into cursive symbols that could be written more quickly. If we compare these to the Gupta numerals above, we can try to see how that evolutionary process might have taken place, but our imagination would be just about all we would have to depend upon since we do not know exactly how the process unfolded.

The Gupta numerals eventually evolved into another form of numerals called the Nagari numerals, and these continued to evolve until the 11th century, at which time they looked like this:[18]

1	2	3	4	5	6	7	8	9	0
९	२	३	८	५	६	७	८	९	०

Note that by this time, the symbol for 0 has appeared! The Mayans in the Americas had a symbol for zero long before this, however, as we shall see later in the chapter.

These numerals were adopted by the Arabs, most likely in the eighth century during Islamic incursions into the northern part of India.[19] It is believed that the Arabs were instrumental in spreading them to other parts of the world, including Spain (see below).

Other examples of variations up to the eleventh century include:

Devangari, eighth century[20]:

West Arab Gobar, tenth century[21]:

Spain, 976 B.C.E.[22]:

Finally, one more graphic[23] shows various forms of these numerals as they developed and eventually converged to the 15th century in Europe.

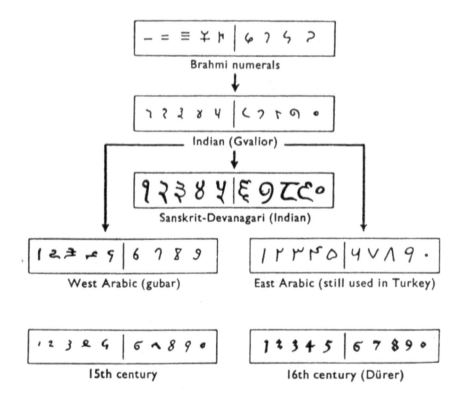

The Positional System

More important than the form of the number symbols is the development of the place value system. Although it is in slight dispute, the earliest known document in which the Indian system displays a positional system dates back to 346 C.E. However, some evidence suggests that they may have actually developed a positional system as far back as the first century C.E.

The Indians were not the first to use a positional system. The Babylonians (as we will see in Chapter 3) used a positional system with 60 as their base. However, there is not much evidence that the Babylonian system had much impact on later numeral systems, except with the Greeks. Also, the Chinese had a base-10 system, probably derived from the use of a counting board[24]. Some believe that the positional system used in India was derived from the Chinese system.

Wherever it may have originated, it appears that around 600 C.E., the Indians abandoned the use of symbols for numbers higher than nine and began to use our familiar system where the position of the symbol determines its overall value.[25] Numerous documents from the seventh century demonstrate the use of this positional system.

Interestingly, the earliest dated inscriptions using the system with a symbol for zero come from Cambodia. In 683, the 605th year of the Saka era is written with three digits and a dot in the middle. The 608th year uses three digits with a modern 0 in the middle.[26] The dot as a symbol for zero also appears in a Chinese work (*Chiu–chih li*). The author of this document gives a strikingly clear description of how the Indian system works:

> *Using the [Indian] numerals, multiplication and division are carried out. Each numeral is written in one stroke. When a number is counted to ten, it is advanced into the higher place. In each vacant place a dot is always put. Thus, the numeral is always denoted in each place. Accordingly, there can be no error in determining the place. With the numerals, calculations are easy...* "[27]

Transmission to Europe

It is not completely known how the system got transmitted to Europe. Traders and travelers of the Mediterranean coast may have carried it there. It is found in a tenth–century Spanish manuscript and may have been introduced to Spain by the Arabs, who invaded the region in 711 C.E. and were there until 1492.

In many societies, a division formed between those who used numbers and calculation for practical, every day business and those who used them for ritualistic purposes or for state business.[28] The former might often use older systems while the latter were inclined to use the newer, more elite written numbers. Competition between the two groups arose and continued for quite some time.

In a 14th century manuscript of Boethius' *The Consolations of Philosophy*, there appears a well-known drawing of two mathematicians. One is a merchant and is using an abacus (the "abacist"). The other is a Pythagorean philosopher (the "algorist") using his "sacred" numbers. They are in a competition that is being judged by the goddess of number. By 1500 C.E., however, the newer symbols and system had won out and has persevered until today. The Seattle Times recently reported that the Hindu–Arabic numeral system has been included in the book *The Greatest Inventions of the Past 2000 Years*.[29]

One question to answer is *why* the Indians would develop such a positional notation. Unfortunately, an answer to that question is not currently known. Some suggest that the system has its origins with the Chinese counting boards. These boards were portable and it is thought that Chinese travelers who passed through India took their boards with them and ignited an idea in Indian mathematics.[30] Others, such as G. G. Joseph propose that it is the Indian fascination with very large numbers that drove them to develop a system whereby these kinds of big numbers could easily be written down. In this theory, the system developed entirely within the Indian mathematical framework without considerable influence from other civilizations.

2.4   The Development and Use of Different Number Bases

Introduction and Basics

During the previous discussions, we have been referring to positional base systems. In this section of the chapter, we will explore exactly what a base system is and what it means if a system is "positional." We will do so by first looking at our own familiar, base-ten system and then deepen our exploration by looking at other possible base systems. In the next part of this section, we will journey back to Mayan civilization and look at their unique base system, which is based on the number 20 rather than the number 10.

A base system is a structure within which we count. The easiest way to describe a base system is to think about our own base-ten system. The base-ten system, which we call the "decimal" system, requires a total of ten different symbols/digits to write any number. They are, of course, 0, 1, 2, …, 9.

The decimal system is also an example of a *positional* base system, which simply means that the position of a digit gives its place value. Not all civilizations had a positional system even though they did have a base with which they worked.

In our base-ten system, a number like 5,783,216 has meaning to us because we are familiar with the system and its places. As we know, there are six ones, since there is a 6 in the ones place. Likewise, there are seven "hundred-thousands," since the 7 resides in that place. Each digit has a value that is explicitly determined by its position within the number. We make a distinction between digit, which is just a symbol such as 5, and a number, which is made up of one or more digits. We can take this number and assign each of its digits a value. One way to do this is with a table, which follows:

5,000,000	$= 5 \times 1,000,000$	$= 5 \times 10^6$	Five million
+700,000	$= 7 \times 100,000$	$= 7 \times 10^5$	Seven hundred thousand
+80,000	$= 8 \times 10,000$	$= 8 \times 10^4$	Eighty thousand
+3,000	$= 3 \times 1000$	$= 3 \times 10^3$	Three thousand
+200	$= 2 \times 100$	$= 2 \times 10^2$	Two hundred
+10	$= 1 \times 10$	$= 1 \times 10^1$	Ten
+6	$= 6 \times 1$	$= 6 \times 10^0$	Six
5,783,216	Five million, seven hundred eighty-three thousand, two hundred sixteen		

From the third column in the table we can see that each place is simply a multiple of ten. Of course, this makes sense given that our base is ten. The digits that are multiplying each place simply tell us how many of that place we have. We are restricted to having at most 9 in any one place before we have to "carry" over to the next place. We cannot, for example, have 11 in the hundreds place. Instead, we would carry 1 to the thousands place and retain 1 in the hundreds place. This comes as no surprise to us since we readily see that 11 hundreds is the same as one-thousand, one-hundred. Carrying is a pretty typical occurrence in a base system.

However, base-ten is not the only option we have. Practically any positive integer greater than or equal to 2 can be used as a base for a number system. Such systems can work just like the decimal system except the number of symbols will be different and each position will depend on the base itself.

Other Bases

For example, let's suppose we adopt a base-five system. The only modern digits we would need for this system are 0,1,2,3 and 4. What are the place values in such a system? To answer that, we start with the ones place, as most base systems do. However, if we were to count in this system, we could only get to four (4) before we had to jump up to the next place. Our base is 5, after all! What is that next place that we would jump to? It would not be tens, since we are no longer in base-ten. We're in a different numerical world. As the base-ten system progresses from 10^0 to 10^1, so does the base-five system moves from 5^0 to $5^1 = 5$. Thus, we move from the ones to the fives.

After the fives, we would move to the 5^2 place, or the twenty-fives. Note that in base-ten, we would have gone from the tens to the hundreds, which is, of course, 10^2.

Let's take an example and build a table. Consider the number 30412 in base five. We will write this as 30412_5 , where the subscript 5 is not part of the number but indicates the base we're using. First off, note that this is NOT the number "thirty thousand, four hundred twelve." We must be careful not to impose the base-ten system on this number. Here's what our table might look like. We will use it to convert this number to our more familiar base-ten system.

	Base 5	This column coverts to base–ten	In Base–Ten
	3×5^4	$= 3 \times 625$	$= 1875$
+	0×5^3	$= 0 \times 125$	$= 0$
+	4×5^2	$= 4 \times 25$	$= 100$
+	1×5^1	$= 1 \times 5$	$= 5$
+	2×5^0	$= 2 \times 1$	$= 2$
		Total	1982

As you can see, the number 30412_5 is equivalent to 1,982 in base–ten. We will say $30412_5 = 1982_{10}$. All of this may seem strange to you, but that's only because you are so used to the only system that you've ever seen.

Example 3

Convert 6234_7 to a base-10 number.

We first note that we are given a base-7 number that we are to convert. Thus, our places will start at the ones (7^0), and then move up to the 7's, 49's (7^2), etc. Here's the breakdown:

	Base 7	Convert	Base 10
	$= 6 \times 7^3$	$= 6 \times 343$	$= 2058$
$+$	$= 2 \times 7^2$	$= 2 \times 49$	$= 98$
$+$	$= 3 \times 7$	$= 3 \times 7$	$= 21$
$+$	$= 4 \times 1$	$= 4 \times 1$	$= 4$
		Total	2181

Thus, $6234_7 = 2181_{10}$.

Try it Now 3
Convert 41065_7 to a base-10 number.

Converting from Base 10 to Other Bases

Converting from an unfamiliar base to the familiar decimal system is not that difficult once you get the hang of it. It's only a matter of identifying each place and then multiplying each digit by the appropriate power. However, going the other direction can be a little trickier. Suppose you have a base-ten number and you want to convert to base-five. Let's start with some simple examples before we get to a more complicated one.

Example 4

Convert twelve to a base-five number.

We can probably easily see that we can rewrite this number as follows:
$$12 = (2 \times 5) + (2 \times 1)$$

Hence, we have two fives and 2 ones. Hence, in base five, we would write twelve as 22_5. Thus, $12_{10} = 22_5$.

Example 5

Convert sixty-nine to a base-five number.

We can see now that we have more than 25, so we rewrite sixty-nine as follows:
$$69 = (2 \times 25) + (3 \times 5) + (4 \times 1)$$

Here, we have two twenty-fives, 3 fives, and 4 ones. Hence, in base five we have 234. Thus, $69_{10} = 234_5$.

Example 6

Convert the base-seven number 3261_7 to base 10.

The powers of 7 are
$7^0 = 1$
$7^1 = 7$
$7^2 = 49$
$7^3 = 343$
Etc...

$3261_7 = (3 \times 343) + (2 \times 49) + (6 \times 7) + (1 \times 1) = 1170_{10}$.
Thus $3261_7 = 1170_{10}$.

Try it Now 4
Convert 143 to a base-5 number.

Try it Now 5
Convert the base-3 number 21021_3 to base 10.

In general, when converting from base-ten to some other base, it is often helpful to determine the highest power of the base that will divide into the given number at least once. In the last example, $5^2 = 25$ is the largest power of five that is present in 69, so that was our starting point. If we had moved to $5^3 = 125$, then 125 would not divide into 69 at least once.

Converting from Base 10 to Base b
1. Find the highest power of the base b that will divide into the given number at least once and then divide.
2. Write down the whole number part, then use the remainder from division in the next step.
3. Repeat step two, dividing by the next highest power of the base b, writing down the whole number part (including 0), and using the remainder in the next step.
4. Continue until the remainder is smaller than the base. This last remainder will be in the "ones" place.
5. Collect all your whole number parts to get your number in base b notation.

Example 7

Convert the base-ten number 348 to base-five.

The powers of five are
$5^0 = 1$
$5^1 = 5$
$5^2 = 25$
$5^3 = 125$
$5^4 = 625$
Etc...

Since 348 is smaller than 625, but bigger than 125, we see that $5^3=125$ is the highest power of five present in 348. So, we divide 125 into 348 to see how many 125's divide into 348:
$348 \div 125 = 2$ with remainder 98

We write down the whole part, 2, and continue with the remainder. There are 98 left over, so we see how many 25's (the next smallest power of five) there are in the remainder:
$98 \div 25 = 3$ with remainder 23

We write down the whole part, 2, and continue with the remainder. There are 23 left over, so we look at the next place, the 5's:
$23 \div 5 = 4$ with remainder 3

This leaves us with 3, which is less than our base, so this number will be in the "ones" place. We are ready to assemble our base-five number:
$348 = (2 \times 5^3) + (3 \times 5^2) + (4 \times 5^1) + (3 \times 1)$

Hence, our base-five number is 2343. We'll say that $348_{10} = 2343_5$.

Example 8

Convert the base-ten number 4,509 to base-seven.

The powers of 7 are

$7^0 = 1$
$7^1 = 7$
$7^2 = 49$
$7^3 = 343$
$7^4 = 2401$
$7^5 = 16807$
Etc…

The highest power of 7 that will divide into 4,509 is $7^4 = 2401$.
With division, we see that it will go in 1 time with a remainder of 2108. So we have 1 in the 7^4 place.

The next power down is $7^3 = 343$, which goes into 2108 six times with a new remainder of 50. So we have 6 in the 7^3 place.

The next power down is $7^2 = 49$, which goes into 50 once with a new remainder of 1. So there is a 1 in the 7^2 place.

The next power down is 7^1 but there was only a remainder of 1, so that means there is a 0 in the 7's place and we still have 1 as a remainder.

That, of course, means that we have 1 in the ones place.

Putting all of this together means that $4,509_{10} = 16101_7$.

$4,509 \div 7^4 =$	1 R 2108	
$2108 \div 7^3 =$	6 R 50	
$50 \div 7^2 =$	1 R 1	
$1 \div 7^1 =$	0 R 1	
$1 \div 7^0 =$	1	
$4,509_{10} = 16101_7$		

Try it Now 6
Convert 657_{10} to a base 4 number.

Try it Now 7
Convert 8377_{10} to a base 8 number.

Another Method for Converting from Base 10 to Other Bases

As you read the solution to this last example and attempted the "Try it Now" problems, you may have had to repeatedly stop and think about what was going on. The fact that you are probably struggling to follow the explanation and reproduce the process yourself is mostly due to the fact that the non-decimal systems are so unfamiliar to you. In fact, the only system that you are probably comfortable with is the decimal system.

As budding mathematicians, you should always be asking questions like "How could I simplify this process?" In general, that is one of the main things that mathematicians do…they look for ways to take complicated situations and make them easier or more familiar. In this section, we will attempt to do that.

To do so, we will start by looking at our own decimal system. What we do may seem obvious and maybe even intuitive but that's the point. We want to find a process that we readily recognize works and makes sense to us in a familiar system and then use it to extend our results to a different, unfamiliar system.

Let's start with the decimal number, 4863_{10}. We will convert this number to base 10. Yeah, I know it's already in base 10, but if you carefully follow what we're doing, you'll see it makes things work out very nicely with other bases later on. We first note that the highest power of 10 that will divide into 4863 at least once is $10^3 = 1000$. *In general, this is the first step in our new process; we find the highest power that a given base that will divide at least once into our given number.*

We now divide 1000 into 4863:

$$4863 \div 1000 = 4.863$$

This says that there are four thousands in 4863 (obviously). However, it also says that there are 0.863 thousands in 4863. This fractional part is our remainder and will be converted to lower powers of our base (10). If we take that decimal and multiply by 10 (since that's the base we're in) we get the following:

$$0.863 \times 10 = 8.63$$

Why multiply by 10 at this point? We need to recognize here that 0.863 thousands is the same as 8.63 hundreds. Think about that until it sinks in.

$$(0.863)(1000) = 863$$
$$(8.63)(100) = 863$$

These two statements are equivalent. So, what we are really doing here by multiplying by 10 is rephrasing or converting from one place (thousands) to the next place down (hundreds).

$$0.863 \times 10 \Rightarrow 8.63$$
$$(\text{Parts of Thousands}) \times 10 \Rightarrow \text{Hundreds}$$

We now have 8 hundreds and a remainder of 0.63 hundreds, which is the same as 6.3 tens. We can do this again with the 0.63 that remains after this first step.

$$0.63 \times 10 \Rightarrow 6.3$$
$$\text{Hundreds} \times 10 \Rightarrow \text{Tens}$$

So, we have six tens and 0.3 tens, which is the same as 3 ones, our last place value.

Now here's the punch line. Let's put all of the together in one place:

$$4863 \div 1000 = \quad ④.863$$
$$0.863 \times 10 = \quad ⑧.63$$
$$0.63 \times 10 = \quad ⑥.3$$
$$0.3 \times 10 = \quad ③.0$$

Note that in each step, the remainder is carried down to the next step and multiplied by 10, the base. Also, at each step, the whole number part, which is circled, gives the digit that belongs in that particular place. What is amazing is that this works for any base! So, to convert from a base 10 number to some other base, *b*, we have the following steps we can follow:

Converting from Base 10 to Base *b*: Another method
1. Find the highest power of the base *b* that will divide into the given number at least once and then divide.
2. Keep the whole number part, and multiply the fractional part by the base *b*.
3. Repeat step two, keeping the whole number part (including 0), carrying the fractional part to the next step until only a whole number result is obtained.
4. Collect all your whole number parts to get your number in base *b* notation.

We will illustrate this procedure with some examples.

Example 9

Convert the base 10 number, 348_{10}, to base 5.

This is actually a conversion that we have done in a previous example. The powers of five are

$5^0 = 1$
$5^1 = 5$
$5^2 = 25$
$5^3 = 125$
$5^4 = 625$
Etc…

The highest power of five that will go into 348 at least once is 5^3. We divide by 125 and then proceed:

$$348 \div 5^3 = ②.784$$

$$0.784 \times 5 = ③.92$$

$$0.92 \times 5 = ④0.6$$

$$0.6 \times 5 = ③.0$$

By keeping all the whole number parts, from top bottom, gives 2343 as our base 5 number. Thus, $2343_5 = 348_{10}$.

We can compare our result with what we saw earlier, or simply check with our calculator, and find that these two numbers really are equivalent to each other.

Example 10

Convert the base 10 number, 3007_{10}, to base 5.

The highest power of 5 that divides at least once into 3007 is $5^4 = 625$. Thus, we have:

$3007 \div 625 = ④.8112$
$0.8112 \times 5 = ④.056$
$0.056 \times 5 = ⓪.28$
$0.28 \times 5 = ①0.4$
$0.4 \times 5 = ②0.0$

This gives us that $3007_{10} = 44012_5$. Notice that in the third line that multiplying by 5 gave us 0 for our whole number part. We don't discard that! The zero tells us that a zero in that place. That is, there are no 5^2's in this number.

This last example shows the importance of using a calculator in certain situations and taking care to avoid clearing the calculator's memory or display until you get to the very end of the process.

Example 11

Convert the base 10 number, 63201_{10}, to base 7.

The powers of 7 are

$7^0 = 1$
$7^1 = 7$
$7^2 = 49$
$7^3 = 343$
$7^4 = 2401$
$7^5 = 16807$
Etc…

The highest power of 7 that will divide at least once into 63201 is 7^5. When we do the initial division on a calculator, we get the following:

$63201 \div 7^5 = 3.760397453$

The decimal part actually fills up the calculators display and we don't know if it terminates at some point or perhaps even repeats down the road. So, if we clear our calculator at this point, we will introduce error that is likely to keep this process from ever ending. To avoid this problem, we leave the result in the calculator and simply subtract 3 from this to get the fractional part all by itself. DO NOT ROUND OFF! Subtraction and then multiplication by seven gives:

$63201 \div 7^5 = ③.760397453$
$0.760397453 \times 7 = ⑤.322782174$
$0.322782174 \times 7 = ②.259475219$
$0.259475219 \times 7 = ①.816326531$

0. 816326531 × 7 = ⑤.714285714

0. 714285714 × 7 = ⑤.000000000

Yes, believe it or not, that last product is exactly 5, *as long as you don't clear anything out on your calculator*. This gives us our final result: $63201_{10} = 352155_7$. If we round, even to two decimal places in each step, clearing our calculator out at each step along the way, we will get a series of numbers that do not terminate, but begin repeating themselves endlessly. (Try it!) We end up with something that doesn't make any sense, at least not in this context. So be careful to use your calculator cautiously on these conversion problems.

Also, remember that if your first division is by 7^5, then you expect to have 6 digits in the final answer, corresponding to the places for 7^5, 7^4, and so on down to 7^0. If you find yourself with more than 6 digits due to rounding errors, you know something went wrong.

Try it Now 8
Convert the base-10 number, 9352_{10}, to base 5.

Try it Now 9
Convert the base-10 number, 1500, to base 3.

Be careful not to clear your calculator on this one. Also, if you're not careful in each step, you may not get all of the digits you're looking for, so move slowly and with caution.

2.5 The Mayan Numeral System

Background

As you might imagine, the development of a base system is an important step in making the counting process more efficient. Our own base-ten system probably arose from the fact that we have 10 fingers (including thumbs) on two hands. This is a natural development. However, other civilizations have had a variety of bases other than ten. For example, the Natives of Queensland used a base–two system, counting as follows: "one, two, two and one, two two's, much." Some Modern South American Tribes have a base-five system counting in this way: "one, two, three, four, hand, hand and one, hand and two," and so on. The Babylonians used a base-60 (sexigesimal) system. In this chapter, we wrap up with a specific example of a civilization that actually used a base system other than 10.

The Mayan civilization is generally dated from 1500 B.C.E to 1700 C.E. The Yucatan Peninsula (see map[31]) in Mexico was the scene for the development of one of the most advanced civilizations of the ancient world. The Mayans had a sophisticated ritual system that was overseen by a priestly class. This class of priests developed a philosophy with time as divine and eternal.[32] The calendar, and calculations related to it, were thus very important to the ritual life of the priestly class, and hence the Mayan people. In fact, much of what we know about this culture comes from their calendar records and astronomy data. Another important source of information on the Mayans is the writings of Father Diego de Landa, who went to Mexico as a missionary in 1549.

1. Chichen Itza
2. Uxmal
3. Tulum
4. Palenque
5. Bonampak, Yaxchilan
6. Tikal
7. Altun Ha
8. Copán

There were two numeral systems developed by the Mayans – one for the common people and one for the priests. Not only did these two systems use different symbols, they also used different base systems. For the

priests, the number system was governed by ritual. The days of the year were thought to be gods, so the formal symbols for the days were decorated heads,[33] like the sample to the left[34] Since the basic calendar was based on 360 days, the priestly numeral system used a mixed base system employing multiples of 20 and 360. This makes for a confusing system, the details of which we will skip.

The Mayan Number System

Instead, we will focus on the numeration system of the "common" people, which used a more consistent base system. As we stated earlier, the Mayans used a base–20 system, called the "vigesimal" system. Like our system, it is positional, meaning that the position of a numeric symbol indicates its place value. In the following table, you can see the place value in its vertical format.[35]

Powers	Base–Ten Value	Place Name
20^7	12,800,000,000	Hablat
20^6	64,000,000	Alau
20^5	3,200,000	Kinchil
20^4	160,000	Cabal
20^3	8,000	Pic
20^2	400	*Bak*
20^1	20	Kal
20^0	1	Hun

In order to write numbers down, there were only three symbols needed in this system. A horizontal bar represented the quantity 5, a dot represented the quantity 1, and a special symbol (thought to be a shell) represented zero. The Mayan system may have been the first to make use of zero as a placeholder/number. The first 20 numbers are shown in the table to the right.[36]

Unlike our system, where the ones place starts on the right and then moves to the left, the Mayan systems places the ones on the <u>bottom</u> of a vertical orientation and moves up as the place value increases.

When numbers are written in vertical form, there should never be more than four dots in a single place. When writing Mayan numbers, every group of five dots becomes one bar. Also, there should never be more than three bars in a single place…four bars would be converted to one dot in the next place up. It's the same as 10 getting converted to a 1 in the next place up when we carry during addition.

Number	Vertical Form	Number	Vertical Form
0	⬭	10	═
1	∘	11	∘ over ═
2	∘ ∘	12	∘ ∘ over ═
3	∘ ∘ ∘	13	∘ ∘ ∘ over ═
4	∘ ∘ ∘ ∘	14	∘ ∘ ∘ ∘ over ═
5	—	15	≡
6	∘ over —	16	∘ over ≡
7	∘ ∘ over —	17	∘ ∘ over ≡
8	∘ ∘ ∘ over —	18	∘ ∘ ∘ over ≡
9	∘ ∘ ∘ ∘ over —	19	∘ ∘ ∘ ∘ over ≡

Example 12

What is the value of this number, which is shown in vertical form?

∘ ∘ ∘

∘ ∘ ∘ over —

Starting from the bottom, we have the ones place. There are two bars and three dots in this place. Since each bar is worth 5, we have 13 ones when we count the three dots in the ones place. Looking to the place value above it (the twenties places), we see there are three dots so we have three twenties.

$\circ\circ\circ$ ← 20's

$\circ\circ\circ$ ← 1's

Hence, we can write this number in base ten as

$$\left(3\times 20^1\right)+\left(13\times 20^0\right)=\left(3\times 20\right)+\left(13\times 1\right)$$
$$=60+13$$
$$=73$$

Example 13

What is the value of the following Mayan number?

This number has 11 in the ones place, zero in the 20's place, and 18 in the $20^2 = 400$'s place. Hence, the value of this number in base-ten is

$18\times400 + 0\times20 + 11\times1 = 7211.$

Try it Now 10
Convert the Mayan number below to base 10.

Example 14

Convert the base-10 number 3575_{10} to Mayan numerals.

This problem is done in two stages. First, we need to convert to a base 20 number. We will do so using the method provided in the last section of the text. The second step is to convert that number to Mayan symbols.

The highest power of 20 that will divide into 3575 is $20^2 = 400$, so we start by dividing that and then proceed from there:

$3575 \div 400 = 8.9375$
$0.9375 \times 20 = 18.75$
$0.75 \times 20 = 15.0$

This means that $3575_{10} = 8,18,15_{20}$

The second step is to convert this to Mayan notation. This number indicates that we have 15 in the ones position. That's three bars at the bottom of the number. We also have 18 in the 20's place, so that's three bars and three dots in the second position. Finally, we have 8 in the 400's place, so that's one bar and three dots on the top. We get the following

Note that in the previous example a new notation was used when we wrote $8,18,15_{20}$. The commas between the three numbers 8, 18, and 15 are now separating place values for us so that we can keep them separate from each other. This use of the comma is slightly different than how they're used in the decimal system. When we write a number in base 10, such as 7,567,323, the commas are used primarily as an aide to read the number easily but they do not separate single place values from each other. We will need this notation whenever the base we use is larger than 10.

Writing numbers with bases bigger than 10
When the base of a number is larger than 10, separate each "digit" with a comma to make the separation of digits clear.

For example, in base 20, to write the number corresponding to $17 \times 20^2 + 6 \times 20^1 + 13 \times 20^0$, we'd write $17,6,13_{20}$.

Try it Now 11
Convert the base 10 number 10553_{10} to Mayan numerals.

Try it Now 12
Convert the base 10 number 5617_{10} to Mayan numerals.

Adding Mayan Numbers

When adding Mayan numbers together, we'll adopt a scheme that the Mayans probably did not use but which will make life a little easier for us.

Example 15

Add, in Mayan, the numbers 37 and 29: [37]

First draw a box around each of the vertical places. This will help keep the place values from being mixed up.

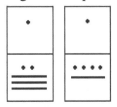

Next, put all of the symbols from both numbers into a single set of places (boxes), and to the right of this new number draw a set of empty boxes where you will place the final sum:

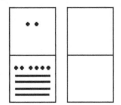

You are now ready to start carrying. Begin with the place that has the lowest value, just as you do with Arabic numbers. Start at the bottom place, where each dot is worth 1. There are six dots, but a maximum of four are allowed in any one place; once you get to five dots, you must convert to a bar. Since five dots make one bar, we draw a bar through five of the dots, leaving us with one dot which is under the four-dot limit. Put this dot into the bottom place of the empty set of boxes you just drew:

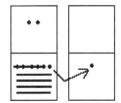

Now look at the bars in the bottom place. There are five, and the maximum number the place can hold is three. *Four bars are equal to one dot in the next highest place.*

Whenever we have four bars in a single place we will automatically convert that to a *dot* in the next place up. We draw a circle around four of the bars and an arrow up to the dots' section of the higher place. At the end of that arrow, draw a new dot. That dot represents

20 just the same as the other dots in that place. Not counting the circled bars in the bottom place, there is one bar left. One bar is under the three-bar limit; put it under the dot in the set of empty places to the right.

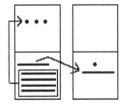

Now there are only three dots in the next highest place, so draw them in the corresponding empty box.

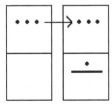

We can see here that we have 3 twenties (60), and 6 ones, for a total of 66. We check and note that 37 + 29 = 66, so we have done this addition correctly. Is it easier to just do it in base–ten? Probably, but that's only because it's more familiar to you. Your task here is to try to learn a new base system and how addition can be done in slightly different ways than what you have seen in the past. Note, however, that the concept of carrying is still used, just as it is in our own addition algorithm.

Try it Now 13
Try adding 174 and 78 in Mayan by first converting to Mayan numbers and then working entirely within that system. Do not add in base-ten (decimal) until the very end when you *check* your work.

2.6 Addition and Subtraction with Other Bases

Addition in Other Bases

As we saw in the previous section with the Mayan numeration system, we can add or subtract in other bases. Below are a series of steps, but, overall, we add as usual while finding its equivalent number from base 10 to the new base.

Adding in base b

1. Rewrite the addition vertically, if not already.
2. Start in the ones place (as usual), but find the number the sum represents in base b.
3. If the sum is larger than base b, then carry over to the b^1 place value.
4. Repeat steps 2 and 3 for the b^2, b^3, ... place values.

Example

Add in base two: $111_{two} + 11_{two}$

Writing this vertically, we get

$$\begin{array}{r} 111_{two} \\ + \ 11_{two} \\ \hline \end{array}$$

Let's add the ones place as usual. If we were in base 10, 1+1=2; 2 in base 10 is equivalent to 0 in base 2 and we carry 1 over to the 2's place. Recall, base two is {0,1}. Hence, the ones place is 0:

$$\begin{array}{r} 11^{+1}1_{two} \\ + \ 1 \ 1_{two} \\ \hline 1 \ 0_{two} \end{array}$$

Adding in the 2's place, 1+1+1=3 in base 10, but 3 is 1 in base 2 and we carry 1 to the 2^2's place:

$$\begin{array}{r} 1^{+1} \ 11_{two} \\ + \ \ \ \ 11_{two} \\ \hline 0 \ 10_{two} \end{array}$$

Adding in the 2^2's place, we get 1+1=2 in base 10, but 2 in base 10 is 0 in base 2 and we carry a 1 over to the 2^3's place:

$$\begin{array}{r} +1 \ 1 \ 11_{two} \\ + \ \ \ \ \ 11_{two} \\ \hline 1 \ 010_{two} \end{array}$$

Hence, $111_{two} + 11_{two} = 1010_{two}$. Note, another way to do this is convert each number to base 10 and add as usual, then convert the result back into base 2.

Example

Add in base five: $44_{five} + 42_{five}$

Let's try this by converting each number to base 10, adding them, then converting the sum back into base 5:

$44_{five} = 4(5) + 4(1) = 24_{ten}$

$42_{five} = 4(5) + 2(1) = 22_{ten}$

Next, we add the base 10 numbers:

$24 + 22 = 46_{ten}$

Converting 46 in base ten into a number in base 5, we use the previous sections' technique and obtain 141_{five}. Thus, 46 in base 10 is equivalent to 141 in base 5.

Subtraction in Other Bases

Subtracting in base *b*
1. Rewrite the subtraction vertically, if not already.
2. Start in the ones place (as usual), but find the number the difference represents in base *b*.
3. If the ones place of the minuend is smaller than the ones place of the subtrahend, then borrow from the place value to the left in that base *b*. Then subtract as usual.
4. Repeat steps 2 and 3 for the b^2, b^3 place values.

Example

Subtract in base 5: 240_{five} - 40_{five}

$$240_5$$
$$- \ 40_5$$

If we take the ones place, 0-0 =0, which is 0 in base 5, then the ones place stays 0.

$$240_5$$
$$- \ 40_5$$
$$\overline{0}$$

Now let's take the 5's place: 4-4=0. Since 0 is in base 5, then the 5's place is 0.

$$240_5$$
$$- \ 40_5$$
$$\overline{00}$$

Let's look at the 5^2's place. Notice, we can just drop the two down and get

240_5
$- \ 40_5$
$\overline{200_5}$

Thus, 240_{five} - 40_{five} = 200_{five}.

What if we have to borrow? We can subtract with borrowing easily in base 10, but what if we wanted to subtract two numbers that included borrowing? Let's see.

Example

Subtract in base 5: 404_{five} - 323_{five}

Rewriting this vertically, we get

404_{five}
-323_{five}

Subtracting in the ones place, we get 4-3=1. Since 1 is in base 5, then the ones place is 1.

404_{five}
-323_{five}
1

Looking at the 5's place, notice that 0 is less than 2 and we have to borrow from the 5^2's place. Recall, we are in base 5, so when we borrow, we still reduce 4 to 3, but we carry 5 since we are in base 5:

$\cancel{4}^3 \ \cancel{0}^5 \ 4_{\text{five}}$
$-3 \quad 2 \ 3_{\text{five}}$
$0 \quad 3 \ 1_{\text{five}}$

Now, we subtract as usual: 5-2 = 3. Since 3 is in base 5, then the 5's place is 3. Subtracting in the 5^2's place, we get 3-3=0; hence, the 5^2's place is 0.

Thus, 404_{five} - 323_{five} = 31_{five}.

Conclusion

In this chapter, we have briefly sketched the development of numbers and our counting system, with the emphasis on the "brief" part. There are numerous sources of information and research that fill many volumes of books on this topic. Unfortunately, we cannot begin to come close to covering all of the information that is out there.

We have only scratched the surface of the wealth of research and information that exists on the development of numbers and counting throughout human history. It is important to note that the system that we use every day is a product of thousands of years of progress and development. It represents contributions by many civilizations and cultures. It does not come down to us from the sky, a gift from the gods. It is not the creation of a textbook publisher. It is indeed as human as we are, as is the rest of mathematics. Behind every symbol, formula and rule there is a human face to be found, or at least sought.

Furthermore, we hope that you now have a basic appreciation for just how interesting and diverse number systems can get. Also, we're pretty sure that you have also begun to recognize that we take our own number system for granted so much that when we try to adapt to other systems or bases, we find ourselves truly having to concentrate and think about what is going on.

Try it Now Answers

1. $1+6\times3+3\times6+2\times12 = 61$ cats.

2. From left to right:
   Cord 1 = 2,162
   Cord 2 = 301
   Cord 3 = 0
   Cord 4 = 2,070

3. $41065_7 = 9994_{10}$
4. $143_{10} = 1033_5$
5. $21021_3 = 196_{10}$
6. $657_{10} = 22101_4$
7. $8377_{10} = 20271_8$
8. $9352_{10} = 244402_5$
9. $1500_{10} = 2001120_3$
10. 1562
11. $10553_{10} = 1,6,7,13_{20}$

12. $5617_{10} = 14,0,17_{20}$. Note that there is a zero in the 20's place, so you'll need to use the appropriate zero symbol in between the ones and 400's places.

13. A sample solution is shown.

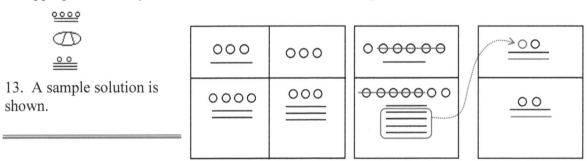

Exercises

Skills

Counting Board and Quipu

1. In the following Peruvian counting board, determine how many of each item is represented. Please show all of your calculations along with some kind of explanation of how you got your answer. Note the key at the bottom of the drawing.

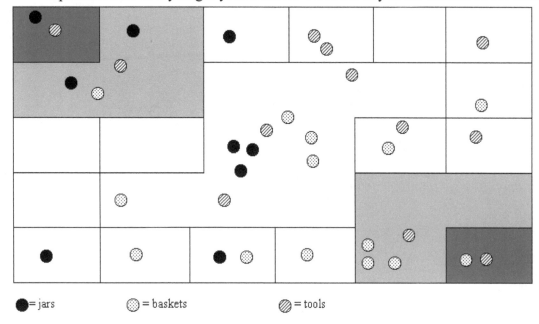

● = jars   ⊚ = baskets   ⊘ = tools

2. Draw a quipu with a main cord that has branches (H cords) that show each of the following numbers on them. (You should produce <u>one</u> drawing for this problem with the cord for part **a** on the left and moving to the right for parts **b** through **d**.)
    a. 232   b. 5065
    c. 23,451   d. 3002

Basic Base Conversions
3.   423 in base 5 to base 10   4. 3044 in base 5 to base 10

5.   387 in base 10 to base 5   6. 2546 in base 10 to base 5

7.   110101 in base 2 to base 10   8. 11010001 in base 2 to base 10

9.   100 in base 10 to base 2   10. 2933 in base 10 to base 2

11. Convert 653 in base 7 to base 10.   12. Convert 653 in base 10 to base 7

13. 3412 in base 5 to base 2   14. 10011011 in base 2 to base 5
(Hint: convert first to base 10 then to the final desired base)

The Caidoz System

Suppose you were to discover an ancient base–12 system made up twelve symbols. Let's call this base system the Caidoz system. Here are the symbols for each of the numbers 0 through 11:

0 = ♈	6 = ♎
1 = ♉	7 = ♏
2 = ♊	8 = ♐
3 = ♋	9 = ♑
4 = ♌	10 = ♒
5 = ♍	11 = ♓

Convert each of the following numbers in Caidoz to base 10

15. ♏♉♒   16. ♐♋♈♓

17. ♎♌♊   18. ♍♉♒♈

Convert the following base 10 numbers to Caidoz, using the symbols shown above.

19. 175   20. 3030

21. 10,000   22. 5507

Mayan Conversions

Convert the following numbers to Mayan notation. Show your calculations used to get your answers.

23. 135   24. 234

25. 360   26. 1,215

27. 10,500   28. 1,100,000

Convert the following Mayan numbers to decimal (base-10) numbers. Show all calculations.

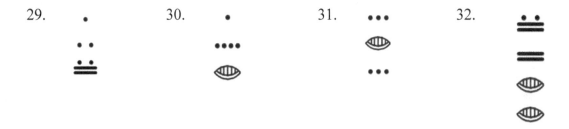

29.

30.

31.

32.

James Bidwell has suggested that Mayan addition was done by "simply combining bars and dots and carrying to the next higher place." He goes on to say, "After the combining of dots and bars, the second step is to exchange every five dots for one bar in the same position." After converting the following base 10 numbers into vertical Maya notation (in base 20, of course), perform the indicated addition:

33. 32 + 11

34. 82 + 15

35. 35 + 148

36. 2412 + 5000

37. 450 + 844

38. 10,000 + 20,000

39. 4,500 + 3,500

40. 130,000 + 30,000

41. Use the fact that the Mayans had a base-20 number system to complete the following multiplication table. The table entries should be in Mayan notation. Remember: Their zero looked like this… ⚬. *Xerox and then cut out the table below, fill it in, and paste it onto your homework assignment if you do not want to duplicate the table with a ruler.* (To think about but not write up: Bidwell claims that only these entries are needed for "Mayan multiplication." What does he mean?)

×	•	••	•••	••••	—	═	≡
•							
••							
•••							
••••							
—							
═							
≡							

Binary and Hexadecimal Conversions

Modern computers operate in a world of "on" and "off" electronic switches, so use a ***binary*** *counting system – base 2, consisting of only two digits: 0 and 1.*

Convert the following binary numbers to decimal (base–10) numbers.

42. 1001

43. 1101

44. 110010

45. 101110

Convert the following base-10 numbers to binary

46. 7

47. 12

48. 36

49. 27

Four binary digits together can represent any base-10 number from 0 to 15. To create a more human-readable representation of binary-coded numbers, hexadecimal numbers, base 16, are commonly used. Instead of using the $8,13,12_{16}$ notation used earlier, the letter A is used to represent the digit 10, B for 11, up to F for 15, so $8,13,12_{16}$ would be written as 8DC.

Convert the following hexadecimal numbers to decimal (base–10) numbers.

50. C3

51. 4D

52. 3A6

53. BC2

Convert the following base-10 numbers to hexadecimal.

54. 152

55. 176

56. 2034

57. 8263

Add or subtract in base 5 or in base 2.

58. $110_{five} - 34_{five}$

62. $10011_{two} + 10000_{two}$

59. $314_{five} - 32_{five}$

63. $20_{five} + 42_{five}$

60. $30_{five} - 12_{five}$

64. $14_{five} + 24_{five}$

61. $110_{two} + 10_{two}$

65. $44_{five} - 34_{five}$

Exploration

66. What are the advantages and disadvantages of bases other than ten.

67. Supposed you are charged with creating a base-15 number system. What symbols would you use for your system and why? Explain with at least two specific examples how you would convert between your base–15 system and the decimal system.

68. Describe an interesting aspect of Mayan civilization that we did not discuss in class. Your findings must come from some source such as an encyclopedia article, or internet site and you must provide reference(s) of the materials you used (either the publishing information or Internet address).

69. For a Papuan tribe in southeast New Guinea, it was necessary to translate the bible passage John 5:5 "And a certain man was there, which had an infirmity 30 and 8 years" into "A man lay ill one man, both hands, five and three years." Based on your own understanding of bases systems (and some common sense), furnish an explanation of the translation. Please use complete sentences to do so. (Hint: To do this problem, I am asking you to think about how base systems work, where they come from, and how they are used. You won't necessarily find an "answer" in readings or such…you'll have to think it through and come up with a reasonable response. Just make sure that you clearly explain why the passage was translated the way that it was.)

70. The Mayan calendar was largely discussed leading up to December 2012. Research how the Mayan calendar works, and how the counts are related to the number based they use.

Endnotes

[1] Eves, Howard; An Introduction to the History of Mathematics, p. 9.

[2] Eves, p. 9.

[3] McLeish, John; The Story of Numbers – How Mathematics Has Shaped Civilization, p. 7.

[4] Bunt, Lucas; Jones, Phillip; Bedient, Jack; The Historical Roots of Elementary Mathematics, p. 2.

[5] http://www.math.buffalo.edu/mad/Ancient-Africa/mad_zaire-uganda.html

[6] Diana, Lind Mae; The Peruvian Quipu in *Mathematics Teacher,* Issue 60 (Oct., 1967), p. 623–28.

[7] Diana, Lind Mae; The Peruvian Quipu in *Mathematics Teacher,* Issue 60 (Oct., 1967), p. 623–28.

[8] http://wiscinfo.doit.wisc.edu/chaysimire/titulo2/khipus/what.htm

[9] Diana, Lind Mae; The Peruvian Quipu in *Mathematics Teacher,* Issue 60 (Oct., 1967), p. 623–28.

[10] http://www.cs.uidaho.edu/~casey931/seminar/quipu.html

[11] http://www-groups.dcs.st-and.ac.uk/~history/HistTopics/Indian_numerals.html

[12] http://www-groups.dcs.st-and.ac.uk/~history/Mathematicians/Al-Biruni.html

[13] http://www-groups.dcs.st-and.ac.uk/~history/HistTopics/Indian_numerals.html

[14] http://www-groups.dcs.st-and.ac.uk/~history/HistTopics/Indian_numerals.html

[15] Ibid

[16] Ibid

[17] Ibid

[18] Ibid

[19] Katz, page 230

[20] Burton, David M., *History of Mathematics, An Introduction*, p. 254–255

[21] Ibid

[22] Ibid

[23] Katz, page 231.

[24] Ibid, page 230

[25] Ibid, page 231.

[26] Ibid, page 232.

[27] Ibid, page 232.

[28] McLeish, p. 18

[29] http://seattletimes.nwsource.com/news/health-science/html98/invs_20000201.html, Seattle Times, Feb. 1, 2000

[30] Ibid, page 232.

[31] http://www.gorp.com/gorp/location/latamer/map_maya.htm

[32] Bidwell, James; Mayan Arithmetic in *Mathematics Teacher*, Issue 74 (Nov., 1967), p. 762–68.

[33] http://www.ukans.edu/~lctls/Mayan/numbers.html

[34] http://www.ukans.edu/~lctls/Mayan/numbers.html

[35] Bidwell

[36] http://www.vpds.wsu.edu/fair_95/gym/UM001.html

[37] http://forum.swarthmore.edu/k12/mayan.math/mayan2.html

Chapter 3: Sets

It is natural for us to classify items into groups, or sets, and consider how those sets overlap with each other. We can use these sets understand relationships between groups, and to analyze survey data.

Table of Contents

3.1 Basics

An art collector might own a collection of paintings, while a music lover might keep a collection of CDs. Any collection of items can form a **set**.

> **Set**
> A **set** is a collection of distinct objects, called **elements** of the set.
>
> A set can be defined by describing the contents, or by listing the elements of the set, enclosed in curly brackets.

Example 1

Some examples of sets defined by describing the contents:
a) The set of all even numbers
b) The set of all books written about travel to Chile

Some examples of sets defined by listing the elements of the set:
a) {1, 3, 9, 12}
b) {red, orange, yellow, green, blue, indigo, purple}

A set simply specifies the contents; order is not important. The set represented by {1, 2, 3} is equivalent to the set {3, 1, 2}.

> **Notation**
> Commonly, we will use a variable to represent a set, to make it easier to refer to that set later.
>
> The symbol ∈ means "is an element of".
>
> A set that contains no elements, { }, is called the **empty set** and is notated ∅.

Example 2

Let $A = \{1, 2, 3, 4\}$

To notate that 2 is element of the set, we'd write $2 \in A$

Sometimes a collection might not contain all the elements of a set. For example, Chris owns three Madonna albums. While Chris's collection is a set, we can also say it is a **subset** of the larger set of all Madonna albums.

> **Subset**
>
> A **subset** of a set A is another set that contains only elements from the set A, but may not contain all the elements of A.
>
> If B is a subset of A, we write $B \subseteq A$.
>
> A **proper subset** is a subset that is not identical to the original set – it contains fewer elements.
>
> If B is a proper subset of A, we write $B \subset A$.

Example 3

Consider these three sets:

A = the set of all even numbers $B = \{2, 4, 6\}$ $C = \{2, 3, 4, 6\}$

Here $B \subset A$ since every element of B is also an even number, so is an element of A.

More formally, we could say $B \subset A$ since if $x \in B$, then $x \in A$.

It is also true that $B \subset C$.

C is not a subset of A, since C contains an element, 3, that is not contained in A

Example 4

Suppose a set contains the plays "Much Ado About Nothing", "MacBeth", and "A Midsummer's Night Dream". What is a larger set this might be a subset of?

There are many possible answers here. One would be the set of plays by Shakespeare. This is also a subset of the set of all plays ever written. It is also a subset of all British literature.

Try it Now 1
The set $A = \{1, 3, 5\}$. What is a larger set this might be a subset of?

3.2 Union, Intersection, and Complement

Commonly sets interact. For example, you and a new roommate decide to have a house party, and you both invite your circle of friends. At this party, two sets are being combined, though it might turn out that there are some friends that were in both sets.

Union, Intersection, and Complement
The **union** of two sets contains all the elements contained in either set (or both sets).
The union is notated $A \cup B$.
More formally, $x \in A \cup B$ if $x \in A$ or $x \in B$ (or both).

The **intersection** of two sets contains only the elements that are in both sets.
The intersection is notated $A \cap B$.
More formally, $x \in A \cap B$ if $x \in A$ and $x \in B$.

The **complement** of a set A contains everything that is *not* in the set A.
The complement is notated A', or A^c, or sometimes $\sim A$.

Example 5

Consider the sets: $A = \{$red, green, blue$\}$ $B = \{$red, yellow, orange$\}$
$C = \{$red, orange, yellow, green, blue, purple$\}$

a) Find $A \cup B$

The union contains all the elements in either set: $A \cup B = \{$red, green, blue, yellow, orange$\}$
Notice we only list red once.

b) Find $A \cap B$

The intersection contains all the elements in both sets: $A \cap B = \{$red$\}$

c) Find $A^c \cap C$

Here we're looking for all the elements that are *not* in set A and are also in C.
$A^c \cap C = \{$orange, yellow, purple$\}$

Try it Now 2
Using the sets from the previous example, find $A \cup C$ and $B^c \cap A$

Notice that in the example above, it would be hard to just ask for A^c, since everything from the color fuchsia to puppies and peanut butter are included in the complement of the set. For this reason, complements are usually only used with intersections, or when we have a universal set in place.

> **Universal Set**
> A **universal set** is a set that contains all the elements we are interested in. This would have to be defined by the context.
>
> A complement is relative to the universal set, so A^c contains all the elements in the universal set that are not in A.

Example 6
___

a) If we were discussing searching for books, the universal set might be all the books in the library.
b) If we were grouping your Facebook friends, the universal set would be all your Facebook friends.
c) If you were working with sets of numbers, the universal set might be all whole numbers, all integers, or all real numbers

Example 7
___

Suppose the universal set is U = all whole numbers from 1 to 9. If $A = \{1, 2, 4\}$, then

$A^c = \{3, 5, 6, 7, 8, 9\}$.

As we saw earlier with the expression $A^c \cap C$, set operations can be grouped together. Grouping symbols can be used like they are with arithmetic – to force an order of operations.

Example 8
___

Suppose $H = \{$cat, dog, rabbit, mouse$\}$, $F = \{$dog, cow, duck, pig, rabbit$\}$
$W = \{$duck, rabbit, deer, frog, mouse$\}$

a) Find $(H \cap F) \cup W$

We start with the intersection: $H \cap F = \{$dog, rabbit$\}$
Now we union that result with W: $(H \cap F) \cup W = \{$dog, duck, rabbit, deer, frog, mouse$\}$

b) Find $H \cap (F \cup W)$

We start with the union: $F \cup W = \{$dog, cow, rabbit, duck, pig, deer, frog, mouse$\}$
Now we intersect that result with H: $H \cap (F \cup W) = \{$dog, rabbit, mouse$\}$

c) Find $(H \cap F)^c \cap W$

We start with the intersection: $H \cap F = \{$dog, rabbit$\}$
Now we want to find the elements of W that are *not* in $H \cap F$
$(H \cap F)^c \cap W = \{$duck, deer, frog, mouse$\}$

3.3 Venn Diagrams

To visualize the interaction of sets, John Venn in 1880 thought to use overlapping circles, building on a similar idea used by Leonhard Euler in the 18th century. These illustrations now called **Venn Diagrams**.

Venn Diagram A Venn diagram represents each set by a circle, usually drawn inside of a containing box representing the universal set. Overlapping areas indicate elements common to both sets.

Basic Venn diagrams can illustrate the interaction of two or three sets.

Example 9

Create Venn diagrams to illustrate $A \cup B$, $A \cap B$, and $A^c \cap B$.

$A \cup B$ contains all elements in *either* set.

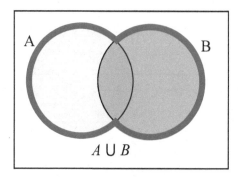

$A \cap B$ contains only those elements in both sets – in the overlap of the circles.

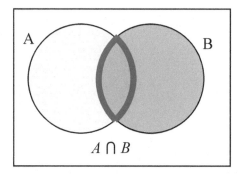

A^c will contain all elements *not* in the set A. $A^c \cap B$ will contain the elements in set B that are not in set A.

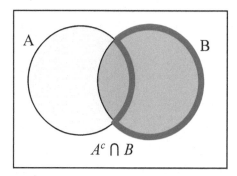

Example 10

Use a Venn diagram to illustrate $(H \cap F)^c \cap W$.

We'll start by identifying everything in the set $H \cap F$:

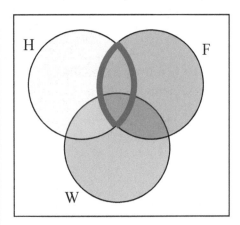

Now, $(H \cap F)^c \cap W$ will contain everything *not* in the set identified above that is also in set W.

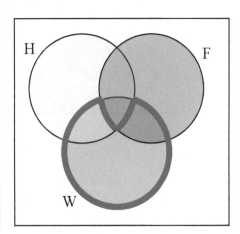

Example 11

Create an expression to represent the outlined part of the Venn diagram shown.

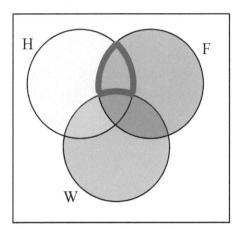

The elements in the outlined set *are* in sets H and F, but are not in set W. So, we could represent this set as $H \cap F \cap W^c$.

Try it Now 3

Create an expression to represent the outlined portion of the Venn diagram shown

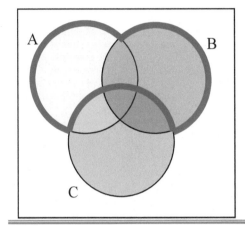

3.4 Cardinality

Often times we are interested in the number of items in a set or subset. This is called the cardinality of the set.

> **Cardinality**
> The number of elements in a set is the cardinality of that set.
>
> The cardinality of the set A is often notated as $|A|$ or n(A).

Example 12

Let $A = \{1, 2, 3, 4, 5, 6\}$ and $B = \{2, 4, 6, 8\}$. What is the cardinality of B? $A \cup B$, $A \cap B$?

The cardinality of B is 4, since there are 4 elements in the set.
The cardinality of $A \cup B$ is 7, since $A \cup B = \{1, 2, 3, 4, 5, 6, 8\}$, which contains 7 elements.
The cardinality of $A \cap B$ is 3, since $A \cap B = \{2, 4, 6\}$, which contains 3 elements.

Example 13

What is the cardinality of P = the set of English names for the months of the year?

The cardinality of this set is 12, since there are 12 months in the year.

Sometimes we may be interested in the cardinality of the union or intersection of sets, but not know the actual elements of each set. This is common in surveying.

Example 14

A survey asks 200 people "What beverage do you drink in the morning", and offers choices
- Tea only
- Coffee only
- Both coffee and tea

Suppose 20 report tea only, 80 report coffee only, 40 report both. How many people drink tea in the morning? How many people drink neither tea or coffee?

This question can most easily be answered by creating a Venn diagram. We can see that we can find the people who drink tea by adding those who drink only tea to those who drink both: 60 people.

We can also see that those who drink neither are those not contained in the any of the three other groupings, so we can count those by subtracting from the cardinality of the universal set, 200.
$200 - 20 - 80 - 40 = 60$ people who drink neither.

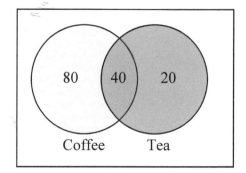

Example 15

A survey asks: Which online services have you used in the last month?
- Twitter
- Facebook
- Have used both

The results show 40% of those surveyed have used Twitter, 70% have used Facebook, and 20% have used both. How many people have used neither Twitter or Facebook?

Let T be the set of all people who have used Twitter, and F be the set of all people who have used Facebook. Notice that while the cardinality of F is 70% and the cardinality of T is 40%, the cardinality of $F \cup T$ is not simply 70% + 40%, since that would count those who use both services twice. To find the cardinality of $F \cup T$, we can add the cardinality of F and the cardinality of T, then subtract those in intersection that we've counted twice. In symbols,
$$n(F \cup T) = n(F) + n(T) - n(F \cap T)$$
$$n(F \cup T) = 70\% + 40\% - 20\% = 90\%$$

Now, to find how many people have not used either service, we're looking for the cardinality of $(F \cup T)^c$. Since the universal set contains 100% of people and the cardinality of $F \cup T = 90\%$, the cardinality of $(F \cup T)^c$ must be the other 10%.

The previous example illustrated two important properties.

Cardinality Properties
$n(A \cup B) = n(A) + n(B) - n(A \cap B)$
$n(A^c) = n(U) - n(A)$

Notice that the first property can also be written in an equivalent form by solving for the cardinality of the intersection:

$$n(A \cap B) = n(A) + n(B) - n(A \cup B)$$

Example 16

Fifty students were surveyed, and asked if they were taking a social science (SS), humanities (HM) or a natural science (NS) course the next quarter.

21 were taking a SS course	26 were taking a HM course
19 were taking a NS course	9 were taking SS and HM
7 were taking SS and NS	10 were taking HM and NS
3 were taking all three	7 were taking none

How many students are only taking a SS course?

It might help to look at a Venn diagram.
From the given data, we know that there are
3 students in region e and
7 students in region h.

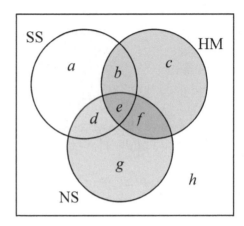

Since 7 students were taking a SS and NS course, we
know that n(d) + n(e) = 7. Since we know there are 3
students in region 3, there must be
7 – 3 = 4 students in region d.

Similarly, since there are 10 students taking HM and
NS, which includes regions e and f, there must be
10 – 3 = 7 students in region f.

Since 9 students were taking SS and HM, there must be 9 – 3 = 6 students in region b.

Now, we know that 21 students were taking a SS course. This includes students from regions
a, b, d, and e. Since we know the number of students in all but region a, we can determine
that 21 – 6 – 4 – 3 = 8 students are in region a.

8 students are taking only a SS course.

Try it Now 4
One hundred fifty people were surveyed and asked if they believed in UFOs, ghosts, and
Bigfoot.

      43 believed in UFOs   44 believed in ghosts
      25 believed in Bigfoot   10 believed in UFOs and ghosts
      8 believed in ghosts and Bigfoot   5 believed in UFOs and Bigfoot
      2 believed in all three

How many people surveyed believed in at least one of these things?

Try it Now Answers
1. There are several answers: The set of all odd numbers less than 10. The set of all odd
numbers. The set of all integers. The set of all real numbers.

2. $A \cup C$ = {red, orange, yellow, green, blue purple}
$B^c \cap A$ = {green, blue}

3. $A \cup B \cap C^c$

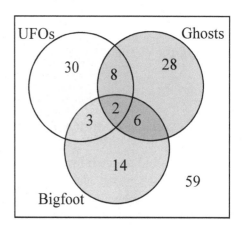

4. Starting with the intersection of all three circles, we work our way out. Since 10 people believe in UFOs and Ghosts, and 2 believe in all three, that leaves 8 that believe in only UFOs and Ghosts. We work our way out, filling in all the regions. Once we have, we can add up all those regions, getting 91 people in the union of all three sets. This leaves 150 – 91 = 59 who believe in none.

Exercises

1. List out the elements of the set "The letters of the word Mississippi."

2. List out the elements of the set "Months of the year."

3. Write a verbal description of the set {3, 6, 9}.

4. Write a verbal description of the set {a, e, i, o, u}.

5. Is {1, 3, 5} a subset of the set of odd integers?

6. Is {A, B, C} a subset of the set of letters of the alphabet?

For problems 7-12, consider the sets below, and indicate if each statement is true or false.

A = {1, 2, 3, 4, 5}   B = {1, 3, 5}   C = {4, 6}   U = {numbers from 0 to 10}

7. $3 \in B$ 8. $5 \in C$ 9. $B \subset A$ 10. $C \subset A$ 11. $C \subset B$ 12. $C \subset D$

Using the sets from above, and treating U as the Universal set, find each of the following:

13. $A \cup B$ 14. $A \cup C$ 15. $A \cap C$ 16. $B \cap C$ 17. A^c 18. B^c

Let D = {b, a, c, k},   E = {t, a, s, k},   F = {b, a, t, h}. Using these sets, find the following:

19. $D^c \cap E$ 20. $F^c \cap D$ 21. $(D \cap E) \cup F$ 22. $D \cap (E \cup F)$

23. $(F \cap E)^c \cap D$ 24. $(D \cup E)^c \cap F$

Create a Venn diagram to illustrate each of the following:

25. $(F \cap E) \cup D$ 26. $(D \cup E)^c \cap F$

27. $(F^c \cap E^c) \cap D$ 28. $(D \cup E) \cup F$

Write an expression for the shaded region.

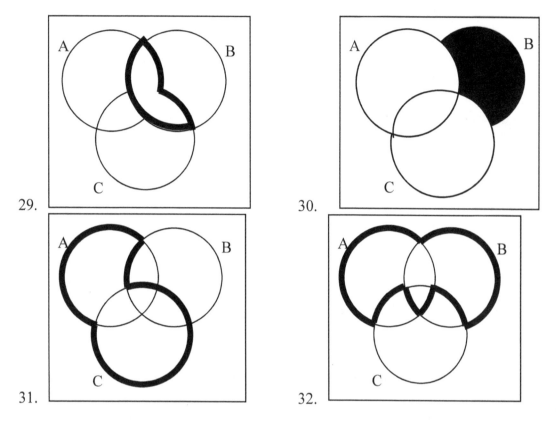

29.

30.

31.

32.

Let A = {1, 2, 3, 4, 5}   B = {1, 3, 5}   C = {4, 6}. Find the cardinality of the given set.

33. n(*A*)   34. n(*B*)   35. n(*A* ∪ *C*)   36. n(*A* ∩ *C*)

The Venn diagram here shows the cardinality of each set. Use this in 37-40 to find the cardinality of given set.

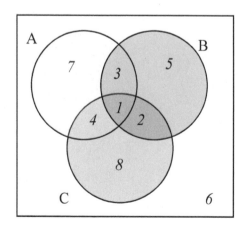

37. n(*A* ∩ *C*) 38. n(*B* ∪ *C*) 39. n(*A* ∩ *B* ∩ *C*^c)   40. n(*A* ∩ *B*^c ∩ *C*)

41. If n(G) = 20, n(H) = 30, n(*G* ∩ *H*) = 5, find n(*G* ∪ *H*)

42. If n(G) = 5, n(H) = 8, n(*G* ∩ *H*) = 4, find n(*G* ∪ *H*)

43. A survey was given asking whether they watch movies at home from Netflix, Redbox, or a video store. Use the results to determine how many people use Redbox.

     52 only use Netflix   62 only use Redbox

     24 only use a video store   16 use only a video store and Redbox

     48 use only Netflix and Redbox   30 use only a video store and Netflix

     10 use all three   25 use none of these

44. A survey asked buyers whether color, size, or brand influenced their choice of cell phone. The results are below. How many people were influenced by brand?

     5 only said color   8 only said size

     16 only said brand   20 said only color and size

     42 said only color and brand 53 said only size and brand

     102 said all three   20 said none of these

45. Use the given information to complete a Venn diagram, then determine: a) how many students have seen exactly one of these movies, and b) how many had seen only *Star Wars*.

     18 had seen *The Matrix*   (M)   24 had seen *Star Wars (SW)*

     20 had seen *Lord of the Rings (LotR)*   10 had seen *M* and *SW*

     14 had seen *LotR* and *SW*   12 had seen *M* and *LotR*

     6 had seen all three

46. A survey asked people what alternative transportation modes they use. Using the data to complete a Venn diagram, then determine: a) what percent of people only ride the bus, and b) how many people don't use any alternate transportation.

     30% use the bus   20% ride a bicycle

     25% walk   5% use the bus and ride a bicycle

     10% ride a bicycle and walk   12% use the bus and walk

     2% use all three

Chapter 4: Logic

Logic is, basically, the study of valid reasoning. When searching the internet, we use Boolean logic – terms like "and" and "or" – to help us find specific web pages that fit in the sets we are interested in. After exploring this form of logic, we will look at logical arguments and how we can determine the validity of a claim.

Table of Contents

4.1   Boolean Logic

We can often classify items as belonging to sets. If you went the library to search for a book and they asked you to express your search using unions, intersections, and complements of sets, that would feel a little strange. Instead, we typically using words like "and", "or", and "not" to connect our keywords together to form a search. These words, which form the basis of **Boolean logic**, are directly related to our set operations.

> **Boolean Logic**
> Boolean logic combines multiple statements that are either true or false into an expression that is either true or false.
>
> In connection to sets, a search is true if the element is part of the set.

Suppose *M* is the set of all mystery books, and *C* is the set of all comedy books. If we search for "mystery", we are looking for all the books that are an element of the set *M*; the search is true for books that are in the set.

When we search for "mystery *and* comedy", we are looking for a book that is an element of both sets, in the intersection. If we were to search for "mystery *or* comedy", we are looking for a book that is a mystery, a comedy, or both, which is the union of the sets. If we searched for "*not* comedy", we are looking for any book in the library that is not a comedy, the complement of the set *C*.

> **Connection to Set Operations**
> A and B elements in the intersection $A \cap B$
> A or B elements in the union $A \cup B$
> Not A elements in the complement A^c

Notice here that *or* is not exclusive. This is a difference between the Boolean logic use of the word and common everyday use. When your significant other asks "do you want to go to the park or the movies?" they usually are proposing an exclusive choice – one option or the other, but not both. In Boolean logic, the *or* is not exclusive – more like being asked at a restaurant "would you like fries or a drink with that?" Answering "both, please" is an acceptable answer.

Example 1

Suppose we are searching a library database for Mexican universities. Express a reasonable search using Boolean logic.

We could start with the search "Mexico *and* university", but would be likely to find results for the U.S. state New Mexico. To account for this, we could revise our search to read:
Mexico *and* university *not* "New Mexico"

In most internet search engines, it is not necessary to include the word *and*; the search engine assumes that if you provide two keywords you are looking for both. In Google's search, the keyword *or* has be capitalized as OR, and a negative sign in front of a word is used to indicate *not*. Quotes around a phrase indicate that the entire phrase should be looked for. The search from the previous example on Google could be written:
Mexico university -"New Mexico"

Example 2

Describe the numbers that meet the condition:
        even and less than 10 and greater than 0

The numbers that satisfy all three requirements are {2, 4, 6, 8}

Sometimes statements made in English can be ambiguous. For this reason, Boolean logic uses parentheses to show precedent, just like in algebraic order of operations.

Example 3

The English phrase "Go to the store and buy me eggs and bagels or cereal" is ambiguous; it is not clear whether the requestors is asking for eggs always along with either bagels or cereal, or whether they're asking for either the combination of eggs and bagels, or just cereal.

For this reason, using parentheses clarifies the intent:
Eggs and (bagels or cereal) means   Option 1: Eggs and bagels, Option 2: Eggs and cereal
(Eggs and bagels) or cereal means   Option 1: Eggs and bagels, Option 2: Cereal

Example 4

Describe the numbers that meet the condition:

> odd number and less than 20 and greater than 0 and (multiple of 3 or multiple of 5)

The first three conditions limit us to the set {1, 3, 5, 7, 9, 11, 13, 15, 17, 19}

The last grouped conditions tell us to find elements of this set that are also either a multiple of 3 or a multiple of 5. This leaves us with the set {3, 5, 9, 15}

Notice that we would have gotten a very different result if we had written

> (odd number and less than 20 and greater than 0 and multiple of 3) or multiple of 5

The first grouped set of conditions would give {3, 9, 15}. When combined with the last condition, though, this set expands without limits:

{3, 5, 9, 15, 20, 25, 30, 35, 40, 45, ...}

Be aware that when a string of conditions is written without grouping symbols, it is often interpreted from the left to right, resulting in the latter interpretation.

4.2   Conditionals

Beyond searching, Boolean logic is commonly used in spreadsheet applications like Excel to do conditional calculations. A **statement** is something that is either true or false. A statement like 3 < 5 is true; a statement like "a rat is a fish" is false. A statement like "x < 5" is true for some values of *x* and false for others. When an action is taken or not depending on the value of a statement, it forms a **conditional**.

Statements and Conditionals
A **statement** is either true or false.
A **conditional** is a compound statement of the form
    "if *p* then *q*"   or   "if *p* then *q*, else *s*".

Example 5

In common language, an example of a conditional statement would be "If it is raining, then we'll go to the mall. Otherwise we'll go for a hike."

The statement "If it is raining" is the condition – this may be true or false for any given day. If the condition is true, then we will follow the first course of action, and go to the mall. If the condition is false, then we will use the alternative, and go for a hike.

Example 6

As mentioned earlier, conditional statements are commonly used in spreadsheet applications like Excel. If Excel, you can enter an expression like

=IF(A1<2000, A1+1, A1*2)

Notice that after the IF, there are three parts. The first part is the condition, and the second two are calculations. Excel will look at the value in cell A1 and compare it to 2000. If that condition is true, then the first calculation is used, and 1 is added to the value of A1 and the result is stored. If the condition is false, then the second calculation is used, and A1 is multiplied by 2 and the result is stored.

In other words, this statement is equivalent to saying "If the value of A1 is less than 2000, then add 1 to the value in A1. Otherwise, multiple A1 by 2"

Example 7

The expression =IF(A1 > 5, 2*A1, 3*A1) is used. Find the result if A1 is 3, and the result if A1 is 8.

This is equivalent to saying
If A1 >5, then calculate 2*A1. Otherwise, calculate 3*A1

If A1 is 3, then the condition is false, since 3 > 5 is not true, so we do the alternate action, and multiple by 3, giving 3*3 = 9

If A1 is 8, then the condition is true, since 8 > 5, so we multiply the value by 2, giving 2*8=16

Example 8

An accountant needs to withhold 15% of income for taxes if the income is below $30,000, and 20% of income if the income is $30,000 or more. Write an expression that would calculate the amount to withhold.

Our conditional needs to compare the value to 30,000. If the income is less than 30,000, we need to calculate 15% of the income: 0.15*income. If the income is more than 30,000, we need to calculate 20% of the income: 0.20*income.

In words, we could write "If income < 30,000, then multiply by 0.15, otherwise multiply by 0.20". In Excel, we would write:

=IF(A1<30000, 0.15*A1, 0.20*A1)

As we did earlier, we can create more complex conditions by using the operators *and*, *or*, and *not* to join simpler conditions together.

Example 9

A parent might say to their child "if you clean your room and take out the garbage, then you can have ice cream."

Here, there are two simpler conditions:
1) The child cleaning her room
2) The child taking out the garbage

Since these conditions were joined with *and*, then the combined conditional will only be true if both simpler conditions are true; if either chore is not completed then the parent's condition is not met.

Notice that if the parent had said "if you clean your room *or* take out the garbage, then you can have ice cream", then the child would only need to complete one chore to meet the condition.

To create the condition "A1 < 3000 and A1 > 100" in Excel, you would need to enter "AND(A1<3000, A1>100)". Likewise, for the condition "A1=4 or A1=6" you would enter "OR(A1=4, A1=6)."

Example 10

In a spreadsheet, cell A1 contains annual income, and A2 contains number of dependents. A certain tax credit applies if someone with no dependents earns less than $10,000 and has no dependents, or if someone with dependents earns less than $20,000. Write a rule that describes this.

There are two ways the rule is met:

income is less than 10,000 *and* dependents is 0, *or*
income is less than 20,000 *and* dependents is not 0.

Informally, we could write these as
(A1 < 10000 *and* A2 = 0) *or* (A1 < 20000 *and* A2 > 0)

Notice that the A2 > 0 condition is actually redundant and not necessary, since we'd only be considering that *or* case if the first pair of conditions were not met. So this could be simplified to
(A1 < 10000 *and* A2 = 0) *or* (A1 < 20000)

In Excel's format, we'd write
= IF (OR(AND(A1 < 10000, A2 = 0), A1 < 20000), "you qualify", "you don't qualify")

4.3   Truth Tables

Because complex Boolean statements can get tricky to think about, we can create a **truth table** to keep track of what truth values for the simple statements make the complex statement true and false

---

Truth table
A table showing what the resulting truth value of a complex statement is for all the possible truth values for the simple statements.

Example 11

Suppose you're picking out a new couch, and your significant other says "get a sectional *or* something with a chaise".

This is a complex statement made of two simpler conditions: "is a sectional", and "has a chaise". For simplicity, let's use S to designate "is a sectional", and C to designate "has a chaise". The condition S is true if the couch is a sectional.
A truth table for this would look like this:

S	C	S or C
T	T	T
T	F	T
F	T	T
F	F	F

In the table, T is used for true, and F for false. In the first row, if S is true and C is also true, then the complex statement "S or C" is true. This would be a sectional that also has a chaise, which meets our desire.

Remember also that *or* in logic is not exclusive; if the couch has both features, it does meet the condition.

To shorthand our notation further, we're going to introduce some symbols that are commonly used for *and*, *or*, and *not*.

Symbols
The symbol $\wedge$ is used for *and*: A and B is notated $A \wedge B$.
The symbol $\vee$ is used for *or*: A or B is notated $A \vee B$
The symbol $\sim$ is used for *not*:   not A is notated $\sim A$

You can remember the first two symbols by relating them to the shapes for the union and intersection. $A \wedge B$ would be the elements that exist in both sets, in $A \cap B$. Likewise, $A \vee B$ would be the elements that exist in either set, in $A \cup B$.

In the previous example, the truth table was really just summarizing what we already know about how the *or* statement work. The truth tables for the basic *and*, *or*, and *not* statements are shown below.

Basic truth tables

A	B	$A \wedge B$
T	T	T
T	F	F
F	T	F
F	F	F

A	B	$A \vee B$
T	T	T
T	F	T
F	T	T
F	F	F

A	$\sim A$
T	F
F	T

Truth tables really become useful when analyzing more complex Boolean statements.

Example 12
_____

Create a truth table for the statement $A \wedge \sim(B \vee C)$.

It helps to work from the inside out when creating truth tables, and create tables for intermediate operations. We start by listing all the possible truth value combinations for A, B, and C. Notice how the first column contains 4 Ts followed by 4 Fs, the second column contains 2 Ts, 2 Fs, then repeats, and the last column alternates. This pattern ensures that all combinations are considered. Along with those initial values, we'll list the truth values for the innermost expression, $B \vee C$.

A	B	C	$B \vee C$
T	T	T	T
T	T	F	T
T	F	T	T
T	F	F	F
F	T	T	T
F	T	F	T
F	F	T	T
F	F	F	F

Next, we can find the negation of $B \vee C$, working off the $B \vee C$ column we just created.

A	B	C	$B \vee C$	$\sim(B \vee C)$
T	T	T	T	F
T	T	F	T	F
T	F	T	T	F
T	F	F	F	T
F	T	T	T	F
F	T	F	T	F
F	F	T	T	F
F	F	F	F	T

Finally, we find the values of *A and* $\sim(B \vee C)$

A	B	C	$B \lor C$	$\sim(B \lor C)$	$A \land \sim(B \lor C)$
T	T	T	T	F	F
T	T	F	T	F	F
T	F	T	T	F	F
T	F	F	F	T	T
F	T	T	T	F	F
F	T	F	T	F	F
F	F	T	T	F	F
F	F	F	F	T	F

It turns out that this complex expression is only true in one case: if A is true, B is false, and C is false.

Try it Now 1
Create a truth table for this statement: $(\sim A \land B) \lor \sim B$

When we discussed conditions earlier, we discussed the type where we take an action based on the value of the condition. We are now going to talk about a more general version of a conditional, sometimes called an **implication**.

> **Implications**
> Implications are logical conditional sentences stating that a statement p, called the antecedent, implies a consequence q.
>
> Implications are commonly written as $p \rightarrow q$

Implications are similar to the conditional statements we looked at earlier; $p \rightarrow q$ is typically written as "if p then q", or "p therefore q". The difference between implications and conditionals is that conditionals we discussed earlier suggest an action – if the condition is true, then we take some action as a result. Implications are a logical statement that suggest that the consequence must logically follow if the antecedent is true.

Example 13

The English statement "If it is raining, then there are clouds is the sky" is a logical implication. It is a valid argument because if the antecedent "it is raining" is true, then the consequence "there are clouds in the sky" must also be true.

Notice that the statement tells us nothing of what to expect if it is not raining. If the antecedent is false, then the implication becomes irrelevant.

Example 14

A friend tells you that "if you upload that picture to Facebook, you'll lose your job". There are four possible outcomes:

1) You upload the picture and keep your job

2) You upload the picture and lose your job
3) You don't upload the picture and keep your job
4) You don't upload the picture and lose your job

There is only one possible case where your friend was lying – the first option where you upload the picture and keep your job. In the last two cases, your friend didn't say anything about what would happen if you didn't upload the picture, so you can't conclude their statement is invalid, even if you didn't upload the picture and still lost your job.

In traditional logic, an implication is considered valid (true) as long as there are no cases in which the antecedent is true and the consequence is false. It is important to keep in mind that symbolic logic cannot capture all the intricacies of the English language.

Truth values for implications

p	q	$p \to q$
T	T	T
T	F	F
F	T	T
F	F	T

Example 15

Construct a truth table for the statement $(m \land \sim p) \to r$

We start by constructing a truth table for the antecedent.

m	p	$\sim p$	$m \land \sim p$
T	T	F	F
T	F	T	T
F	T	F	F
F	F	T	F

Now we can build the truth table for the implication

m	p	$\sim p$	$m \land \sim p$	r	$(m \land \sim p) \to r$
T	T	F	F	T	T
T	F	T	T	T	T
F	T	F	F	T	T
F	F	T	F	T	T
T	T	F	F	F	T
T	F	T	T	F	F
F	T	F	F	F	T
F	F	T	F	F	T

In this case, when m is true, p is false, and r is false, then the antecedent $m \wedge \sim p$ will be true but the consequence false, resulting in an invalid implication; every other case gives a valid implication.

For any implication, there are three related statements, the **converse**, the **inverse**, and the **contrapositive**.

Related Statements		
The original implication is	"if p then q"	$p \rightarrow q$
The converse is:	"if q then p"	$q \rightarrow p$
The inverse is	"if not p then not q"	$\sim p \rightarrow \sim q$
The contrapositive is	"if not q then not p"	$\sim q \rightarrow \sim p$

Example 16

Consider again the valid implication "If it is raining, then there are clouds in the sky".

The converse would be "If there are clouds in the sky, it is raining." This is certainly not always true.

The inverse would be "If it is not raining, then there are not clouds in the sky." Likewise, this is not always true.

The contrapositive would be "If there are not clouds in the sky, then it is not raining." This statement is valid, and is equivalent to the original implication.

Looking at truth tables, we can see that the original conditional and the contrapositive are logically equivalent, and that the converse and inverse are logically equivalent.

p	q	Implication $p \rightarrow q$	Converse $q \rightarrow p$	Inverse $\sim p \rightarrow \sim q$	Contrapositive $\sim q \rightarrow \sim p$
T	T	T	T	T	T
T	F	F	T	T	F
F	T	T	F	F	T
F	F	T	T	T	T

Equivalent

Equivalence
A conditional statement and its contrapositive are logically equivalent.
The converse and inverse of a statement are logically equivalent.

4.4   Arguments

A logical argument is a claim that a set of premises support a conclusion. There are two general types of arguments: inductive and deductive arguments.

Argument types
An **inductive** argument uses a collection of specific examples as its premises and uses them to propose a general conclusion.

A **deductive** argument uses a collection of general statements as its premises and uses them to propose a specific situation as the conclusion.

Example 17

The argument "when I went to the store last week I forgot my purse, and when I went today I forgot my purse. I always forget my purse when I go the store" is an inductive argument.

The premises are:
I forgot my purse last week
I forgot my purse today

The conclusion is:
I always forget my purse

Notice that the premises are specific situations, while the conclusion is a general statement. In this case, this is a fairly weak argument, since it is based on only two instances.

Example 18

The argument "every day for the past year, a plane flies over my house at 2 p.m. A plane will fly over my house every day at 2 p.m." is a stronger inductive argument, since it is based on a larger set of evidence.

Evaluating inductive arguments
An inductive argument is never able to prove the conclusion true, but it can provide either weak or strong evidence to suggest it may be true.

Many scientific theories, such as the big bang theory, can never be proven. Instead, they are inductive arguments supported by a wide variety of evidence. Usually in science, an idea is considered a hypothesis until it has been well tested, at which point it graduates to being considered a theory. The commonly known scientific theories, like Newton's theory of gravity, have all stood up to years of testing and evidence, though sometimes they need to be adjusted based on new evidence. For gravity, this happened when Einstein proposed the theory of general relativity.

A deductive argument is more clearly valid or not, which makes them easier to evaluate.

> **Evaluating deductive arguments**
> A deductive argument is considered valid if all the premises are true, and the conclusion follows logically from those premises. In other words, the premises are true, and the conclusion follows necessarily from those premises.

Example 19

The argument "All cats are mammals and a tiger is a cat, so a tiger is a mammal" is a valid deductive argument.

The premises are:
All cats are mammals
A tiger is a cat

The conclusion is:
A tiger is a mammal

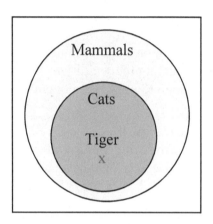

Both the premises are true. To see that the premises must logically lead to the conclusion, one approach would be use a Venn diagram. From the first premise, we can conclude that the set of cats is a subset of the set of mammals. From the second premise, we are told that a tiger lies within the set of cats. From that, we can see in the Venn diagram that the tiger also lies inside the set of mammals, so the conclusion is valid.

> **Analyzing arguments with Venn diagrams[1]**
> To analyze an argument with a Venn diagram
> 1) Draw a Venn diagram based on the premises of the argument
> 2) If the premises are insufficient to determine what determine the location of an element, indicate that.
> 3) The argument is valid if it is clear that the conclusion must be true

Example 20

---

[1] Technically, these are Euler circles or Euler diagrams, not Venn diagrams, but for the sake of simplicity we'll continue to call them Venn diagrams.

Premise: All firefighters know CPR
Premise: Jill knows CPR
Conclusion: Jill is a firefighter

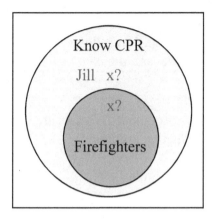

From the first premise, we know that firefighters all lie inside the set of those who know CPR. From the second premise, we know that Jill is a member of that larger set, but we do not have enough information to know if she also is a member of the smaller subset that is firefighters.

Since the conclusion does not necessarily follow from the premises, this is an invalid argument, regardless of whether Jill actually is a firefighter.

It is important to note that whether or not Jill is actually a firefighter is not important in evaluating the validity of the argument; we are only concerned with whether the premises are enough to prove the conclusion.

Try it Now 2
Determine the validity of this argument:
Premise: No cows are purple
Premise: Fido is not a cow
Conclusion: Fido is purple

In addition to these categorical style premises of the form "all ___", "some ____", and "no ____", it is also common to see premises that are implications.

Example 21

Premise: If you live in Seattle, you live in Washington.
Premise: Marcus does not live in Seattle
Conclusion: Marcus does not live in Washington

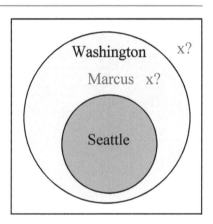

From the first premise, we know that the set of people who live in Seattle is inside the set of those who live in Washington. From the second premise, we know that Marcus does not lie in the Seattle set, but we have insufficient information to know whether or not Marcus lives in Washington or not. This is an invalid argument.

Example 22

Consider the argument "You are a married man, so you must have a wife."

This is an invalid argument, since there are, at least in parts of the world, men who are married to other men, so the premise not insufficient to imply the conclusion.

Some arguments are better analyzed using truth tables.

Example 23

Consider the argument
Premise: If you bought bread, then you went to the store
Premise: You bought bread
Conclusion: You went to the store

While this example is hopefully fairly obviously a valid argument, we can analyze it using a truth table by representing each of the premises symbolically. We can then look at the implication that the premises together imply the conclusion. If the truth table is a tautology (always true), then the argument is valid.

We'll get B represent "you bought bread" and S represent "you went to the store". Then the argument becomes:
Premise: $B \rightarrow S$
Premise: B
Conclusion: S

To test the validity, we look at whether the combination of both premises implies the conclusion; is it true that $[(B \rightarrow S) \wedge B] \rightarrow S$?

B	S	$B \rightarrow S$	$(B \rightarrow S) \wedge B$	$[(B \rightarrow S) \wedge B] \rightarrow S$
T	T	T	T	T
T	F	F	F	T
F	T	T	F	T
F	F	T	F	T

Since the truth table for $[(B \rightarrow S) \wedge B] \rightarrow S$ is always true, this is a valid argument.

Try it Now 3
Determine if the argument is valid:
Premise: If I have a shovel I can dig a hole.
Premise: I dug a hole
Conclusion: Therefore, I had a shovel

Analyzing arguments using truth tables
To analyze an argument with a truth table:
1. Represent each of the premises symbolically
2. Create a conditional statement, joining all the premises with and to form the antecedent, and using the conclusion as the consequent.
3. Create a truth table for that statement. If it is always true, then the argument is valid.

Example 24

Premise: If I go to the mall, then I'll buy new jeans
Premise: If I buy new jeans, I'll buy a shirt to go with it
Conclusion: If I got to the mall, I'll buy a shirt.

Let M = I go to the mall, J = I buy jeans, and S = I buy a shirt.
The premises and conclusion can be stated as:
Premise: $M \rightarrow J$
Premise: $J \rightarrow S$
Conclusion: $M \rightarrow S$

We can construct a truth table for $[(M{\rightarrow}J) \wedge (J{\rightarrow}S)] \rightarrow (M{\rightarrow}S)$

M	J	S	$M{\rightarrow}J$	$J{\rightarrow}S$	$(M{\rightarrow}J) \wedge (J{\rightarrow}S)$	$M{\rightarrow}S$	$[(M{\rightarrow}J) \wedge (J{\rightarrow}S)] \rightarrow (M{\rightarrow}S)$
T	T	T	T	T	T	T	T
T	T	F	T	F	F	F	T
T	F	T	F	T	F	T	T
T	F	F	F	T	F	F	T
F	T	T	T	T	T	T	T
F	T	F	T	F	F	T	T
F	F	T	T	T	T	T	T
F	F	F	T	T	T	T	T

From the truth table, we can see this is a valid argument.

The previous problem is an example of a syllogism.

Syllogism
A syllogism is an implication derived from two others, where the consequence of one is the antecedent to the other.   The general form of a syllogism is:
Premise: $p \rightarrow q$
Premise: $q \rightarrow r$
Conclusion: $p \rightarrow r$

This is sometime called the transitive property for implication.

Example 25

Premise: If I work hard, I'll get a raise.
Premise: If I get a raise, I'll buy a boat.
Conclusion: If I don't buy a boat, I must not have worked hard.

If we let W = working hard, R = getting a raise, and B = buying a boat, then we can represent our argument symbolically:
Premise $H \rightarrow R$
Premise $R \rightarrow B$
Conclusion: $\sim B \rightarrow \sim H$

We could construct a truth table for this argument, but instead, we will use the notation of the contrapositive we learned earlier to note that the implication $\sim B \rightarrow \sim H$ is equivalent to the implication $H \rightarrow B$.   Rewritten, we can see that this conclusion is indeed a logical syllogism derived from the premises.

Try it Now 4

Is this argument valid?

Premise: If I go to the party, I'll be really tired tomorrow.
Premise: If I go to the party, I'll get to see friends.
Conclusion: If I don't see friends, I won't be tired tomorrow.

Lewis Carroll, author of *Alice in Wonderland*, was a math and logic teacher, and wrote two books on logic. In them, he would propose premises as a puzzle, to be connected using syllogisms.

Example 26

Solve the puzzle. In other words, find a logical conclusion from these premises.
All babies are illogical.
Nobody is despised who can manage a crocodile.
Illogical persons are despised.

Let B = is a baby, D = is despised, I = is illogical, and M = can manage a crocodile.
Then we can write the premises as:
$B \rightarrow I$
$M \rightarrow \sim D$
$I \rightarrow D$

From the first and third premises, we can conclude that $B \rightarrow D$; that babies are despised. Using the contrapositive of the second premised, $D \rightarrow \sim M$, we can conclude that $B \rightarrow \sim M$; that babies cannot manage crocodiles.

While silly, this is a logical conclusion from the given premises.

4.5   Logical Fallacies in Common Language

In the previous discussion, we saw that logical arguments can be invalid when the premises are not true, when the premises are not sufficient to guarantee the conclusion, or when there are invalid chains in logic. There are a number of other ways in which arguments can be invalid, a sampling of which are given here.

> **Ad hominem**
> An ad hominem argument attacks the person making the argument, ignoring the argument itself.

Example 27

"Jane says that whales aren't fish, but she's only in the second grade, so she can't be right."

Here the argument is attacking Jane, not the validity of her claim, so this is an ad hominem argument.

Example 28

"Jane says that whales aren't fish, but everyone knows that they're really mammals – she's so stupid."

This certainly isn't very nice, but it is *not* ad hominem since a valid counterargument is made along with the personal insult.

> **Appeal to ignorance**
> This type of argument assumes something it true because it hasn't been proven false.

Example 29

"Nobody has proven that photo isn't Bigfoot, so it must be Bigfoot."

> **Appeal to authority**
> These arguments attempt to use the authority of a person to prove a claim. While often authority can provide strength to an argument, problems can occur when the person's opinion is not shared by other experts, or when the authority is irrelevant to the claim.

Example 30

"A diet high in bacon can be healthy – Doctor Atkins said so."

Here, an appeal to the authority of a doctor is used for the argument. This generally would provide strength to the argument, except that the opinion that eating a diet high in saturated fat runs counter to general medical opinion. More supporting evidence would be needed to justify this claim.

Example 31

"Jennifer Hudson lost weight with Weight Watchers, so their program must work."

Here, there is an appeal to the authority of a celebrity. While her experience does provide evidence, it provides no more than any other person's experience would.

Appeal to consequence
An appeal to consequence concludes that a premise is true or false based on whether the consequences are desirable or not.

Example 32

"Humans will travel faster than light: faster-than-light travel would be beneficial for space travel."

False dilemma
A false dilemma argument falsely frames an argument as an "either or" choice, without allowing for additional options.

Example 33

"Either those lights in the sky were an airplane or aliens. There are no airplanes scheduled for tonight, so it must be aliens."

This argument ignores the possibility that the lights could be something other than an airplane or aliens.

Circular reasoning
Circular reasoning is an argument that relies on the conclusion being true for the premise to be true.

Example 34

"I shouldn't have gotten a C in that class; I'm an A student!"

In this argument, the student is claiming that because they're an A student, though shouldn't have gotten a C. But because they got a C, they're not an A student.

Straw man
A straw man argument involves misrepresenting the argument in a less favorable way to make it easier to attack.

Example 35

"Senator Jones has proposed reducing military funding by 10%. Apparently, he wants to leave us defenseless against attacks by terrorists"

Here the arguer has represented a 10% funding cut as equivalent to leaving us defenseless, making it easier to attack.

Post hoc (post hoc ergo propter hoc)
A post hoc argument claims that because two things happened sequentially, then the first must have caused the second.

Example 36

"Today I wore a red shirt, and my football team won! I need to wear a red shirt every time they play to make sure they keep winning."

> **Correlation implies causation**
> Similar to post hoc, but without the requirement of sequence, this fallacy assumes that just because two things are related one must have caused the other. Often there is a third variable not considered.

Example 37

"Months with high ice cream sales also have a high rate of deaths by drowning. Therefore, ice cream must be causing people to drown."

This argument is implying a causal relation, when really both are more likely dependent on the weather; that ice cream and drowning are both more likely during warm summer months.

Try it Now 5
Identify the logical fallacy in each of the arguments

a. Only an untrustworthy person would run for office. The fact that politicians are untrustworthy is proof of this.
b. Since the 1950s, both the atmospheric carbon dioxide level and obesity levels have increased sharply. Hence, atmospheric carbon dioxide causes obesity.
c. The oven was working fine until you started using it, so you must have broken it.
d. You can't give me a D in the class – I can't afford to retake it.
e. The senator wants to increase support for food stamps. He wants to take the taxpayers' hard-earned money and give it away to lazy people. This isn't fair so we shouldn't do it.

Try it Now Answers
1.

A	*B*	~*A*	~*A* ∧ *B*	~*B*	(~*A* ∧ *B*) ∨ ~*B*
T	T	F	F	F	F
T	F	F	F	T	T
F	T	T	T	F	T
F	F	T	F	T	T

2. Since no cows are purple, we know there is no overlap between the set of cows and the set of purple things. We know Fido is not in the cow set, but that is not enough to conclude that Fido is in the purple things set.

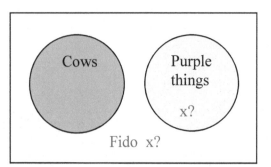

3. Let S: have a shovel, D: dig a hole

The first premise is equivalent to S→D. The second premise is D. The conclusion is S. We are testing [(S→D) ∧ D] → S

S	*D*	*S→D*	*(S→D) ∧ D*	*[(S→D) ∧ D] → S*
T	T	T	T	T
T	F	F	F	T
F	T	T	T	F
F	F	T	F	T

This is not a tautology, so this is an invalid argument.

4. Letting P = go to the party, T = being tired, and F = seeing friends, then we can represent this argument as P:

Premise: P → T
Premise: P → F
Conclusion: ~F → ~T

We could rewrite the second premise using the contrapositive to state ~F → ~P, but that does not allow us to form a syllogism. If we don't see friends, then we didn't go the party, but that is not sufficient to claim I won't be tired tomorrow. Maybe I stayed up all night watching movies.

5.
a. Circular
b. Correlation does not imply causation
c. Post hoc
d. Appeal to consequence
e. Straw man

Exercises

1. Consider the set of whole numbers from 1 to 10, inclusive. List the numbers that meet the condition: *less than seven or odd*

2. An accountant needs to withhold 13% of income for taxes if the income is below $40,000, and 20% of income if the income is $40,000 or more. The income amount is in cell **A1**. Write a spreadsheet expression that would calculate the amount to withhold.

3. Complete the truth table for the statement $\sim P \wedge \sim S$.

4. Complete the truth table for the statement $\sim Q \to S$.

5. For the statement $Q \to R$, identify the inverse, converse, contrapositive, and original statement.

    _____ $\sim Q \to \sim R$ A. Inverse
    _____ $\sim R \to \sim Q$ B. Contrapositive
    _____ $R \to Q$ C. Converse
    _____ $Q \to R$ D. Statement

6. Complete the truth table for the statement (A $\vee$ B) $\wedge$ C.

7. Complete the truth table for the statement $\sim$(A $\vee$ B) $\wedge$ C.

8. Complete the truth table for the implication $\sim$(A $\wedge$ B) $\to$ C.

9. Determine if the conclusion follows logically from the premises. Is this a valid or invalid argument?

    Premise: No loving persons are thoughtless persons.
    Premise: No loving persons are aggressive people.
    Conclusion: No aggressive people are thoughtless persons.

10. Determine if the conclusion follows logically from the premises. Is this a valid or invalid argument?

    Premise: All sensations are ideas of external material things.
    Premise: Some ideas are not sensations.
    Conclusion: Some ideas are not ideas of external material things.

11. Determine if the conclusion follows logically from the premises. Is this a valid or invalid argument?

Premise: All complex ideas are works of the mind.
Premise: All relations are works of the mind.
Conclusion: All relations are complex ideas.

12. Determine if the conclusion follows logically from the premises. Is this a valid or invalid argument?

   Premise: If an animal is a rodent, then it is a mammal.
   Premise: Rats are rodents.
   Conclusion: Rats are mammals.

13. Determine if the conclusion follows logically from the premises. Is this a valid or invalid argument?

   Premise: If it has an engine, I can fix it.
   Premise: Cars have engines.
   Conclusion: I can fix cars.

14. Determine if the conclusion follows logically from the premises. Is this a valid or invalid argument?

   Premise: If I upgrade my computer, then it will run faster.
   Premise: If my computer runs after, I will be more productive.
   Conclusion: If I upgrade my computer, then I will be more productive.

15. Categorize the following logical fallacy.

   "It was his fault, Officer. You can tell by the kind of car I'm driving and by my clothes that I am a good citizen and would not lie. Look at the rattletrap he is driving, and look at how he is dressed. You can't believe anything that a dirty, long-haired hippie like that might tell you. Search his car; he probably has pot in it."

16. Categorize the following logical fallacy.

   My client is an integral part of this community. If he is sent to prison not only will this city suffer but also, he will be most missed by his family. You surely cannot find it in your hearts to reach any other verdict than "not guilty."

17. Categorize the following logical fallacy.

   "Old man Brown claims that he saw a flying saucer in his farm, but he never got beyond the fourth grade in school and can hardly read or write. He is completely ignorant of what scientists have written on the subject, so his report cannot possibly be true."

Chapter 5: Measurement

Table of Contents

In this chapter, we discuss measurement. We've seen some measurement and geometry in Chapter 1 with problem solving, but, in this chapter, we discuss the concepts in more depth.

5.1 U.S. Customary Measurement System

Learning Objective(s)
1 Define units of length and convert from one to another.
2 Perform arithmetic calculations on units of length.
3 Solve application problems involving units of length.

Introduction

Measurement is a number that describes the size or amount of something. You can measure many things like length, area, capacity, weight, temperature and time. In the United States, two main systems of measurement are used: the **metric system** and the **U.S. customary measurement system**. This topic addresses the measurement of length using the U.S. customary measurement system.

Suppose you want to purchase tubing for a project, and you see two signs in a hardware store: *$1.88 for 2 feet* of tubing and *$5.49 for 3 yards* of tubing. If both types of tubing will work equally well for your project, which is the better price? You need to know about two **units of measurement**, yards and feet, in order to determine the answer.

Units of Length

Objective 1

Length is the distance from one end of an object to the other end, or from one object to another. For example, the length of a letter-sized piece of paper is 11 inches. The system for measuring length in the United States is based on the four customary units of length: **inch**, **foot**, **yard**, and **mile**. Below are examples to show measurement in each of these units.

Unit	Description	Image
Inch/Inches	Some people donate their hair to be made into wigs for cancer patients who have lost hair as a result of treatment. One company requires hair donations to be at least 8 inches long.	
	Frame size of a bike: the distance from the center of the crank to the top of the seat tube. Frame size is usually measured in inches. This frame is 16 inches.	
Foot/Feet	Rugs are typically sold in standard lengths. One typical size is a rug that is 8 feet wide and 11 feet long. This is often described as an 8 by 11 rug.	
Yard/Yards	Soccer fields vary some in their size. An official field can be any length between 100 and 130 yards.	

Mile/Miles	A marathon is 26.2 miles long. One marathon route is shown in the map to the right.	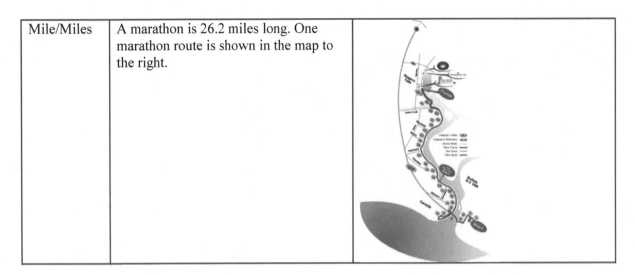

You can use any of these four U.S. customary measurement units to describe the length of something, but it makes more sense to use certain units for certain purposes. For example, it makes more sense to describe the length of a rug in feet rather than miles, and to describe a marathon in miles rather than inches.

You may need to convert between units of measurement. For example, you might want to express your height using feet and inches (5 feet 4 inches) or using only inches (64 inches). You need to know the unit equivalents in order to make these conversions between units.

The table below shows equivalents and conversion factors for the four customary units of measurement of length.

Unit Equivalents	Conversion Factors (longer to shorter units of measurement)	Conversion Factors (shorter to longer units of measurement)
1 foot = 12 inches	$\dfrac{12 \text{ inches}}{1 \text{ foot}}$	$\dfrac{1 \text{ foot}}{12 \text{ inches}}$
1 yard = 3 feet	$\dfrac{3 \text{ feet}}{1 \text{ yard}}$	$\dfrac{1 \text{ yard}}{3 \text{ feet}}$
1 mile = 5,280 feet	$\dfrac{5,280 \text{ feet}}{1 \text{ mile}}$	$\dfrac{1 \text{ mile}}{5,280 \text{ feet}}$

Note that each of these conversion factors is a ratio of equal values, so each conversion factor equals 1. Multiplying a measurement by a conversion factor does not change the size of the measurement at all since it is the same as multiplying by 1; it just changes the units that you are using to measure.

Converting Between Units of Length

You can use the conversion factors to convert a measurement, such as feet, to another type of measurement, such as inches.

Note that there are many more inches for a measurement than there are feet for the same measurement, as feet is a longer unit of measurement. You could use the conversion factor $\frac{12 \text{ inches}}{1 \text{ foot}}$.

If a length is measured in feet, and you'd like to convert the length to yards, you can think, "I am converting from a shorter unit to a longer one, so the length in yards will be less than the length in feet." You could use the conversion factor $\frac{1 \text{ yard}}{3 \text{ feet}}$.

If a distance is measured in miles, and you want to know how many feet it is, you can think, "I am converting from a longer unit of measurement to a shorter one, so the number of feet would be greater than the number of miles." You could use the conversion factor $\frac{5,280 \text{ feet}}{1 \text{ mile}}$.

You can use the **factor label method** to convert a length from one unit of measure to another using the conversion factors. In the factor label method, you multiply by unit fractions to convert a measurement from one unit to another. Study the example below to see how the factor label method can be used to convert $3\frac{1}{2}$ feet into an equivalent number of inches.

Example	
Problem	**How many inches are in $3\frac{1}{2}$ feet?**
$3\frac{1}{2} \text{feet} = \underline{\quad}$ inches	Begin by reasoning about your answer. Since a foot is longer than an inch, this means the answer would be greater than $3\frac{1}{2}$.
$3\frac{1}{2}\text{feet} \cdot \dfrac{12 \text{ inches}}{1 \text{ foot}} = \underline{\quad}$ inches	Find the conversion factor that compares inches and feet, with "inches" in the numerator, and multiply.
$\dfrac{7 \text{ feet}}{2} \cdot \dfrac{12 \text{ inches}}{1 \text{ foot}} = \underline{\quad}$ inches	Rewrite the mixed number as an improper fraction before multiplying.
$\dfrac{7 \ \cancel{\text{feet}}}{2} \cdot \dfrac{12 \text{ inches}}{1 \ \cancel{\text{foot}}} = \underline{\quad}$ inches	You can reduce similar units when they appear in the numerator *and* the denominator. So here, reduce the similar units "feet" and "foot." This eliminates this unit from the problem.

$$\frac{7}{2} \cdot \frac{12 \text{ inches}}{1} = \underline{\quad} \text{ inches}$$

$$\frac{7 \cdot 12 \text{ inches}}{2 \cdot 1} = \underline{\quad} \text{ inches} \qquad \text{Rewrite as multiplication of numerators and denominators.}$$

$$\frac{84 \text{ inches}}{2} = \underline{\quad} \text{ inches} \qquad \text{Multiply.}$$

$$\frac{84 \text{ inches}}{2} = 42 \text{ inches} \qquad \text{Divide.}$$

Answer

There are 42 inches in $3\frac{1}{2}$ feet.

Notice that by using the factor label method you can reduce the units out of the problem, just as if they were numbers. You can only reduce if the unit being reduced is in both the numerator and denominator of the fractions you are multiplying.

In the problem above, you reduced *feet* and *foot* leaving you with *inches*, which is what you were trying to find.

$$\frac{7 \; \cancel{\text{feet}}}{2} \cdot \frac{12 \text{ inches}}{1 \; \cancel{\text{foot}}} = \underline{\quad} \text{ inches}$$

What if you had used the wrong conversion factor?

$$\frac{7 \text{ feet}}{2} \cdot \frac{1 \text{ foot}}{12 \text{ inches}} =$$

You could not reduce the feet because the unit is not the same in *both* the numerator and the denominator. So if you complete the computation, you would still have both feet and inches in the answer and no conversion would take place.

Here is another example of a length conversion using the factor label method.

Example		
Problem	**How many yards is 7 feet?**	
	7 feet = ___ yards	Start by reasoning about the size of your answer. Since a yard is longer than a foot, there will be fewer yards. So your answer will be less than 7.

$$7 \text{ feet} \cdot \frac{1 \text{ yard}}{3 \text{ feet}} = \underline{\quad} \text{ yards}$$

Find the conversion factor that compares feet and yards, with yards in the numerator.

$$\frac{7 \text{ feet}}{1} \cdot \frac{1 \text{ yard}}{3 \text{ feet}} = \underline{\quad} \text{ yards}$$

Rewrite the whole number as a fraction in order to multiply.

$$\frac{7 \cancel{\text{ feet}}}{1} \cdot \frac{1 \text{ yard}}{3 \cancel{\text{ feet}}} = \underline{\quad} \text{ yards}$$

Reduce the similar units "feet" and "feet" leaving only yards.

$$\frac{7}{1} \cdot \frac{1 \text{ yard}}{3} = \underline{\quad} \text{ yards}$$

$$\frac{7 \cdot 1 \text{ yard}}{1 \cdot 3} = \underline{\quad} \text{ yards}$$

Multiply.

$$\frac{7 \text{ yards}}{3} = 2\frac{1}{3} \text{ yards}$$

Divide, and write as a mixed number.

Answer

$$7 \text{ feet equals } 2\frac{1}{3} \text{ yards.}$$

Note that if the units do not reduce to give you the answer you are trying to find, you may not have used the correct conversion factor.

Try it Now 1

How many feet are in $2\frac{1}{2}$ miles?

Applying Unit Conversions

Objective 2

Objective 3

There are times when you will need to perform computations on measurements that are given in different units. For example, consider the tubing problem given earlier. You must decide which of the two options is a better price, and you have to compare prices given in different unit measurements.

In order to compare, you need to convert the measurements into one single, common unit of measurement. To be sure you have made the computation accurately, think about whether the unit you are converting to is smaller or larger than the number you have. Its relative size will tell you whether the number you are trying to find is greater or lesser than the given number.

Example
Problem An interior decorator needs border trim for a home she is wallpapering. She needs 15 feet of border trim for the living room, 30 feet of border trim for the bedroom, and 26 feet of

	border trim for the dining room. How many yards of border trim does she need?
15 feet + 30 feet + 26 feet = 71 feet	You need to find the total length of border trim that is needed for all three rooms in the house. Since the measurements for each room are given in feet, you can add the numbers.
71 feet = ___ yards	How many yards is 71 feet? Reason about the size of your answer. Since a yard is longer than a foot, there will be fewer yards. Expect your answer to be less than 71.
$\dfrac{71 \text{ feet}}{1} \cdot \dfrac{1 \text{ yard}}{3 \text{ feet}} =$ ___ yards	Use the conversion factor $\dfrac{1 \text{ yard}}{3 \text{ feet}}$.
$\dfrac{71 \text{ \cancel{feet}}}{1} \cdot \dfrac{1 \text{ yard}}{3 \text{ \cancel{feet}}} =$ ___ yards	Since "feet" is in the numerator and denominator, you can reduce this unit.
$\dfrac{71}{1} \cdot \dfrac{1 \text{ yard}}{3} =$ ___ yards	
$\dfrac{71 \cdot 1 \text{ yard}}{1 \cdot 3} =$ ___ yards	Multiply.
$\dfrac{71 \text{ yards}}{3} =$ ___ yards	
$\dfrac{71 \text{ yards}}{3} = 23\dfrac{2}{3}$ yards	Divide, and write as a mixed number.
Answer The interior decorator needs $23\dfrac{2}{3}$ yards of border trim.	

The next example uses the factor label method to solve a problem that requires converting from miles to feet.

Example	
Problem	**Two runners were comparing how much they had trained earlier that day. Jo said, "According to my pedometer, I ran 8.3 miles." Alex said, "That's a little more than what I ran. I ran 8.1 miles." How many more feet did Jo run than Alex?**

8.3 miles – 8.1 miles = 0.2 mile | You need to find the difference between the distance Jo ran and the distance Alex ran. Since both distances are given in the same unit, you can subtract and keep the unit the same.

$$0.2 \text{ mile} = \frac{2}{10} \text{ mile}$$

$$\frac{2}{10} \text{ mile} = \underline{\quad} \text{ feet}$$

Since the problem asks for the difference in *feet*, you must convert from miles to feet. How many feet is 0.2 mile? Reason about the size of your answer. Since a mile is longer than a foot, the distance when expressed as feet will be a number greater than 0.2.

$$\frac{2 \text{ mile}}{10} \cdot \frac{5{,}280 \text{ feet}}{1 \text{ mile}} = \underline{\quad} \text{ feet}$$

Use the conversion factor $\dfrac{5{,}280 \text{ feet}}{1 \text{ mile}}$.

$$\frac{2 \cancel{\text{ mile}}}{10} \cdot \frac{5{,}280 \text{ feet}}{1 \cancel{\text{ mile}}} = \underline{\quad} \text{ feet}$$

$$\frac{2}{10} \cdot \frac{5{,}280 \text{ feet}}{1} = \underline{\quad} \text{ feet}$$

$$\frac{2 \cdot 5{,}280 \text{ feet}}{10 \cdot 1} = \underline{\quad} \text{ feet} \quad \text{Multiply.}$$

$$\frac{10{,}560 \text{ feet}}{10} = \underline{\quad} \text{ feet}$$

$$\frac{10{,}560 \text{ feet}}{10} = 1{,}056 \text{ feet} \quad \text{Divide.}$$

Answer Jo ran 1,056 feet further than Alex.

Now let's revisit the question from earlier.

Example
Problem **You are walking through a hardware store and notice two sales on tubing.** - **3 yards of Tubing A costs \$5.49.** - **Tubing B sells for \$1.88 for 2 feet.** **Either tubing is acceptable for your project. Which tubing is less expensive?**

Tubing A	3 yards = $5.49	Find the unit price for each tubing. This will make it easier to compare.
	$\dfrac{\$5.49 \div 3}{3 \text{ yards} \div 3} = \dfrac{\$1.83}{1 \text{ yard}}$	Find the cost per yard of Tubing A by dividing the cost of 3 yards of the tubing by 3.
Tubing B	2 feet = $1.88	Tubing B is sold by the foot. Find the cost per foot by dividing $1.88 by 2 feet.
	$\dfrac{\$1.88 \div 2}{2 \text{ feet} \div 2} = \dfrac{\$0.94}{1 \text{ foot}}$	

$\dfrac{\$0.94}{1 \text{ foot}} \cdot \dfrac{3 \text{ feet}}{1 \text{ yard}} = \dfrac{\$ \underline{\quad}}{\underline{\quad} \text{ yard}}$	To compare the prices, you need to have the same unit of measure.
$\dfrac{\$0.94}{1 \text{ foot}} \cdot \dfrac{3 \text{ feet}}{1 \text{ yard}} = \dfrac{\$2.82}{1 \text{ yard}}$	Use the conversion factor $\dfrac{3 \text{ feet}}{1 \text{ yard}}$, reduce and
$2.82 per yard	multiply.
Tubing A: $1.83 per yard Tubing B: $2.82 per yard	Compare prices for 1 yard of each tubing.

Answer   Tubing A is less expensive than Tubing B.

In the problem above, you could also have found the price per foot for each kind of tubing and compared the unit prices of each per foot.

Try it Now 2

A fence company is measuring a rectangular area in order to install a fence around its perimeter. If the length of the rectangular area is 130 yards and the width is 75 feet, what is the total length of the distance to be fenced?

Summary

The four basic units of measurement that are used in the U.S. customary measurement system are: inch, foot, yard, and mile. Typically, people use yards, miles, and sometimes feet to describe long distances. Measurement in inches is common for shorter objects or lengths.

You need to convert from one unit of measure to another if you are solving problems that include measurements involving more than one type of measurement. Each of the units can be converted to one of the other units using the table of equivalents, the conversion factors, and/or the factor label method shown in this topic.

Try it Now Answers

1. 13,200 feet; there are 5,280 feet in a mile, so multiply $2\frac{1}{2}$ by 5,280 to get 13,200 feet.

2. 930 feet; 130 yards is equivalent to 390 feet. To find the perimeter, add length + length + width + width: 390 feet + 390 feet + 75 feet + 75 feet = 930 feet.

Weight

Learning Objective(s)
1 Define units of weight and convert from one to another.
2 Perform arithmetic calculations on units of weight.
3 Solve application problems involving units of weight.

Introduction

When you mention how heavy or light an object is, you are referring to its weight. In the U.S. customary system of measurement, weight is measured in ounces, pounds, and tons. Like other units of measurement, you can convert between these units and you sometimes need to do this to solve problems.

In 2010, the post office charges $0.44 to mail something that weighs an ounce or less. The post office charges $0.17 for each additional ounce, or fraction of an ounce, of weight. How much will it cost to mail a package that weighs two pounds three ounces? To answer this question, you need to understand the relationship between ounces and pounds.

Units of Weight

Objective 1

You often use the word **weight** to describe how heavy or light an object or person is. Weight is measured in the U.S. customary system using three units: ounces, pounds, and tons. An **ounce** is the smallest unit for measuring weight, a **pound** is a larger unit, and a **ton** is the largest unit.

Whales are some of the largest animals in the world. Some species can reach weights of up to 200 tons- that's equal to 400,000 pounds.	
Meat is a product that is typically sold by the pound. One pound of ground beef makes about four hamburger patties.	

Ounces are used to measure lighter objects. A stack of 11 pennies is equal to about one ounce.		

You can use any of the customary measurement units to describe the weight of something, but it makes more sense to use certain units for certain purposes. For example, it makes more sense to describe the weight of a human being in pounds rather than tons. It makes more sense to describe the weight of a car in tons rather than ounces.

$$1 \text{ pound} = 16 \text{ ounces}$$

$$1 \text{ ton} = 2{,}000 \text{ pounds}$$

Converting Between Units of Weight

Four ounces is a typical serving size of meat. Since meat is sold by the pound, you might want to convert the weight of a package of meat from pounds to ounces in order to determine how many servings are contained in a package of meat.

The weight capacity of a truck is often provided in tons. You might need to convert pounds into tons if you are trying to determine whether a truck can safely transport a big shipment of heavy materials.

The table below shows the unit conversions and conversion factors that are used to make conversions between customary units of weight.

Unit Equivalents	Conversion Factors (heavier to lighter units of measurement)	Conversion Factors (lighter to heavier units of measurement)
1 pound = 16 ounces	$\dfrac{16 \text{ ounces}}{1 \text{ pound}}$	$\dfrac{1 \text{ pound}}{16 \text{ ounces}}$
1 ton = 2000 pounds	$\dfrac{2000 \text{ pounds}}{1 \text{ ton}}$	$\dfrac{1 \text{ ton}}{2000 \text{ pounds}}$

You can use the *factor label method* to convert one customary unit of weight to another customary unit of weight. This method uses conversion factors, which allow you to "reduce" units to end up with your desired unit of measurement.

Each of these conversion factors is a ratio of equal values, so each conversion factor equals 1. Multiplying a measurement by a conversion factor does not change the size of the measurement at all, since it is the same as multiplying by 1. It just changes the units that you are using to measure it in.

Two examples illustrating the factor label method are shown below.

Example

| Problem | **How many ounces are in $2\frac{1}{4}$ pounds?** |

$2\frac{1}{4}$ pounds = ___ ounces

Begin by reasoning about your answer. Since a pound is heavier than an ounce, expect your answer to be a number greater than $2\frac{1}{4}$.

$2\frac{1}{4}$ pounds $\cdot \dfrac{16 \text{ ounces}}{1 \text{ pound}} =$ ___ ounces

Multiply by the conversion factor that relates ounces and pounds: $\dfrac{16 \text{ ounces}}{1 \text{ pound}}$.

$\dfrac{9 \text{ pounds}}{4} \cdot \dfrac{16 \text{ ounces}}{1 \text{ pound}} =$ ___ ounces

Write the mixed number as an improper fraction.

$\dfrac{9 \text{ \sout{pounds}}}{4} \cdot \dfrac{16 \text{ ounces}}{1 \text{ \sout{pound}}} =$ ___ ounces

The common unit "pound" can be reduced because it appears in both the numerator and denominator.

$\dfrac{9}{4} \cdot \dfrac{16 \text{ ounces}}{1} =$ ___ ounces

Multiply and simplify.

$\dfrac{9 \cdot 16 \text{ ounces}}{4 \cdot 1} =$ ___ ounces

$\dfrac{144 \text{ ounces}}{4} =$ ___ ounces

$\dfrac{144 \text{ ounces}}{4} = 36$ ounces

Answer   There are 36 ounces in $2\frac{1}{4}$ pounds.

Example
Problem	**How many tons is 6,500 pounds?**

6,500 pounds = ___ tons   Begin by reasoning about your answer. Since a ton is heavier than a pound, expect your answer to be a number less than 6,500.

$6{,}500 \text{ pounds} \cdot \dfrac{1 \text{ ton}}{2{,}000 \text{ pounds}} =$ ___ tons   Multiply by the conversion factor that relates tons to pounds: $\dfrac{1 \text{ ton}}{2{,}000 \text{ pounds}}$.

$\dfrac{6{,}500 \text{ pounds}}{1} \cdot \dfrac{1 \text{ ton}}{2{,}000 \text{ pounds}} =$ ___ tons

$\dfrac{6{,}500 \text{ pounds}}{1} \cdot \dfrac{1 \text{ ton}}{2{,}000 \text{ pounds}} =$ ___ tons   Apply the Factor Label method.

$\dfrac{6{,}500}{1} \cdot \dfrac{1 \text{ ton}}{2{,}000} =$ ___ tons   Multiply and simplify.

$\dfrac{6{,}500 \text{ tons}}{2{,}000} =$ ___ tons

$\dfrac{6{,}500 \text{ tons}}{2{,}000} = 3\dfrac{1}{4} \text{ tons}$

Answer   6,500 pounds is equal to $3\dfrac{1}{4}$ tons.

Try it Now 1

How many pounds is 72 ounces?

Applying Unit Conversions

Objective 2

Objective 3

There are times when you need to perform calculations on measurements that are given in different units. To solve these problems, you need to convert one of the measurements to the same unit of measurement as the other measurement.

Think about whether the unit you are converting to is smaller or larger than the unit you are converting from. This will help you be sure that you are making the right computation. You can use the factor label method to make the conversion from one unit to another.

Here is an example of a problem that requires converting between units.

Example

Problem   **A municipal trash facility allows a person to throw away a maximum of 30 pounds of trash per week. Last week, 140 people threw away the maximum allowable trash. How many tons of trash did this equal?**

$$140 \cdot 30 \text{ pounds} = 4{,}200 \text{ pounds}$$

Determine the total trash for the week expressed in pounds.
If 140 people each throw away 30 pounds, you can find the total by multiplying.

$$4{,}200 \text{ pounds} = \underline{\quad} \text{ tons}$$

Then convert 4,200 pounds to tons. Reason about your answer. Since a ton is heavier than a pound, expect your answer to be a number less than 4,200.

$$\frac{4{,}200 \text{ pounds}}{1} \cdot \frac{1 \text{ ton}}{2{,}000 \text{ pounds}} = \underline{\quad} \text{ tons}$$

Find the conversion factor appropriate for the situation:
$$\frac{1 \text{ ton}}{2{,}000 \text{ pounds}}.$$

$$\frac{4{,}200 \;\cancel{\text{pounds}}}{1} \cdot \frac{1 \text{ ton}}{2{,}000 \;\cancel{\text{pounds}}} = \underline{\quad} \text{ tons}$$

$$\frac{4{,}200}{1} \cdot \frac{1 \text{ ton}}{2{,}000} = \underline{\quad} \text{ tons}$$

$$\frac{4{,}200 \cdot 1 \text{ ton}}{1 \cdot 2{,}000} = \underline{\quad} \text{ tons}$$

Multiply and simplify.

$$\frac{4{,}200 \text{ ton}}{2{,}000} = \underline{\quad} \text{ tons}$$

$$\frac{4{,}200 \text{ ton}}{2{,}000} = 2\frac{1}{10} \text{ tons}$$

Answer   The total amount of trash generated is $2\frac{1}{10}$ tons.

Let's revisit the post office problem that was posed earlier. We can use unit conversion to solve this problem.

Example

Problem **The post office charges \$0.44 to mail something that weighs an ounce or less. The charge for each additional ounce, or fraction of an ounce, of weight is \$0.17. At this rate, how much will it cost to mail a package that weighs 2 pounds 3 ounces?**

2 pounds 3 ounces = ___ ounces   Since the pricing is for ounces, convert the weight of the package from pounds and ounces into just ounces.

$$\frac{2 \text{ pounds}}{1} \cdot \frac{16 \text{ ounces}}{\text{pound}} = \text{\_\_\_ ounces}$$ First use the factor label method to convert 2 pounds to ounces.

$$\frac{2 \text{ } \cancel{\text{pounds}}}{1} \cdot \frac{16 \text{ ounces}}{\cancel{\text{pound}}} = \text{\_\_\_ ounces}$$

$$\frac{2}{1} \cdot \frac{16 \text{ ounces}}{1} = 32 \text{ ounces}$$ 2 pounds = 32 ounces.

32 ounces + 3 ounces = 35 ounces   Add the additional 3 ounces to find the weight of the package. The package weighs 35 ounces. There are 34 additional ounces, since 35 − 1 = 34.

\$0.44 + \$0.17(34)   Apply the pricing formula. \$0.44 for the first ounce and \$0.17 for each

\$0.44 + \$5.78   additional ounce.

\$0.44 + \$5.78 = \$6.22

Answer   It will cost \$6.22 to mail a package that weighs 2 pounds 3 ounces.

Try it Now 2

The average weight of a northern Bluefin tuna is 1,800 pounds. The average weight of a great white shark is $2\frac{1}{2}$ tons. On average, how much more does a great white shark weigh, in pounds, than a northern Bluefin tuna?

Summary

In the U.S. customary system of measurement, weight is measured in three units: ounces, pounds, and tons. A pound is equivalent to 16 ounces, and a ton is equivalent to 2,000 pounds. While an object's weight can be described using any of these units, it is typical to describe very heavy objects using tons and very light objects using an ounce. Pounds are used to describe the weight of many objects and people. Often, in order to compare the weights of two objects or people or to solve problems involving weight, you must convert from one unit of measurement to another unit of measurement. Using conversion factors with the factor label method is an effective strategy for converting units and solving problems.

Try it Now Answers

1. $4\frac{1}{2}$ pounds; There are 16 ounces in one pound, so 72 ounces $\cdot \dfrac{1\ \text{pound}}{16\ \text{ounces}} = 4\frac{1}{2}$ pounds.

2. 3,200 pounds; $2\frac{1}{2}$ tons = 5,000 pounds. 5,000 pounds – 1,800 pounds = 3,200 pounds.

Capacity

Learning Objective(s)
1 Define units of capacity and convert from one to another.
2 Perform arithmetic calculations on units of capacity.
3 Solve application problems involving units of capacity.

Introduction

Capacity is the amount of liquid (or other pourable substance) that an object can hold when it's full. When a liquid, such as milk, is being described in gallons or quarts, this is a measure of capacity.

Understanding units of capacity can help you solve problems like this: Sven and Johanna were hosting a potluck dinner. They did not ask their guests to tell them what they would be bringing, and three people ended up bringing soup. Erin brought 1 quart, Richard brought 3 pints, and LeVar brought 9 cups. How many cups of soup did they have all together?

Units of Capacity

Objective 1

There are five main units for measuring capacity in the U.S. customary measurement system. The smallest unit of measurement is a **fluid ounce**. "Ounce" is also used as a measure of weight, so it is important to use the word "fluid" with ounce when you are talking about capacity. Sometimes the prefix "fluid" is not used when it is clear from the context that the measurement is capacity, not weight.

The other units of capacity in the customary system are the **cup**, **pint**, **quart**, and **gallon**. The table below describes each unit of capacity and provides an example to illustrate the size of the unit of measurement.

Fluid Ounce	
A unit of capacity equal to $\frac{1}{8}$ of a cup. One fluid ounce of water at 62°F weighs about one ounce. The amount of liquid medicine is often measured in fluid ounces.	

Cup	
A unit equal to 8 fluid ounces. The capacity of a standard measuring cup is one cup.	
Pint	
A unit equal to 16 fluid ounces, or 2 cups. The capacity of a carton of ice cream is often measured in pints.	
Quart	
A unit equal to 32 fluid ounces, or 4 cups. You often see quarts of milk being sold in the supermarket.	
Gallon	
A unit equal to 4 quarts, or 128 fluid ounces. When you fill up your car with gasoline, the price of gas is often listed in dollars per gallon.	

You can use any of these five measurement units to describe the capacity of an object, but it makes more sense to use certain units for certain purposes. For example, it makes more sense to describe the capacity of a swimming pool in gallons and the capacity of an expensive perfume in fluid ounces.

Sometimes you will need to convert between units of measurement. For example, you might want to express 5 gallons of lemonade in cups if you are trying to determine how many 8-fluid ounce servings the amount of lemonade would yield.

The table below shows some of the most common equivalents and conversion factors for the five customary units of measurement of capacity.

Unit Equivalents	Conversion Factors (heavier to lighter units of measurement)	Conversion Factors (lighter to heavier units of measurement)
1 cup = 8 fluid ounces	$\dfrac{1 \text{ cup}}{8 \text{ fluid ounces}}$	$\dfrac{8 \text{ fluid ounces}}{1 \text{ cup}}$
1 pint = 2 cups	$\dfrac{1 \text{ pint}}{2 \text{ cups}}$	$\dfrac{2 \text{ cups}}{1 \text{ pint}}$
1 quart = 2 pints	$\dfrac{1 \text{ quart}}{2 \text{ pints}}$	$\dfrac{2 \text{ pints}}{1 \text{ quart}}$
1 quart = 4 cups	$\dfrac{1 \text{ quart}}{4 \text{ cups}}$	$\dfrac{4 \text{ cups}}{1 \text{ quart}}$
1 gallon = 4 quarts	$\dfrac{1 \text{ gallon}}{4 \text{ quarts}}$	$\dfrac{4 \text{ quarts}}{1 \text{ gallon}}$
1 gallon = 16 cups	$\dfrac{1 \text{ gallon}}{16 \text{ cups}}$	$\dfrac{16 \text{ cups}}{1 \text{ gallon}}$

Converting Between Units of Capacity

Objective **2**

As with converting units of length and weight, you can use the factor label method to convert from one unit of capacity to another. An example of this method is shown below.

Example	
Problem	How many pints is $2\dfrac{3}{4}$ gallons?
$2\dfrac{3}{4}$ gallons = ____ pints	Begin by reasoning about your answer. Since a gallon is larger than a pint, expect the answer in pints to be a number greater than $2\dfrac{3}{4}$.
$\dfrac{11 \text{ gallons}}{4} \cdot \dfrac{4 \text{ quarts}}{1 \text{ gallon}} \cdot \dfrac{2 \text{ pints}}{1 \text{ quart}} =$ ____ pints	The table above does not contain a conversion factor for gallons and pints, so you cannot convert it in one step. However, you can use quarts as an intermediate unit, as shown here. Set up the equation so that two sets of labels reduce—gallons and quarts.

$$\frac{11 \text{ gallons}}{4} \cdot \frac{4 \text{ quarts}}{1 \text{ gallon}} \cdot \frac{2 \text{ pints}}{1 \text{ quart}} = \underline{} \text{ pints}$$

$$\frac{11}{4} \cdot \frac{4}{1} \cdot \frac{2 \text{ pints}}{1} = \underline{} \text{ pints}$$

$$\frac{11 \cdot 4 \cdot 2 \text{ pints}}{4 \cdot 1 \cdot 1} = \underline{} \text{ pints} \qquad \text{Multiply and simplify.}$$

$$\frac{88 \text{ pints}}{4} = 22 \text{ pints}$$

Answer $2\frac{3}{4}$ gallons is 22 pints.

Example

Problem	**How many gallons is 32 fluid ounces?**

32 fluid ounces = $\underline{}$ gallons   Begin by reasoning about your answer. Since gallons is a larger unit than fluid ounces, expect the answer to be less than 32.

$$\frac{32 \text{ fl oz}}{1} \cdot \frac{1 \text{ cup}}{8 \text{ fl oz}} \cdot \frac{1 \text{ pt}}{2 \text{ cups}} \cdot \frac{1 \text{ qt}}{2 \text{ pt}} \cdot \frac{1 \text{ gal}}{4 \text{ qt}} = \underline{} \text{ gal}$$

The table above does not contain a conversion factor for gallons and fluid ounces, so you cannot convert it in one step. Use a series of intermediate units, as shown here.

$$\frac{32 \text{ fl oz}}{1} \cdot \frac{1 \text{ cup}}{8 \text{ fl oz}} \cdot \frac{1 \text{ pt}}{2 \text{ cups}} \cdot \frac{1 \text{ qt}}{2 \text{ pt}} \cdot \frac{1 \text{ gal}}{4 \text{ qt}} = \underline{} \text{ gal}$$

Reduce units that appear in both the numerator and denominator.

$$\frac{32}{1} \cdot \frac{1}{8} \cdot \frac{1}{2} \cdot \frac{1}{2} \cdot \frac{1 \text{ gal}}{4} = \underline{} \text{ gal}$$

$$\frac{32 \cdot 1 \cdot 1 \cdot 1 \cdot 1 \text{ gal}}{1 \cdot 8 \cdot 2 \cdot 2 \cdot 4} = \underline{} \text{ gal} \qquad \text{Multiply and simplify.}$$

$$\frac{32 \text{ gal}}{128} = \frac{1}{4} \text{ gal}$$

Answer	32 fluid ounces is the same as $\dfrac{1}{4}$ gallon.

Try it Now 1

Find the sum of 4 gallons and 2 pints. Express your answer in cups.

Applying Unit Conversions

Objective 3

There are times when you will need to combine measurements that are given in different units. In order to do this, you need to convert first so that the units are the same.

Consider the situation posed earlier in this topic.

Example	
Problem	**Sven and Johanna were hosting a potluck dinner. They did not ask their guests to tell them what they would be bringing, and three people ended up bringing soup. Erin brought 1 quart, Richard brought 3 pints, and LeVar brought 9 cups. How much soup did they have total?**
1 quart + 3 pints + 9 cups	Since the problem asks for the total amount of soup, you must add the three quantities. Before adding, you must convert the quantities to the same unit.
	The problem does not require a particular unit, so you can choose. Cups might be the easiest computation.
1 quart = 4 cups	This is given in the table of equivalents.
$\dfrac{3 \text{ pints}}{1} \cdot \dfrac{2 \text{ cups}}{1 \text{ pint}} = \underline{}$ cups	Use the factor label method to convert pints to cups.
$\dfrac{3 \, \cancel{\text{pints}}}{1} \cdot \dfrac{2 \text{ cups}}{1 \, \cancel{\text{pint}}} = 6$ cups	
4 cups + 6 cups + 9 cups = 19 cups	Add the 3 quantities.
Answer	There are 19 cups of soup for the dinner.

	Example
Problem	**Natasha is making lemonade to bring to the beach. She has two containers. One holds one gallon and the other holds 2 quarts. If she fills both containers, how many cups of lemonade will she have?**

1 gallon + 2 quarts = ___ cups   This problem requires you to find the sum of the capacity of each container and then convert that sum to cups.

4 quarts + 2 quarts = 6 quarts   First, find the sum in quarts. 1 gallon is equal to 4 quarts.

$$\frac{6 \text{ quarts}}{1} \cdot \frac{2 \text{ pints}}{1 \text{ quart}} \cdot \frac{2 \text{ cups}}{1 \text{ pint}} = \underline{} \text{ cups}$$

Since the problem asks for the capacity in cups, convert 6 quarts to cups.

$$\frac{6 \;\cancel{\text{quarts}}}{1} \cdot \frac{2 \;\cancel{\text{pints}}}{1 \;\cancel{\text{quart}}} \cdot \frac{2 \text{ cups}}{1 \;\cancel{\text{pint}}} = \underline{} \text{ cups}$$

Reduce units that appear in both the numerator and denominator.

$$6 \cdot 2 \cdot 2 = 24 \text{ cups}$$

Multiply.

Answer	Natasha will have 24 cups of lemonade.

Another way to work the problem above would be to first change 1 gallon to 16 cups and change 2 quarts to 8 cups. Then add: 16 + 8 = 24 cups.

Try it Now 2

Alan is making chili. He is using a recipe that makes 24 cups of chili. He has a 5-quart pot and a 2-gallon pot and is trying to determine whether the chili will all fit in one of these pots. Which of the pots will fit the chili?

Summary

There are five basic units for measuring capacity in the U.S. customary measurement system. These are the fluid ounce, cup, pint, quart, and gallon. These measurement units are related to one another, and capacity can be described using any of the units. Typically, people use gallons to describe larger quantities and fluid ounces, cups, pints, or quarts to describe smaller quantities. Often, in order to compare or to solve problems involving the amount of liquid in a container, you need to convert from one unit of measurement to another.

Try it Now Answers

1. 68 cups; each gallon has 16 cups, so 4 x 16 = 64 will give you the number of cups in 4 gallons. Each pint has 2 cups, so 2 x 2 = 4 will give you the number of cups in 2 pints. 64 + 4 = 68 cups.
2. The chili will fit into the 2-gallon pot only; 5 quarts = 5 x 4 cups = 20 cups, so 24 cups of chili will not fit into the 5-quart pot. 2 gallons = 32 cups, so 24 cups of chili will fit in this pot.

5.2 Metric Units of Measurement

Learning Objective(s)
1 Describe the general relationship between the U.S. customary units and metric units of length, weight/mass, and volume.
2 Define the metric prefixes and use them to perform basic conversions among metric units.

Introduction

Objective 1

In the United States, both the **U.S. customary measurement system** and the **metric system** are used, especially in medical, scientific, and technical fields. In most other countries, the metric system is the primary system of measurement. If you travel to other countries, you will see that road signs list distances in kilometers and milk is sold in liters. People in many countries use words like "kilometer," "liter," and "milligram" to measure the length, volume, and weight of different objects. These measurement units are part of the metric system.

Unlike the U.S. customary system of measurement, the metric system is based on 10s. For example, a liter is 10 times larger than a deciliter, and a centigram is 10 times larger than a milligram. This idea of "10" is not present in the U.S. customary system—there are 12 inches in a foot, and 3 feet in a yard…and 5,280 feet in a mile!

So, what if you have to find out how many milligrams are in a decigram? Or, what if you want to convert meters to kilometers? Understanding how the metric system works is a good start.

What is Metric?

The metric system uses units such as **meter**, **liter**, and **gram** to measure length, liquid volume, and mass, just as the U.S. customary system uses feet, quarts, and ounces to measure these.

In addition to the difference in the basic units, the metric system is based on factors of 10, and different measures for length include kilometer, meter, decimeter, centimeter, and millimeter. Notice that the word "meter" is part of all of these units.

The metric system also applies the idea that units within the system get larger or smaller by a power of 10. This means that a meter is 100 times larger than a centimeter, and a kilogram is 1,000 times heavier than a gram. You will explore this idea a bit later. For now, notice how this idea of "getting bigger or smaller by 10" is very different than the relationship between units in the U.S. customary system, where 3 feet equals 1 yard, and 16 ounces equals 1 pound.

Length, Mass, and Volume

The table below shows the basic units of the metric system. Note that the names of all metric units follow from these three basic units.

Length	Mass	Volume
basic units		
meter	gram	liter

other units you may see		
kilometer	kilogram	dekaliter
centimeter	centigram	centiliter
millimeter	milligram	milliliter

In the metric system, the basic unit of length is the meter. A meter is slightly larger than a yardstick, or just over three feet.

The basic metric unit of mass is the gram. A regular-sized paperclip has a mass of about 1 gram.

Among scientists, one gram is defined as the mass of water that would fill a 1-centimeter cube. You may notice that the word "mass" is used here instead of "weight." In the sciences and technical fields, a distinction is made between weight and mass. Weight is a measure of the pull of gravity on an object. For this reason, an object's weight would be different if it was weighed on Earth or on the moon because of the difference in the gravitational forces. However, the object's mass would remain the same in both places because mass measures the amount of substance in an object. As long as you are planning on only measuring objects on Earth, you can use mass/weight fairly interchangeably—but it is worth noting that there is a difference!

Finally, the basic metric unit of volume is the liter. A liter is slightly larger than a quart.

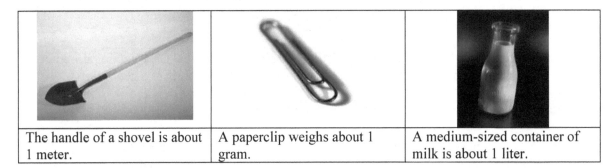

The handle of a shovel is about 1 meter.	A paperclip weighs about 1 gram.	A medium-sized container of milk is about 1 liter.

Though it is rarely necessary to convert between the customary and metric systems, sometimes it helps to have a mental image of how large or small some units are. The table below shows the relationship between some common units in both systems.

	Common Measurements in Customary and Metric Systems
Length	1 centimeter is a little less than half an inch.
	1.6 kilometers is about 1 mile.
	1 meter is about 3 inches longer than 1 yard.
Mass	1 kilogram is a little more than 2 pounds.
	28 grams is about the same as 1 ounce.
Volume	1 liter is a little more than 1 quart.
	4 liters is a little more than 1 gallon.

Prefixes in the Metric System

Objective 2

The metric system is a base-10 system. This means that each successive unit is 10 times larger than the previous one.

The names of metric units are formed by adding a prefix to the basic unit of measurement. To tell how large or small a unit is, you look at the **prefix**. To tell whether the unit is measuring length, mass, or volume, you look at the base.

Prefixes in the Metric System						
kilo-	*hecto-*	*deka-*	meter gram liter	*deci-*	*centi-*	*milli-*
1,000 times **larger** than base unit	100 times **larger** than base unit	10 times **larger** than base unit	base units	10 times **smaller** than base unit	100 times **smaller** than base unit	1,000 times **smaller** than base unit

Using this table as a reference, you can see the following:
- A kilogram is 1,000 times larger than one gram (so 1 kilogram = 1,000 grams).
- A centimeter is 100 times smaller than one meter (so 1 meter = 100 centimeters).
- A dekaliter is 10 times larger than one liter (so 1 dekaliter = 10 liters).

Here is a similar table that just shows the metric units of measurement for mass, along with their size relative to 1 gram (the base unit). The common abbreviations for these metric units have been included as well.

Measuring Mass in the Metric System						
kilogram (kg)	hectogram (hg)	dekagram (dag)	gram (g)	decigram (dg)	centigram (cg)	milligram (mg)
1,000 grams	100 grams	10 grams	gram	0.1 gram	0.01 gram	0.001 gram

Since the prefixes remain constant through the metric system, you could create similar charts for length and volume. The prefixes have the same meanings whether they are attached to the units of length (meter), mass (gram), or volume (liter).

Try it Now 1

Which of the following sets of three units are all metric measurements of length?

A) inch, foot, yard
B) kilometer, centimeter, millimeter
C) kilogram, gram, centigram
D) kilometer, foot, decimeter

Converting Units Up and Down the Metric Scale

Converting between metric units of measure requires knowledge of the metric prefixes and an understanding of the decimal system—that's about it.

For instance, you can figure out how many centigrams are in one dekagram by using the table above. One dekagram is larger than one centigram, so you expect that one dekagram will equal many centigrams.

In the table, each unit is 10 times larger than the one to its immediate right. This means that 1 dekagram = 10 grams; 10 grams = 100 decigrams; and 100 decigrams = 1,000 centigrams. So, 1 dekagram = 1,000 centigrams.

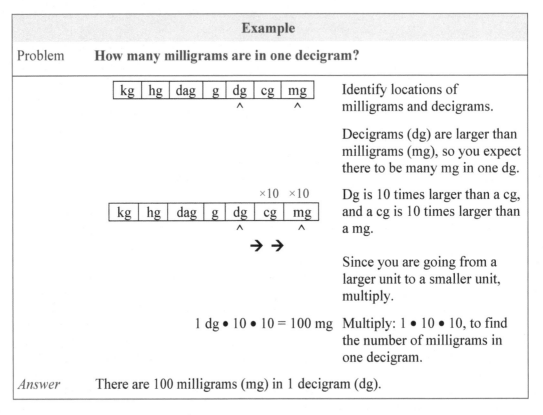

	Example
Problem	**How many milligrams are in one decigram?**

	Identify locations of milligrams and decigrams.
	Decigrams (dg) are larger than milligrams (mg), so you expect there to be many mg in one dg.
	Dg is 10 times larger than a cg, and a cg is 10 times larger than a mg.
	Since you are going from a larger unit to a smaller unit, multiply.
1 dg • 10 • 10 = 100 mg	Multiply: 1 • 10 • 10, to find the number of milligrams in one decigram.
Answer	There are 100 milligrams (mg) in 1 decigram (dg).

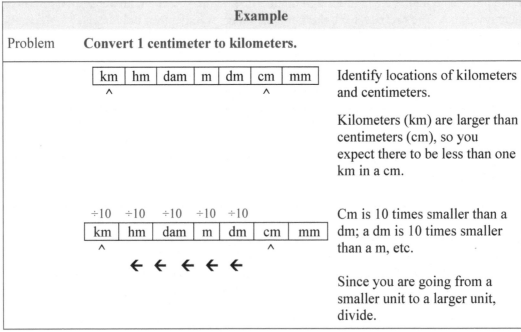

	Example
Problem	**Convert 1 centimeter to kilometers.**

	Identify locations of kilometers and centimeters.
	Kilometers (km) are larger than centimeters (cm), so you expect there to be less than one km in a cm.
	Cm is 10 times smaller than a dm; a dm is 10 times smaller than a m, etc.
	Since you are going from a smaller unit to a larger unit, divide.

> 1 cm ÷ 10 ÷ 10 ÷ 10 ÷ 10 ÷ 10 = 0.00001   Divide: 1 ÷ 10 ÷ 10 ÷ 10 ÷ 10 km ÷ 10, to find the number of kilometers in one centimeter.
>
> *Answer*   1 centimeter (cm) = 0.00001 kilometers (km).

Once you begin to understand the metric system, you can use a shortcut to convert among different metric units. The size of metric units increases tenfold as you go up the metric scale. The decimal system works the same way: a tenth is 10 times larger than a hundredth; a hundredth is 10 times larger than a thousandth, etc. By applying what you know about decimals to the metric system, converting among units is as simple as moving decimal points.

Here is the first problem from above: How many milligrams are in one decigram? You can recreate the order of the metric units as shown below:

$$kg \quad hg \quad dag \quad g \quad \underset{1}{d\,g} \quad \underset{2}{c\,g} \quad mg$$

This question asks you to start with 1 decigram and convert that to milligrams. As shown above, milligrams is two places to the right of decigrams. You can just move the decimal point two places to the right to convert decigrams to milligrams: $1\,dg = 1\underset{1\,2}{00}.\,mg$.

The same method works when you are converting from a smaller to a larger unit, as in the problem: Convert 1 centimeter to kilometers.

$$\underset{5}{k\,m} \quad \underset{4}{h\,m} \quad \underset{3}{d\,am} \quad \underset{2}{m \quad d}\,m \quad \underset{1}{c\,m} \quad mm$$

Note that instead of moving to the right, you are now moving to the left—so the decimal point must do the same: $1\,cm = 0.\underset{5\,4\,3\,2\,1}{00001}\,km$.

Try it Now 2

How many milliliters are in 1 liter?

Summary

The metric system is an alternative system of measurement used in most countries, as well as in the United States. The metric system is based on joining one of a series of prefixes, including kilo-, hecto-, deka-, deci-, centi-, and milli-, with a base unit of measurement, such as meter, liter, or gram. Units in the metric system are all related by a power of 10, which means that each successive unit is 10 times larger than the previous one. This makes converting one metric measurement to another a straightforward process, and is often as simple as moving a decimal point. It is always important, though, to consider the direction of the conversion. If you are converting a smaller unit to a larger unit, then the decimal point has to move to the left (making your number smaller); if you are converting a larger unit to a smaller unit, then the decimal point has to move to the right (making your number larger).

Try it Now Answers

1. B) kilometer, centimeter, millimeter; all of these measurements are from the metric system. They are measurements of length because they all contain the word "meter."
2. 1,000; there are 10 milliliters in a centiliter, 10 centiliters in a deciliter, and 10 deciliters in a liter. Multiply: 10 • 10 • 10, to find the number of milliliters in a liter, 1,000.

Converting within the Metric System

Learning Objective(s)
1 Perform arithmetic calculations on metric units of length, mass, and volume.

Introduction

While knowing the different units used in the metric system is important, the real purpose behind learning the metric system is for you to be able to use these measurement units to calculate the size, mass, or volume of different objects. In practice, it is often necessary to convert one metric measurement to another unit—this happens frequently in the medical, scientific, and technical fields, where the metric system is commonly used.

If you have a prescription for 5,000 mg of medicine, and upon getting it filled, the dosage reads 5 g of medicine, did the pharmacist make a mistake?

For a moment, imagine that you are a pharmacist. You receive three prescriptions for liquid amoxicillin: one calls for 2.5 centiliters, one calls for 0.3 deciliters, and one calls for 450 milliliters. Amoxicillin is stored in the refrigerator in 1 liter, 1 deciliter, and 1 centiliter containers. Which container should you use to ensure you are not wasting any of the unused drug?

To solve this problem, you need to know how to convert from one measurement to another as well as how to add different quantities together. Let's take a look at how to do this.

Converting from Larger to Smaller Units

Objective 1

Converting between measurements in the metric system is simply a matter of identifying the unit that you have, the unit that you want to convert to, and then counting the number of units between them. A basic example of this is shown below.

Example									
Problem	**Convert 1 kilometer to decimeters.**								

	km	hm	dam	m	dm	cm	mm	Identify locations of kilometers and decimeters.
	^				^			

Kilometers (km) are larger than decimeters (dm), so you expect

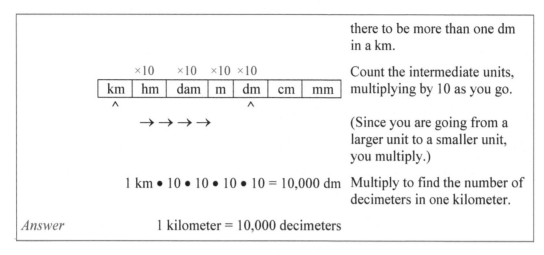

	there to be more than one dm in a km.
×10 ×10 ×10 ×10	Count the intermediate units, multiplying by 10 as you go.
km | hm | dam | m | dm | cm | mm	
→ → → →	(Since you are going from a larger unit to a smaller unit, you multiply.)
1 km • 10 • 10 • 10 • 10 = 10,000 dm	Multiply to find the number of decimeters in one kilometer.
Answer 1 kilometer = 10,000 decimeters	

This problem is straightforward because you are converting 1 kilometer to another unit. The example below shows how you would solve this problem if you were asked to convert 8.2 *kilometers* to *decimeters*. Notice that most steps are the same; the critical difference is that you multiply by 8.2 in the final step.

Example	
Problem	**Convert 8.2 kilometers to decimeters.**
km | hm | dam | m | dm | cm | mm	Identify locations of kilometers and decimeters.
	Kilometers (km) are larger than decimeters (dm), so you expect there to be more than one dm in a km.
×10 ×10 ×10 ×10 km | hm | dam | m | dm | cm | mm	Count the intermediate units, multiplying by 10 as you go.
→ → → →	Since you are going from a larger unit to a smaller unit, multiply.
8.2 km • 10 • 10 • 10 • 10 = 82,000 dm	Multiply to find the number of decimeters in 8.2 kilometers.
Answer 8.2 kilometers = 82,000 decimeters	

You can also apply the rules of base 10 to use the "move the decimal" shortcut method in this example. Notice how decimeters (*dm*) is four places to the right of kilometers (*km*); similarly, you move the decimal point four places to the right when converting 8.2 kilometers to decimeters.

$$k m \quad h m \quad da m \quad m \quad d m \quad cm \quad mm$$

$$\underbrace{\qquad}_{1} \underbrace{\qquad}_{2} \underbrace{\qquad}_{3} \underbrace{\qquad}_{4}$$

$$8.2 \; km = 8.\underset{1 \; 2 \; 3 \; 4}{2000.} \; dm$$

$$8.2 \; km = 82{,}000 \; dm$$

Example	
Problem	**Convert 0.55 liters to centiliters.**

$kl \quad hl \quad dal \quad \underset{1}{\underbrace{l \quad dl}} \; \underset{2}{\underbrace{cl}} \quad ml$	Count two places from liters to centiliters.
$0.55 \, l = 0.\underset{1 \; 2}{55.} \; cl$	In 0.55 l, move the decimal point two places to the right.
$0.55 \, l = 55 \; cl$	
Answer	0.55 liters = 55 centiliters

Try it Now 1

How many dekaliters are in 0.5 deciliters?

Converting from Smaller to Larger Units

You can use similar processes when converting from smaller to larger units. When converting a larger unit to a smaller one, you multiply; when you convert a smaller unit to a larger one, you divide. Here is an example.

Example	
Problem	**Convert 739 centigrams to grams.**

kg | hg | dag | g | dg | cg | mg	Identify locations of centigrams and grams.
	Centigrams (cg) are smaller than grams (g), so you expect there to be less than 739 g in 739 cg.
$\div 10 \; \div 10$ kg | hg | dag | g | dg | cg | mg $\leftarrow \leftarrow$	Count the intermediate units, dividing by 10 as you go. Since you are going from a smaller unit to a larger unit, divide.
$739 \div 10 \div 10 = 7.39 \; g$	Divide to find the number of grams in 739 centigrams.
Answer	739 centigrams = 7.39 grams

Notice that the shortcut method of counting prefixes and moving the decimal the same number of places also works here. Just make sure you are moving the decimal point in the correct direction for the conversion.

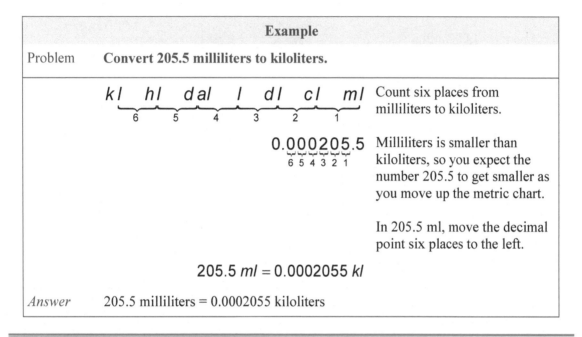

	Example
Problem	**Convert 205.5 milliliters to kiloliters.**

Count six places from milliliters to kiloliters.

Milliliters is smaller than kiloliters, so you expect the number 205.5 to get smaller as you move up the metric chart.

In 205.5 ml, move the decimal point six places to the left.

$$205.5 \ ml = 0.0002055 \ kl$$

Answer   205.5 milliliters = 0.0002055 kiloliters

Try it Now 2

Convert 3,085 milligrams to grams.

Factor Label Method

There is yet another method that you can use to convert metric measurements—the **factor label method**. You used this method when you were converting measurement units within the U.S. customary system.

The factor label method works the same in the metric system; it relies on the use of unit fractions and the reducing of intermediate units. The table below shows some of the **unit equivalents** and **unit fractions** for length in the metric system. (You should notice that all of the unit fractions contain a factor of 10. Remember that the metric system is based on the notion that each unit is 10 times larger than the one that came before it.)

Also, notice that two new prefixes have been added here: mega- (which is very big) and micro- (which is very small).

Unit Equivalents	Conversion Factors	
1 meter = 1,000,000 micrometers	$\dfrac{1\,m}{1,000,000\ \mu m}$	$\dfrac{1,000,000\ \mu m}{1\,m}$
1 meter = 1,000 millimeters	$\dfrac{1\,m}{1,000\ mm}$	$\dfrac{1,000\ mm}{1\,m}$
1 meter = 100 centimeters	$\dfrac{1\,m}{100\ cm}$	$\dfrac{100\ cm}{1\,m}$
1 meter = 10 decimeters	$\dfrac{1\,m}{10\ dm}$	$\dfrac{10\ dm}{1\,m}$
1 dekameter = 10 meters	$\dfrac{1\,dam}{10\ m}$	$\dfrac{10\ m}{1\,dam}$
1 hectometer = 100 meters	$\dfrac{1\,hm}{100\ m}$	$\dfrac{100\ m}{1\,hm}$
1 kilometer = 1,000 meters	$\dfrac{1\,km}{1,000\ m}$	$\dfrac{1,000\ m}{1\,km}$
1 megameter = 1,000,000 meters	$\dfrac{1\,Mm}{1,000,000\ m}$	$\dfrac{1,000,000\ m}{1\,Mm}$

When applying the factor label method in the metric system, be sure to check that you are not skipping over any intermediate units of measurement.

Example	
Problem	**Convert 7,225 centimeters to meters.**

$$7{,}225\ cm = \underline{\quad}\ m$$

Meters is larger than centimeters, so you expect your answer to be less than 7,225.

$$\frac{7{,}225\ cm}{1}\cdot\frac{1\,m}{100\ cm}=\underline{\quad}\ m$$

Using the factor label method, write 7,225 cm as a fraction and use unit fractions to convert it to m.

$$\frac{7{,}225\ \cancel{cm}}{1}\cdot\frac{1\,m}{100\ \cancel{cm}}=\underline{\quad}\ m$$

Reduce similar units, multiply, and simplify.

$$\frac{7{,}225}{1}\cdot\frac{1\,m}{100}=\frac{7{,}225}{100}\,m$$

$$\frac{7{,}225\,m}{100}=72.25\,m$$

| Answer | 7,225 centimeters = **72.25** meters |

Try it Now 3

Using whichever method you prefer, convert 32.5 kilometers to meters.

Now that you have seen how to convert among metric measurements in multiple ways, let's revisit the problem posed earlier.

Example	
Problem	**If you have a prescription for 5,000 mg of medicine, and upon getting it filled, the dosage reads 5 g of medicine, did the pharmacist make a mistake?**
	5,000 mg = ___ g? Need to convert mg to g.

$$\frac{5,000\ mg}{1} \cdot \frac{1g}{1,000\ mg} = \underline{\quad} g$$

$$\frac{5,000\ \cancel{mg}}{1} \cdot \frac{1g}{1,000\ \cancel{mg}} = \underline{\quad} g$$

$$\frac{5,000 \cdot 1g}{1 \cdot 1,000} = \frac{5,000g}{1,000}$$

$$\frac{5,000g}{1,000} = 5\ g$$

| Answer | 5 g = 5,000 mg, so the pharmacist did not make a mistake. |

Summary

To convert among units in the metric system, identify the unit that you have, the unit that you want to convert to, and then count the number of units between them. If you are going from a larger unit to a smaller unit, you multiply by 10 successively. If you are going from a smaller unit to a larger unit, you divide by 10 successively. The factor label method can also be applied to conversions within the metric system. To use the factor label method, you multiply the original measurement by unit fractions; this allows you to represent the original measurement in a different measurement unit.

Try it Now Answers

1. 0.005 dekaliters; one deciliter is 100 times smaller than a dekaliter, so you move the decimal point two places to the left to convert 0.5 deciliters to 0.005 dekaliters.

2. 3.085 grams; one gram is 1,000 times larger than a milligram, so you can move the decimal point in 3,085 three places to the left.

3. 32,500 meters; to find the number of m in 32.5 km, you can set up the following equation: $\frac{32.5 \, km}{1} \cdot \frac{1,000 \, m}{1 \, km} = \frac{32,500 \, m}{1}$. The km units reduce, leaving the answer in meters.

Using Metric Conversions to Solve Problems

Learning Objective(s)
1 Solve application problems involving metric units of length, mass, and volume.

Introduction

Learning how to solve real-world problems using metric conversions is as important as learning how to do the conversions themselves. Mathematicians, scientists, nurses, and even athletes are often confronted with situations where they are presented with information using metric measurements, and must then make informed decisions based on that data.

To solve these problems effectively, you need to understand the context of a problem, perform conversions, and then check the reasonableness of your answer. Do all three of these steps and you will succeed in whatever measurement system you find yourself using.

Understanding Context and Performing Conversions

Objective 1

The first step in solving any real-world problem is to understand its context. This will help you figure out what kinds of solutions are reasonable (and the problem itself may give you clues about what types of conversions are necessary). Here is an example.

Example	
Problem	**In the Summer Olympic Games, athletes compete in races of the following lengths: 100 meters, 200 meters, 400 meters, 800 meters, 1500 meters, 5000 meters and 10,000 meters. If a runner were to run in all these races, how many kilometers would he run?**
	$\begin{array}{r} 10,000 \\ 5,000 \\ 1,500 \\ 800 \\ 400 \\ 200 \\ + \quad 100 \\ \hline 18,000 \end{array}$ To figure out how many kilometers he would run, you need to first add all of the lengths of the races together and *then* convert that measurement to kilometers.
	$\dfrac{18,000 \, m}{1} \cdot \dfrac{1 \, km}{1,000 \, m} = \underline{\quad} \, km$ Use the factor label method and unit fractions to convert

from meters to kilometers.

$$\frac{18,000 \cancel{m}}{1} \cdot \frac{1\,km}{1,000\,\cancel{m}} = \underline{\quad}\ km$$ Reduce, multiply, and solve.

$$\frac{18,000}{1} \cdot \frac{1\,km}{1,000} = \frac{18,000\,km}{1,000}$$

$$\frac{18,000\,km}{1,000} = 18\,km$$

Answer   The runner would run 18 kilometers.

This may not be likely to happen (a runner would have to be quite an athlete to compete in all of these races) but it is an interesting question to consider. The problem required you to find the total distance that the runner would run (in kilometers). The example showed how to add the distances, in meters, and then convert that number to kilometers.

An example with a different context, but still requiring conversions, is shown below.

Example
Problem **One bottle holds 295 dl while another one holds 28,000 ml. What is the difference in capacity between the two bottles?**

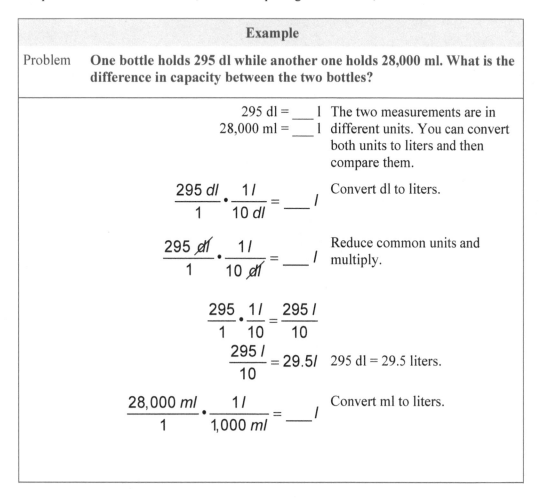

$$295\ dl = \underline{\quad}\ l$$
$$28,000\ ml = \underline{\quad}\ l$$
The two measurements are in different units. You can convert both units to liters and then compare them.

$$\frac{295\ dl}{1} \cdot \frac{1\,l}{10\ dl} = \underline{\quad}\ l$$ Convert dl to liters.

$$\frac{295\ \cancel{dl}}{1} \cdot \frac{1\,l}{10\ \cancel{dl}} = \underline{\quad}\ l$$ Reduce common units and multiply.

$$\frac{295}{1} \cdot \frac{1\,l}{10} = \frac{295\ l}{10}$$

$$\frac{295\ l}{10} = 29.5\,l$$ 295 dl = 29.5 liters.

$$\frac{28,000\ ml}{1} \cdot \frac{1\,l}{1,000\ ml} = \underline{\quad}\ l$$ Convert ml to liters.

$$\frac{28,000 \ \cancel{ml}}{1} \cdot \frac{1 \ l}{1,000 \ \cancel{ml}} = \underline{\quad} \ l$$

$$\frac{28,000}{1} \cdot \frac{1 \ l}{1,000} = \frac{28,000 \ l}{1,000}$$

$$\frac{28,000 \ l}{1,000} = 28 \ l \qquad 28,000 \ ml = 28 \ \text{liters}$$

29.5 liters − 28 liters = 1.5 liters   The question asks for "difference in capacity" between the bottles.

Answer   There is a difference in capacity of 1.5 liters between the two bottles.

This problem asked for the difference between two quantities. The easiest way to find this is to convert one quantity so that both quantities are measured in the same unit, and then subtract one from the other.

Try it Now 1

One boxer weighs in at 85 kg. He is 80 dag heavier than his opponent. How much does his opponent weigh?

Checking your Conversions

Sometimes it is a good idea to check your conversions using a second method. This usually helps you catch any errors that you may make, such as using the wrong unit fractions or moving the decimal point the wrong way.

Example
Problem   **A two-liter bottle contains 87 centiliters of oil and 4.1 deciliters of water. How much more liquid is needed to fill the bottle?**
$87 \ cl + 4.1 \ dl + \underline{\quad} = 2 \ l$ You are looking for the amount of liquid needed to fill the bottle. Convert both measurements to liters and then solve the problem.
$87 \ cl = \underline{\quad} \ l$ Convert 87 cl to liters.
$\dfrac{87 \ cl}{1} \cdot \dfrac{1 \ l}{100 \ cl} = \underline{\quad} \ l$

$$\frac{87 \, \cancel{cl}}{1} \cdot \frac{1 \, l}{100 \, \cancel{cl}} = \underline{\quad} \, l$$

$$\frac{87}{1} \cdot \frac{1 \, l}{100} = \frac{87 \, l}{100}$$

$$\frac{87 \, l}{100} = 0.87l$$

$$4.1 \, dl = \underline{\quad} \, l \qquad \text{Convert 4.1 dl to liters.}$$

$$\frac{4.1 \, dl}{1} \cdot \frac{1 \, l}{10 \, dl} = \underline{\quad} \, l$$

$$\frac{4.1 \, \cancel{dl}}{1} \cdot \frac{1 \, l}{10 \, \cancel{dl}} = \underline{\quad} \, l$$

$$\frac{4.1}{1} \cdot \frac{1 \, l}{10} = \frac{4.1 \, l}{10}$$

$$\frac{4.1 \, l}{10} = 0.41l$$

$$87 \, cl + 4.1 \, dl + \underline{\quad} = 2 \, l \qquad \text{Subtract to find how much} \\ \text{more liquid is needed to fill the}$$

$$0.87 \text{ liter} + 0.41 \text{ liter} + \underline{\quad} = 2 \text{ liters} \quad \text{bottle.}$$

$$\text{·s} - 0.87 \text{ liter} - 0.41 \text{ liter} = 0.72 \text{ liter}$$

Answer   The amount of liquid needed to fill the bottle is 0.72 liter.

Having come up with the answer, you could also check your conversions using the quicker "move the decimal" method, shown below.

Example
Problem   **A two-liter bottle contains 87 centiliters of oil and 4.1 deciliters of water. How much more liquid is needed to fill the bottle?**
$87 \, cl + 4.1 \, dl + \underline{\quad} = 2 \, l$ You are looking for the amount of liquid needed to fill the bottle. $87 \, cl = \underline{\quad} \, l$ Convert 87 cl to liters.

kl   hl   dal   l   dl   cl   ml   On the chart, l is two places to
                 ‿‿‿‿‿‿   the left of cl.
                  2   1

                          .87. cl   Move the decimal point two
                          ‿‿‿   places to the left in 87 cl.
                          2 1

                   87 cl = 0.87 l

                   4.1 dl = ___ l   Convert 4.1 dl to liters.

kl   hl   dal   l   dl   cl   ml   On the chart, l is one place to
                ‿‿‿‿   the left of dl.
                   1

                          .4.1 dl   Move the decimal point one
                          ‿   place to the left in 4.1 dl.
                          1

                   4.1 dl = 0.41 l

        87 cl + 4.1 dl + ___ = 2 l   Subtract to find how much
                                     more liquid is needed to fill the
   0.87 liter + 0.41 liter + ___ = 2 liters   bottle.

   2 liters – 0.87 liter – 0.41 liter = 0.72 liter

Answer   The amount of liquid needed to fill the bottle is 0.72 liter.

The initial answer checks out—0.72 liter of liquid is needed to fill the bottle. Checking one conversion with another method is a good practice for catching any errors in scale.

Summary

Understanding the context of real-life application problems is important. Look for words within the problem that help you identify what operations are needed, and then apply the correct unit conversions. Checking your final answer by using another conversion method (such as the "move the decimal" method, if you have used the factor label method to solve the problem) can cut down on errors in your calculations.

Try it Now Answers

1. 84.2 kg; 80 dag = 0.8 kg, and 85 – 0.8 = 84.2

5.3 Temperature Scales

Learning Objective(s)
1 State the freezing and boiling points of water on the Celsius and Fahrenheit temperature scales.
2 Convert from one temperature scale to the other, using conversion formulas.

Introduction

Turn on the television any morning and you will see meteorologists talking about the day's weather forecast. In addition to telling you what the weather conditions will be like (sunny, cloudy, rainy, muggy), they also tell you the day's forecast for high and low temperatures. A hot summer day may reach 100° in Philadelphia, while a cool spring day may have a low of 40° in Seattle.

If you have been to other countries, though, you may notice that meteorologists measure heat and cold differently outside of the United States. For example, a TV weatherman in San Diego may forecast a high of 89°, but a similar forecaster in Tijuana, Mexico—which is only 20 miles south—may look at the same weather pattern and say that the day's high temperature is going to be 32°. What's going on here?

The difference is that the two countries use different temperature scales. In the United States, temperatures are usually measured using the **Fahrenheit** scale, while most countries that use the metric system use the **Celsius** scale to record temperatures. Learning about the different scales—including how to convert between them—will help you figure out what the weather is going to be like, no matter which country you find yourself in.

Measuring Temperature on Two Scales

Objective 1

Fahrenheit and Celsius are two different scales for measuring temperature.

A thermometer measuring a temperature of 22° Celsius is shown here.

On the Celsius scale, water freezes at 0° and boils at 100°.

If the United States were to adopt the Celsius scale, forecast temperatures would rarely go below -30° or above 45°. (A temperature of -18° may be forecast for a cold winter day in Michigan, while a temperature of 43° may be predicted for a hot summer day in Arizona.)

Most office buildings maintain an indoor temperature between 18°C and 24°C to keep employees comfortable.

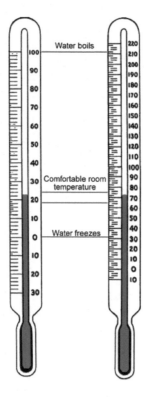

A thermometer measuring a temperature of 72° Fahrenheit is shown here.

On the Fahrenheit scale, water freezes at 32° and boils at 212°.

In the United States, forecast temperatures measured in Fahrenheit rarely go below -20° or above 120°. (A temperature of 0° may be forecast for a cold winter day in Michigan, while a temperature of 110° may be predicted for a hot summer day in Arizona.)

Most office buildings maintain an indoor temperature between 65°F and 75°F to keep employees comfortable.

Celsius   **Fahrenheit**

Try it Now 1

A cook puts a thermometer into a pot of water to see how hot it is. The thermometer reads 132°, but the water is not boiling yet. Which temperature scale is the thermometer measuring?

Converting Between the Scales

Objective 2

By looking at the two thermometers shown, you can make some general comparisons between the scales. For example, many people tend to be comfortable in outdoor temperatures between 50°F and 80°F (or between 10°C and 25°C). If a meteorologist predicts an average temperature of 0°C (or 32°F), then it is a safe bet that you will need a winter jacket.

Sometimes, it is necessary to convert a Celsius measurement to its exact Fahrenheit measurement or vice versa. For example, what if you want to know the temperature of your child in Fahrenheit, and the only thermometer you have measures temperature in Celsius measurement? Converting temperature between the systems is a straightforward process as long as you use the formulas provided below.

Temperature Conversion Formulas

To convert a Fahrenheit measurement to a Celsius measurement, use this formula.

$$C = \frac{5}{9}(F - 32)$$

To convert a Celsius measurement to a Fahrenheit measurement, use this formula.

$$F = \frac{9}{5}C + 32$$

How were these formulas developed? They came from comparing the two scales. Since the freezing point is 0° in the Celsius scale and 32° on the Fahrenheit scale, we subtract 32 when converting from Fahrenheit to Celsius, and add 32 when converting from Celsius to Fahrenheit.

There is a reason for the fractions $\frac{5}{9}$ and $\frac{9}{5}$, also. There are 100 degrees between the freezing (0°) and boiling points (100°) of water on the Celsius scale and 180 degrees between the similar points (32° and 212°) on the Fahrenheit scale. Writing these two scales as a ratio, $\frac{F°}{C°}$, gives $\frac{180°}{100°} = \frac{180° \div 20}{100° \div 20} = \frac{9}{5}$. If you flip the ratio to be $\frac{C°}{F°}$, you get $\frac{100°}{180°} = \frac{100° \div 20}{180° \div 20} = \frac{5}{9}$. Notice how these fractions are used in the conversion formulas.

The example below illustrates the conversion of Celsius temperature to Fahrenheit temperature, using the boiling point of water, which is 100° C.

Example	
Problem	**The boiling point of water is 100°C. What temperature does water boil at in the Fahrenheit scale?**
$F = \frac{9}{5}C + 32$	A Celsius temperature is given. To convert it to the Fahrenheit scale, use the formula at the left.
$F = \frac{9}{5}(100) + 32$	Substitute 100 for *C* and multiply.
$F = \frac{900}{5} + 32$	

$$F = \frac{900 \div 5}{5 \div 5} + 32$$ Simplify $\frac{900}{5}$ by dividing numerator and denominator by 5.

$$F = \frac{180}{1} + 32$$

$$F = 212$$ Add $180 + 32$.

Answer   The boiling point of water is 212°F.

Example
Problem   **Water freezes at 32°F. On the Celsius scale, what temperature is this?**

$$C = \frac{5}{9}(F - 32)$$	A Fahrenheit temperature is given. To convert it to the Celsius scale, use the formula at the left.
$$C = \frac{5}{9}(32 - 32)$$	Substitute 32 for F and subtract.
$$C = \frac{5}{9}(0)$$	Any number multiplied by 0 is 0.
$$C = 0$$	

Answer   The freezing point of water is 0°C.

The two previous problems used the conversion formulas to verify some temperature conversions that were discussed earlier—the boiling and freezing points of water. The next example shows how these formulas can be used to solve a real-world problem using different temperature scales.

Example
Problem   **Two scientists are doing an experiment designed to identify the boiling point of an unknown liquid. One scientist gets a result of 120°C; the other gets a result of 250°F. Which temperature is higher and by how much?**

What is the difference between 120°C and 250°F?	One temperature is given in °C, and the other is given in °F. To find the difference between them, we need to measure them on the same scale.
$$F = \frac{9}{5}C + 32$$	Use the conversion formula to convert 120°C to °F.

(You could convert 250°F to °C instead; this is explained in the text after this example.)

$$F = \frac{9}{5}(120) + 32$$

Substitute 120 for C.

$$F = \frac{1080}{5} + 32$$

Multiply.

$$F = \frac{1080 \div 5}{5 \div 5} + 32$$

Simplify $\dfrac{1080}{5}$ by dividing numerator and denominator by 5.

$$F = \frac{216}{1} + 32$$

Add $216 + 32$.

$$F = 248$$

You have found that 120°C = 248°F.

$$250°F - 248°F = 2°F$$

To find the difference between 248°*F* and 250°F, subtract.

Answer   250°F is the higher temperature by 2°F.

You could have converted 250°F to °C instead, and then found the difference in the two measurements. (Had you done it this way, you would have found that 250°F = 121.1°C, and that 121.1°C is 1.1°C higher than 120°C.) Whichever way you choose, it is important to compare the temperature measurements within the same scale, and to apply the conversion formulas accurately.

Try it Now 2

Tatiana is researching vacation destinations, and she sees that the average summer temperature in Barcelona, Spain is around 26°C. What is the average temperature in degrees Fahrenheit?

Summary

Temperature is often measured in one of two scales: the Celsius scale and the Fahrenheit scale. A Celsius thermometer will measure the boiling point of water at 100° and its freezing point at 0°; a Fahrenheit thermometer will measure the same events at 212° for the boiling point of water and 32° as its freezing point. You can use conversion formulas to convert a measurement made in one scale to the other scale.

Try it Now Answers

1. Fahrenheit; water boils at 212° on the Fahrenheit scale, so a measurement of 132° on a Fahrenheit scale is legitimate for hot (but non-boiling) water.

2. 79°F; Tatiana can find the Fahrenheit equivalent by solving the equation $F = \frac{9}{5}(26) + 32$. The result is 78.8°F, which rounds to 79°F.

Unit Recap

5.1 Length

The four basic units of measurement that are used in the U.S. customary measurement system are: inch, foot, yard, and mile. Typically, people use yards, miles, and sometimes feet to describe long distances. Measurement in inches is common for shorter objects or lengths.

You need to convert from one unit of measure to another if you are solving problems that include measurements involving more than one type of measurement. Each of the units can be converted to one of the other units using the table of equivalents, the conversion factors, and/or the factor label method shown in this topic.

5.1 Weight

In the U.S. customary system of measurement, weight is measured in three units: ounces, pounds, and tons. A pound is equivalent to 16 ounces, and a ton is equivalent to 2,000 pounds. While an object's weight can be described using any of these units, it is typical to describe very heavy objects using tons and very light objects using an ounce. Pounds are used to describe the weight of many objects and people. Often, in order to compare the weights of two objects or people or to solve problems involving weight, you must convert from one unit of measurement to another unit of measurement. Using conversion factors with the factor label method is an effective strategy for converting units and solving problems.

5.1 Capacity

There are five basic units for measuring capacity in the U.S. customary measurement system. These are the fluid ounce, cup, pint, quart, and gallon. These measurement units are related to one another, and capacity can be described using any of the units. Typically, people use gallons to describe larger quantities and fluid ounces, cups, pints, or quarts to describe smaller quantities. Often, in order to compare or to solve problems involving the amount of liquid in a container, you need to convert from one unit of measurement to another.

5.2 The Metric System

The metric system is an alternative system of measurement used in most countries, as well as in the United States. The metric system is based on joining one of a series of prefixes, including kilo-, hecto-, deka-, deci-, centi-, and milli-, with a base unit of measurement, such as meter, liter, or gram. Units in the metric system are all related by a power of 10, which means that each successive unit is 10 times larger than the previous one. This makes converting one metric measurement to another a straightforward process, and is often as simple as moving a decimal point. It is always important, though, to consider the direction of the conversion. If you are converting a smaller unit to a larger unit, then the decimal point has to move to the left (making your number smaller); if you are converting a larger unit to a smaller unit, then the decimal point has to move to the right (making your number larger).

5.2 Converting within the Metric System

To convert among units in the metric system, identify the unit that you have, the unit that you want to convert to, and then count the number of units between them. If you are going from a larger unit to a smaller unit, you multiply by 10 successively. If you are going from a smaller unit to a larger unit, you divide by 10 successively. The factor label method can also be applied to conversions within the metric system. To use the factor label method, you multiply the original measurement by unit fractions; this allows you to represent the original measurement in a different measurement unit.

5.2 Using Metric Conversion to Solve Problems

Understanding the context of real-life application problems is important. Look for words within the problem that help you identify what operations are needed, and then apply the correct unit conversions. Checking your final answer by using another conversion method (such as the "move the decimal" method, if you have used the factor label method to solve the problem) can cut down on errors in your calculations.

5.3 Temperature Scales

Temperature is often measured in one of two scales: the Celsius scale and the Fahrenheit scale. A Celsius thermometer will measure the boiling point of water at 100° and its freezing point at 0°; a Fahrenheit thermometer will measure the same events at 212° for the boiling point of water and 32° as its freezing point. You can use conversion formulas to convert a measurement made in one scale to the other scale.

Glossary

capacity	The amount of liquid (or other pourable substance) that an object can hold when it's full.
Celsius	A measure of temperature commonly used in countries that use the metric system. On the Celsius scale, water freezes at 0° and boils at 100°.
cup	A unit of capacity equal to 8 fluid ounces.
factor label method	One method of converting a measurement from one unit of measurement to another unit of measurement. In this method, you multiply the original measurement by unit fractions containing different units of measurement to obtain the new unit of measurement.
Fahrenheit	A measure of temperature commonly used in the United States. On the Fahrenheit scale, water freezes at 32° F and boils at 212° F.
fluid ounce	A unit of capacity equal to $\frac{1}{8}$ of a cup. One fluid ounce of water at 62°F weighs about one ounce.
foot	A unit for measuring length in the U.S. customary measurement system. 1 foot = 12 inches
gallon **gram** **inch**	A unit equal to 4 quarts, or 128 fluid ounces. The base unit of mass in the Metric system. A unit for measuring length in the U.S. customary measurement system. 1 foot = 12 inches
length	The distance from one end to the other or the distance from one point to another.
liter	The base unit of volume in the Metric system.
measurement	The use of standard units to find out the size or quantity of items such as length, width, height, mass, weight, volume, temperature or time.
meter	The base unit of length in the Metric system.
metric system	A widely-used system of measurement that is based on the decimal system and multiples of 10.
mile	A unit for measuring length in the U.S. customary measurement system. 1 mile = 5,280 feet or 1,760 yards.
ounce	A unit for measuring weight in the U.S. customary measurement system. 16 ounces = 1 pound.
pint	A unit of capacity equal to 16 fluid ounces, or 2 cups.

pound	A unit for measuring weight in the U.S. customary measurement system. 16 ounces = 1 pound.
prefix	A short set of letters that denote the size of measurement units in the Metric System. Metric prefixes include centi-, milli-, kilo-, and hecto-.
quart	A unit of capacity equal to 32 fluid ounces, or 4 cups.
ton	A unit for measuring the weight of heavier items in the U.S. customary measurement system. 1 ton = 2,000 pounds.
U.S. customary measurement system	The most common system of measurement used in the United States. It is based on English measurement systems of the 18th century.
unit equivalents	Statements of equivalence between measurement units within a system or in comparison to another system of units. For example, 1 foot = 12 inches or 1 inch = 2.54 centimeters are both examples of unit equivalents.
unit fractions	A fraction where the numerator and denominator are equal amounts, as in $\frac{1kg}{1000g}$ or $\frac{12 inches}{1 foot}$. Unit fractions serve to help with conversions in the Factor Label method.
unit of measurement	A standard amount or quantity. For example, an inch is a unit of measurement.
weight	A mathematical description of how heavy an object is.
yard	A unit for measuring length in the U.S. customary measurement system. 1 yard = 3 feet or 36 inches.

To see these and all other available Instructor Resources, visit the NROC Network.

Exercises

For 1-26, convert the measurements. *Round your answer up to 3 decimal places, if needed.*

1. 5 yards = _____ feet
2. 6 yards = _____ inches
3. 108 inches = _____ yards
4. 257 miles = _____ feet
5. 253 miles = _____ feet
6. 8 pounds = _____ ounces
7. 370 ounces = _____ ton
8. 417 ounces = _____ ton
9. 112 cups = _____ gallons
10. $8\frac{1}{2}$ gallons = _____ cups
11. 7 quarts = _____ fluid ounces
12. 9 fluid ounces = _____ quarts
13. 8 meters = _____ centimeters
14. 4580 milliliters = _____ liters
15. 3520 milliliters = _____ liters
16. 2 millimeters = _____ meters
17. 171 centimeters = _____ meters
18. 5,180 milliliters = _____ liters
19. 2,610 grams = _____ kilograms
20. 2 kilograms = _____ grams
21. 58.3 millimeters = _____ centimeters
22. 0.3429 kilograms = _____ milligrams
23. $14°C =$ _____ $°F$
24. $41°F =$ _____ $°C$
25. $12°C =$ _____ $°F$
26. $99°F =$ _____ $°C$

Exploration

27. Apollo Spas services 281 hot tubs. If each hot tub needs 105 milliliters of muriatic acid, how many liters of acid are needed for all of the hot tubs?
28. The photo sharing site Flickr had 2.7 billion photos in June 2012. Create a comparison to understand this number by assuming each picture is about 2 megabytes in size, and comparing to the data stored on other media like DVDs, iPods, or flash drives.
29. In June 2012, Twitter was reporting 400 million tweets per day. Each tweet can consist of up to 140 characters (letter, numbers, etc.). Create a comparison to help understand the amount of tweets in a year by imagining each character was a drop of water and comparing to filling something up.
30. During the landing of the Mars Science Laboratory *Curiosity*, it was reported that the signal from the rover would take 14 minutes to reach earth. Radio signals travel at the speed of light, about 186,000 miles per second. How far was Mars from Earth when *Curiosity* landed?
31. It is estimated that a driver takes, on average, 1.5 seconds from seeing an obstacle to reacting by applying the brake or swerving. How far will a car traveling at 60 miles per hour travel (in feet) before the driver reacts to an obstacle?

Chapter 6: Geometry

Table of Contents

In this chapter, we discuss basic shapes and solids of geometry in addition to calculating perimeter, volume, and areas of these shapes and solids. We, also, use these concepts for application problems involving geometry.

6.1   Basic Geometric Concepts and Figures

> **Learning Objective(s)**
> 1 Identify and define points, lines, line segments, rays and planes.
> 2 Classify angles as acute, right, obtuse, or straight.

Introduction

You use geometric terms in everyday language, often without thinking about it. For example, any time you say "walk along this line" or "watch out, this road quickly angles to the left" you are using geometric terms to make sense of the environment around you. You use these terms flexibly, and people generally know what you are talking about.

In the world of mathematics, each of these geometric terms has a specific definition. It is important to know these definitions—as well as how different figures are constructed—to become familiar with the language of geometry. Let's start with a basic geometric figure: the plane.

Figures on a Plane

Objective 1

A **plane** is a flat surface that continues forever (or, in mathematical terms, infinitely) in every direction. It has two dimensions: length and width.

You can visualize a plane by placing a piece of paper on a table. Now imagine that the piece of paper stays perfectly flat and extends as far as you can see in two directions, left-to-right and front-to-back. This gigantic piece of paper gives you a sense of what a geometric plane is like: it continues infinitely in two directions. (Unlike the piece of paper example, though, a geometric plane has no height.)

A plane can contain a number of geometric figures. The most basic geometric idea is a **point**, which has no dimensions. A point is simply a location on the plane. It is represented by a dot. Three points that don't lie in a straight line will determine a plane.

The image below shows four points, labeled *A*, *B*, *C*, and *D*.

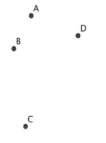

Figure 1 A set of points

Two points on a plane determine a line. A **line** is a one-dimensional figure that is made up of an infinite number of individual points placed side by side. In geometry, all lines are assumed to be straight; if they bend they are called a curve. A line continues infinitely in two directions.

Below is line *AB* or, in geometric notation, $\overleftrightarrow{AB}$. The arrows indicate that the line keeps going forever in the two directions. This line could also be called line *BA*. While the order of the points does not matter for a line, it is customary to name the two points in alphabetical order.

The image below shows the points *A* and *B* and the line $\overleftrightarrow{AB}$.

Figure 2 Line AB

Example	
Problem	**Name the line shown in red.**
	The red line goes through the points *C* and *F*, so the line is $\overleftrightarrow{CF}$.
Answer	$\overleftrightarrow{CF}$

There are two more figures to consider. The section between any two points on a line is called a **line segment**. A line segment can be very long, very short, or somewhere in between. The difference between a line and a line segment is that the line segment has two endpoints and a line goes on forever. A line segment is denoted by its two endpoints, as in $\overline{CD}$.

Figure 3 Line segment CD

A **ray** has one endpoint and goes on forever in one direction. Mathematicians name a ray with notation like $\overrightarrow{EF}$, where point E is the endpoint and F is a point on the ray. When naming a ray, we always say the endpoint first. Note that $\overrightarrow{FE}$ would have the endpoint at F, and continue through E, which is a different ray than $\overrightarrow{EF}$, which would have an endpoint at E, and continue through F.

The term "ray" may be familiar because it is a common word in English. "Ray" is often used when talking about light. While a ray of light resembles the geometric term "ray," it does not go on forever, and it has some width. A geometric ray has no width; only length.

Below is an image of ray *EF* or $\overrightarrow{EF}$. Notice that the end point is *E*.

Figure 4 Ray EF

Example

Problem   **Identify each line and line segment in the picture below.**

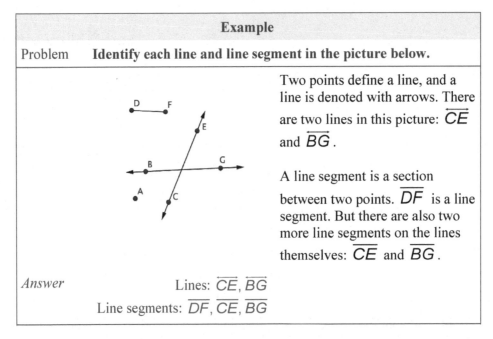

Two points define a line, and a line is denoted with arrows. There are two lines in this picture: $\overleftrightarrow{CE}$ and $\overleftrightarrow{BG}$.

A line segment is a section between two points. $\overline{DF}$ is a line segment. But there are also two more line segments on the lines themselves: $\overline{CE}$ and $\overline{BG}$.

Answer   Lines: $\overleftrightarrow{CE}, \overleftrightarrow{BG}$

Line segments: $\overline{DF}, \overline{CE}, \overline{BG}$

Example

Problem   **Identify each point and ray in the picture below.**

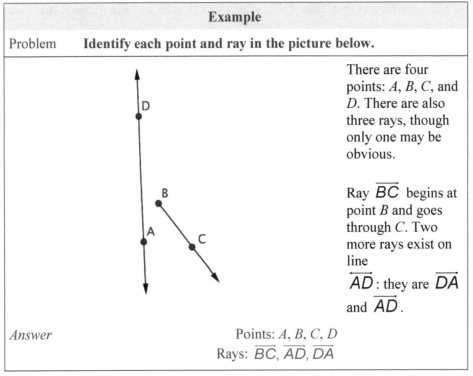

There are four points: *A*, *B*, *C*, and *D*. There are also three rays, though only one may be obvious.

Ray $\overrightarrow{BC}$ begins at point *B* and goes through *C*. Two more rays exist on line $\overleftrightarrow{AD}$: they are $\overrightarrow{DA}$ and $\overrightarrow{AD}$.

Answer   Points: *A*, *B*, *C*, *D*

Rays: $\overrightarrow{BC}, \overrightarrow{AD}, \overrightarrow{DA}$

Try it Now 1

Which of the following is *not* represented in the image below?

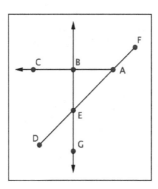

A) *BG*
B) *BA*
C) $\overline{DF}$
D) *AC*

Angles

Lines, line segments, points, and rays are the building blocks of other figures. For example, two rays with a common endpoint make up an **angle**. The common endpoint of the angle is called the **vertex**.

The angle *ABC* is shown below. This angle can also be called $\angle ABC$, $\angle CBA$ or simply $\angle B$. When you are naming angles, be careful to include the vertex (here, point *B*) as the middle letter.

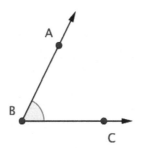

The image below shows a few angles on a plane. Notice that the label of each angle is written "point-vertex-point," and the geometric notation is in the form $\angle ABC$.

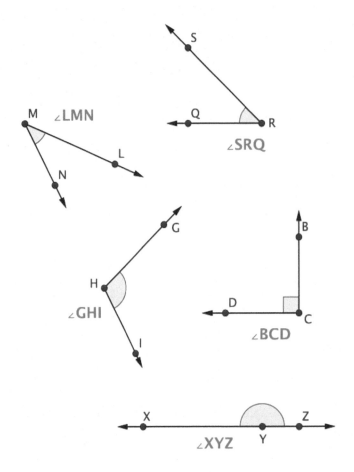

Sometimes angles are very narrow; sometimes they are very wide. When people talk about the "size" of an angle, they are referring to the arc between the two rays. The length of the rays has nothing to do with the size of the angle itself. Drawings of angles will often include an arc (as shown above) to help the reader identify the correct 'side' of the angle.

Think about an analog clock face. The minute and hour hands are both fixed at a point in the middle of the clock. As time passes, the hands rotate around the fixed point, making larger and smaller angles as they go. The length of the hands does not impact the angle that is made by the hands.

An angle is measured in degrees, represented by the symbol °. A circle is defined as having 360°. (In skateboarding and basketball, "doing a 360" refers to jumping and doing one complete body rotation.

A **right angle** is any degree that measures exactly 90°. This represents exactly one-quarter of the way around a circle. Rectangles contain exactly four right angles. A corner mark is often used to denote a right angle, as shown in right angle DCB below.

Angles that are between 0° and 90° (smaller than right angles) are called **acute angles**. Angles that are between 90° and 180° (larger than right angles and less than 180°) are called **obtuse angles**. And an angle that measures exactly 180° is called a **straight angle** because it forms a straight line.

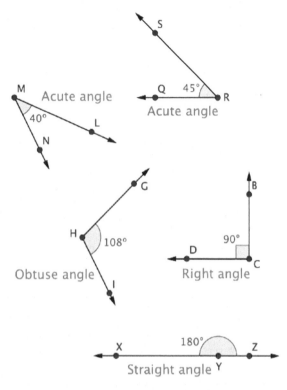

Figure 5 Examples of Angles

Example
Problem   **Label each angle below as acute, right, or obtuse.**

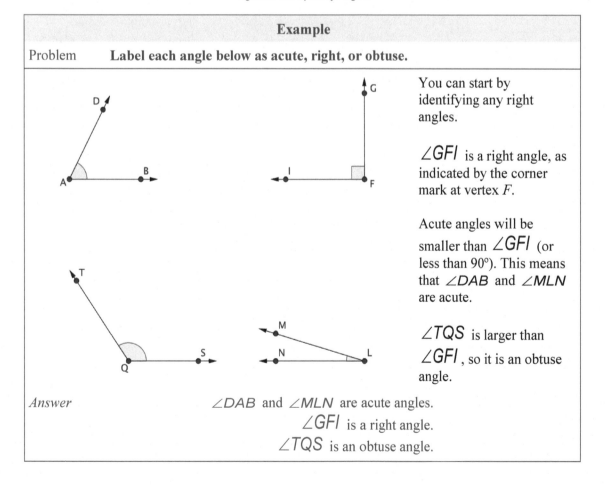

You can start by identifying any right angles.

$\angle GFI$ is a right angle, as indicated by the corner mark at vertex *F*.

Acute angles will be smaller than $\angle GFI$ (or less than 90°). This means that $\angle DAB$ and $\angle MLN$ are acute.

$\angle TQS$ is larger than $\angle GFI$, so it is an obtuse angle.

Answer $\angle DAB$ and $\angle MLN$ are acute angles.
$\angle GFI$ is a right angle.
$\angle TQS$ is an obtuse angle.

Example	
Problem	**Identify each point, ray, and angle in the picture below.**

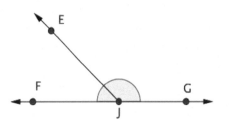

Begin by identifying each point in the figure. There are 4: *E*, *F*, *G*, and *J*.

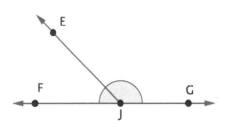

Now find rays. A ray begins at one point, and then continues through another point towards infinity (indicated by an arrow). Three rays start at point *J*: $\overrightarrow{JE}$, $\overrightarrow{JF}$, and $\overrightarrow{JG}$. But also notice that a ray could start at point *F* and go through *J* and *G*, and another could start at point *G* and go through *J* and *F*. These rays can be represented by $\overrightarrow{GF}$ and $\overrightarrow{FG}$.

Finally, look for angles. $\angle EJG$ is obtuse, $\angle EJF$ is acute, and $\angle FJG$ is straight. (Don't forget those straight angles!)

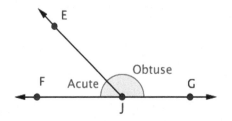

Answer

Points: *E*, *F*, *G*, *J*

Rays: $\overrightarrow{JE}$, $\overrightarrow{JG}$, $\overrightarrow{JF}$, $\overrightarrow{GF}$, $\overrightarrow{FG}$

Angles: $\angle EJG$, $\angle EJF$, $\angle FJG$

Try it Now 2

Identify the acute angles in the given image:

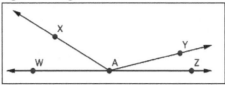

Measuring Angles with a Protractor

Learning how to measure angles can help you become more comfortable identifying the difference between angle measurements. For instance, how is a 135° angle different from a 45° angle?

Measuring angles requires a **protractor**, which is a semi-circular tool containing 180 individual hash marks. Each hash mark represents 1°. (Think of it like this: a circle is 360°, so a semi-circle is 180°.) To use the protractor, do the following three steps:

Step 1. Line up the vertex of the angle with the dot in the middle of the flat side (bottom) of the protractor,

Step 2. Align one side of the angle with the line on the protractor that is at the zero-degree mark, and

Step 3. Look at the curved section of the protractor to read the measurement.

The example below shows you how to use a protractor to measure the size of an angle.

Example
Problem   **Use a protractor to measure the angle shown below.**

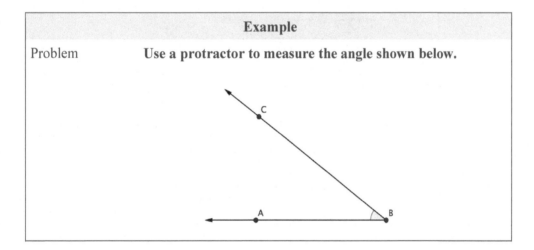

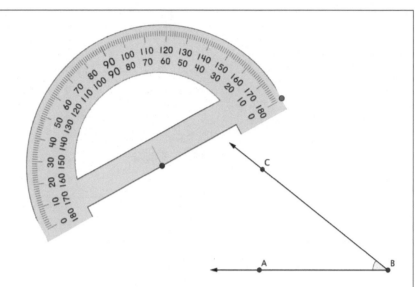

Use a protractor to measure the angle.

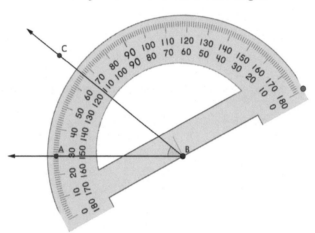

Align the blue dot on the protractor with the vertex of the angle you
want to measure.

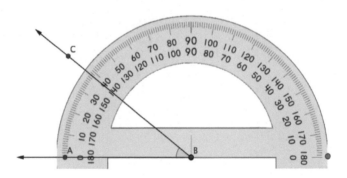

Rotate the protractor around the vertex of the angle until the side of the
angle is aligned with the 0 degree mark of the protractor.

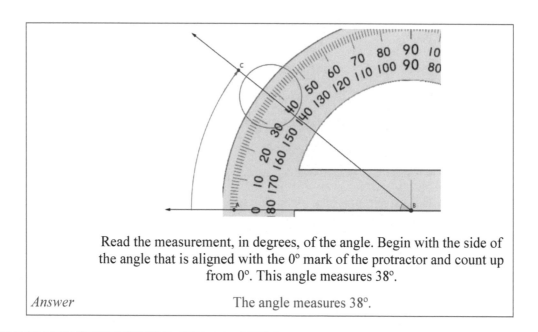

Read the measurement, in degrees, of the angle. Begin with the side of the angle that is aligned with the 0° mark of the protractor and count up from 0°. This angle measures 38°.

Answer   The angle measures 38°.

Try it Now 3

What is the measurement of the angle shown below?

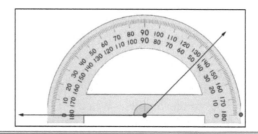

Summary

Geometry begins with simple concepts like points, lines, segments, rays, etc. and expands with angles. As we can see from this section, there are multiple types of angles and several ways to measure them. The most accurate way of measuring an angle is using a protractor. When we put angles together, we obtain geometric shapes and solids, which we discuss in future sections. Next, we discuss lines, and using properties to obtain measures of angles.

Try it Now Answers

1. *BA*; this image does not show any ray that begins at point *B* and goes through point *A*.

2. ∠*WAX* and ∠*YAZ*; both ∠*WAX* and ∠*YAZ* are acute angles.

3. 135°; this protractor is aligned correctly, and the correct measurement is 135°.

6.1.1 Properties of Angles

Learning Objective(s)
1 Identify parallel and perpendicular lines.
2 Find measures of angles.
3 Identify complementary and supplementary angles.

Introduction

Imagine two separate and distinct lines on a plane. There are two possibilities for these lines: they will either intersect at one point, or they will never intersect. When two lines intersect, four angles are formed. Understanding how these angles relate to each other can help you figure out how to measure them, even if you only have information about the size of one angle.

Parallel and Perpendicular

Objective 1

Parallel lines are two or more lines that never intersect. Likewise, parallel line segments are two line segments that never intersect even if the line segments were turned into lines that continued forever. Examples of parallel line segments are all around you, in the two sides of this page and in the shelves of a bookcase. When you see lines or structures that seem to run in the same direction, never cross one another, and are always the same distance apart, there's a good chance that they are parallel.

Perpendicular lines are two lines that intersect at a 90° (right) angle. And perpendicular line segments also intersect at a 90° (right) angle. You can see examples of perpendicular lines everywhere as well—on graph paper, in the crossing pattern of roads at an intersection, to the colored lines of a plaid shirt. In our daily lives, you may be happy to call two lines perpendicular if they merely seem to be at right angles to one another. When studying geometry, however, you need to make sure that two lines intersect at a 90° angle before declaring them to be perpendicular.

The image below shows some parallel and perpendicular lines. The geometric symbol for parallel is ||, so you can show that $AB \parallel CD$. Parallel lines are also often indicated by the marking >> on each line (or just a single > on each line). Perpendicular lines are indicated by the symbol ⊥, so you can write $\overrightarrow{WX} \perp \overrightarrow{YZ}$.

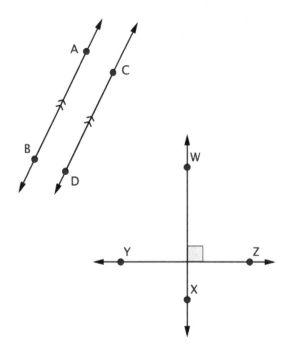

If two lines are parallel, then any line that is perpendicular to one line will also be perpendicular to the other line. Similarly, if two lines are both perpendicular to the same line, then those two lines are parallel to each other. Let's take a look at one example and identify some of these types of lines.

Example	
Problem	**Identify a set of parallel lines and a set of perpendicular lines in the image below.**

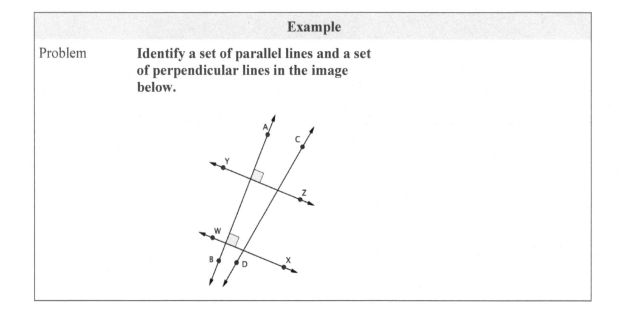

Parallel lines never meet, and perpendicular lines intersect at a right angle.

$\overleftrightarrow{AB}$ and $\overleftrightarrow{CD}$ do not intersect in this image, but if you imagine extending both lines, they will intersect soon. So, they are neither parallel nor perpendicular.

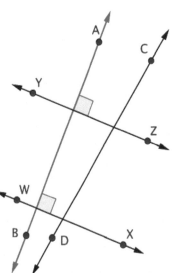

$\overleftrightarrow{AB}$ is perpendicular to both $\overleftrightarrow{WX}$ and $\overleftrightarrow{YZ}$, as indicated by the right-angle marks at the intersection of those lines.

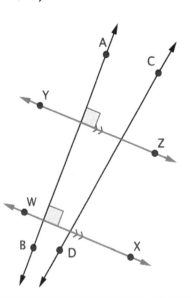

Since $\overleftrightarrow{AB}$ is perpendicular to both lines, then $\overleftrightarrow{WX}$ and $\overleftrightarrow{YZ}$ are parallel.

Answer	$\overrightarrow{WX} \parallel \overrightarrow{YZ}$
	$\overline{AB} \perp \overrightarrow{WX}, \ \overline{AB} \perp \overrightarrow{YZ}$

Try it Now 1

Which statement most accurately represents the image below?

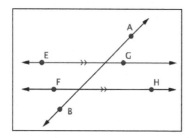

A) $EF \parallel GH$
B) $AB \perp EG$
C) $FH \parallel EG$
D) $AB \parallel FH$

Finding Angle Measurements   Objective 2

Understanding how parallel and perpendicular lines relate can help you figure out the measurements of some unknown angles. To start, all you need to remember is that perpendicular lines intersect at a 90° angle, and that a straight angle measures 180°.

The measure of an angle such as $\angle A$ is written as $m\angle A$. Look at the example below. How can you find the measurements of the unmarked angles?

	Example
Problem	**Find the measurement of $\angle IJF$.**
	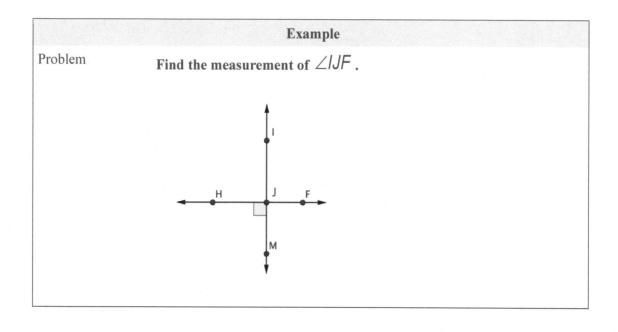

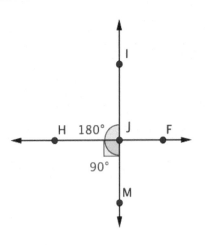

Only one angle, $\angle HJM$, is marked in the image. Notice that it is a right angle, so it measures 90°.

$\angle HJM$ is formed by the intersection of lines $\overleftrightarrow{IM}$ and $\overleftrightarrow{HF}$. Since $\overleftrightarrow{IM}$ is a line, $\angle IJM$ is a straight angle measuring 180°.

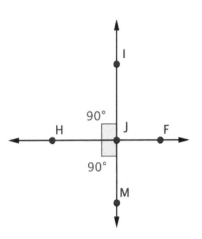

You can use this information to find the measurement of $\angle HJI$:

$$m\angle HJM + m\angle HJI = m\angle IJM$$
$$90° + m\angle HJI = 180°$$
$$m\angle HJI = 90°$$

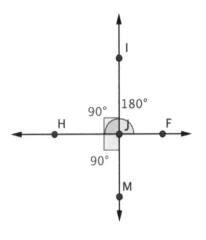

Now use the same logic to find the measurement of $\angle IJF$.

$\angle IJF$ is formed by the intersection of lines $\overleftrightarrow{IM}$ and $\overleftrightarrow{HF}$. Since $\overleftrightarrow{HF}$ is a line, $\angle HJF$ will be a straight angle measuring 180°.

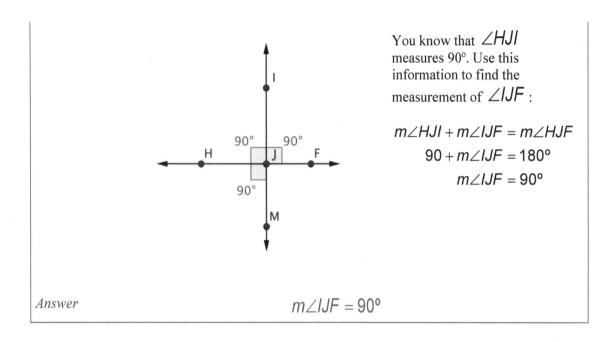

You know that $\angle HJI$ measures 90°. Use this information to find the measurement of $\angle IJF$:

$$m\angle HJI + m\angle IJF = m\angle HJF$$
$$90 + m\angle IJF = 180°$$
$$m\angle IJF = 90°$$

Answer $m\angle IJF = 90°$

In this example, you may have noticed that angles $\angle HJI$, $\angle IJF$, and $\angle HJM$ are all right angles. (If you were asked to find the measurement of $\angle FJM$, you would find that angle to be 90°, too.) This is what happens when two lines are perpendicular—the four angles created by the intersection are all right angles.

Not all intersections happen at right angles, though. In the example below, notice how you can use the same technique as shown above (using straight angles) to find the measurement of a missing angle.

Example

Problem   **Find the measurement of $\angle DAC$.**

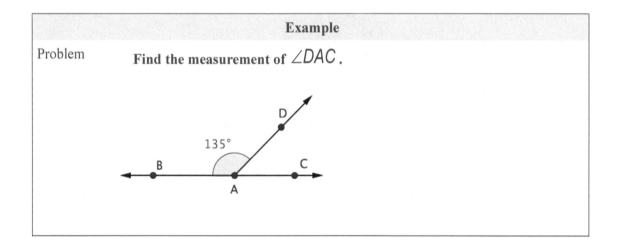

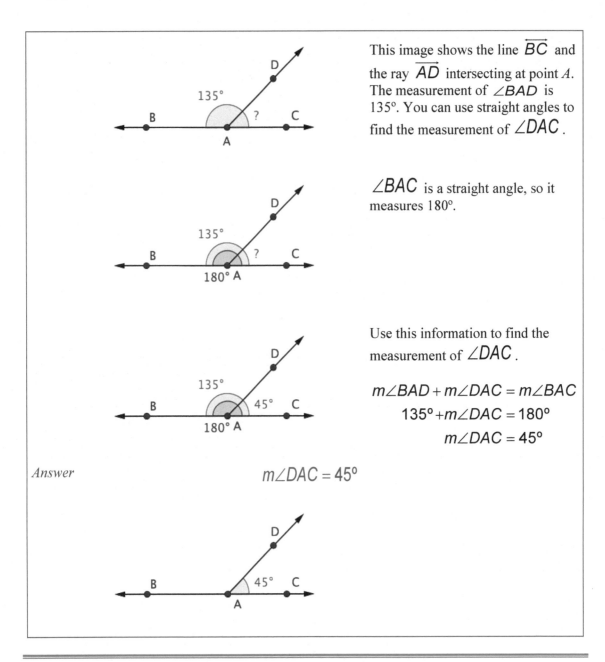

This image shows the line $\overleftrightarrow{BC}$ and the ray $\overrightarrow{AD}$ intersecting at point A. The measurement of $\angle BAD$ is 135°. You can use straight angles to find the measurement of $\angle DAC$.

$\angle BAC$ is a straight angle, so it measures 180°.

Use this information to find the measurement of $\angle DAC$.

$$m\angle BAD + m\angle DAC = m\angle BAC$$
$$135° + m\angle DAC = 180°$$
$$m\angle DAC = 45°$$

Answer $m\angle DAC = 45°$

Try it Now 2

Find the measurement of $\angle CAD$.

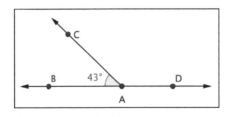

Supplementary and Complementary

In the example above, $m\angle BAC$ and $m\angle DAC$ add up to 180°. Two angles whose measures add up to 180° are called **supplementary angles**. There's also a term for two angles whose measurements add up to 90°, they are called **complementary angles**.

One way to remember the difference between the two terms is that *"c*orner" and *"c*omplementary" each begin with *c* (a 90° angle looks like a corner), while *s*traight and *"s*upplementary" each begin with *s* (a straight angle measures 180°).

If you can identify supplementary or complementary angles within a problem, finding missing angle measurements is often simply a matter of adding or subtracting.

Example
Problem   **Two angles are supplementary. If one of the angles measures 48°, what is the measurement of the other angle?**

$$m\angle A + m\angle B = 180°$$	Two supplementary angles make up a straight angle, so the measurements of the two angles will be 180°.
$48° + m\angle B = 180°$ $m\angle B = 180° - 48°$ $m\angle B = 132°$	You know the measurement of one angle. To find the measurement of the second angle, subtract 48° from 180°.
Answer The measurement of the other angle is 132°.	

Example
Problem   **Find the measurement of $\angle AXZ$.**

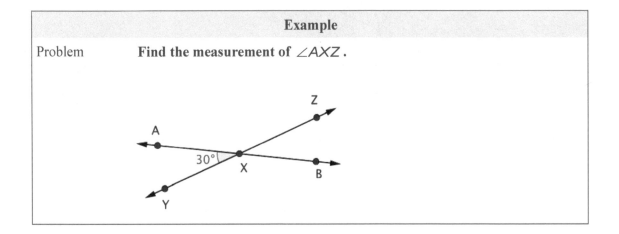

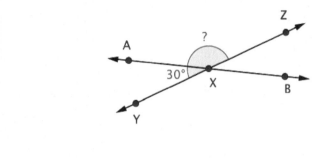

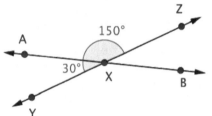

This image shows two intersecting lines, $\overrightarrow{AB}$ and $\overrightarrow{YZ}$. They intersect at point X, forming four angles.

Angles $\angle AXY$ and $\angle AXZ$ are supplementary because together they make up the straight angle $\angle YXZ$.

Use this information to find the measurement of $\angle AXZ$.

$$m\angle AXY + m\angle AXZ = m\angle YXZ$$
$$30° + m\angle AXZ = 180°$$
$$m\angle AXZ = 150°$$

Answer $m\angle AXZ = 150°$

Example

Problem   **Find the measurement of $\angle BAC$.**

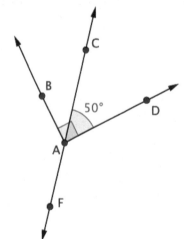

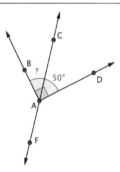

This image shows the line $\overleftrightarrow{CF}$ and the rays $\overrightarrow{AB}$ and $\overrightarrow{AD}$, all intersecting at point A. Angle $\angle BAD$ is a right angle.

Angles $\angle BAC$ and $\angle CAD$ are complementary, because together they create $\angle BAD$.

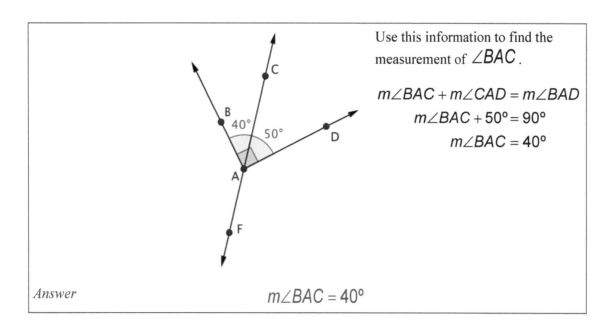

Use this information to find the measurement of $\angle BAC$.

$$m\angle BAC + m\angle CAD = m\angle BAD$$
$$m\angle BAC + 50° = 90°$$
$$m\angle BAC = 40°$$

Answer $m\angle BAC = 40°$

Example

Problem   **Find the measurement of $\angle CAD$.**

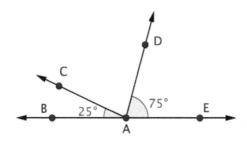

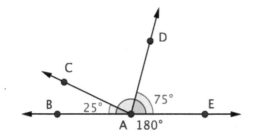

You know the measurements of two angles here: $\angle CAB$ and $\angle DAE$. You also know that $m\angle BAE = 180°$.

Use this information to find the measurement of $\angle CAD$.

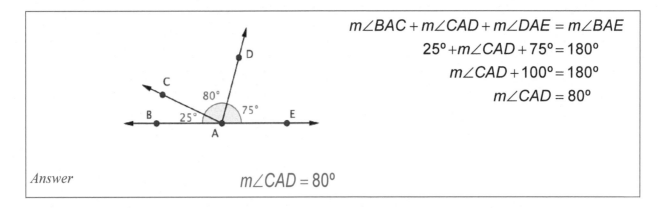

$$m\angle BAC + m\angle CAD + m\angle DAE = m\angle BAE$$
$$25° + m\angle CAD + 75° = 180°$$
$$m\angle CAD + 100° = 180°$$
$$m\angle CAD = 80°$$

Answer $m\angle CAD = 80°$

Try it Now 3

Which pair of angles is complementary?

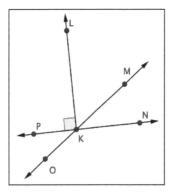

A) $\angle PKO$ and $\angle MKN$

B) $\angle PKO$ and $\angle PKM$

C) $\angle LKP$ and $\angle LKN$

D) $\angle LKM$ and $\angle MKN$

Summary

Parallel lines do not intersect, while perpendicular lines cross at a 90° angle. Two angles whose measurements add up to 180° are said to be supplementary, and two angles whose measurements add up to 90° are said to be complementary. For most pairs of intersecting lines, all you need is the measurement of one angle to find the measurements of all other angles formed by the intersection.

Try it Now Answers

1. C) $FH \parallel EG$; both EG and FH are marked with $>>$ on each line, and those markings mean they are parallel.

2. 137°; $\angle BAD$ is a straight angle measuring 180°. Since $\angle BAC$ measures 43°, the measure of $\angle CAD$ must be $180° - 43° = 137°$.

3. D) ∠*LKM* and ∠*MKN* ; the measurements of two complementary angles will add up to 90°. ∠*LKP* is a right angle, so ∠*LKN* must be a right angle as well. ∠*LKM* + ∠*MKN* = ∠*LKN* , so ∠*LKM* and ∠*MKN* are complementary.

6.1.2 Triangles

> **Learning Objective(s)**
> 1 Identify equilateral, isosceles, scalene, acute, right, and obtuse triangles.
> 2 Identify whether triangles are similar, congruent, or neither.
> 3 Identify corresponding sides of congruent and similar triangles.
> 4 Find the missing measurements in a pair of similar triangles.
> 5 Solve application problems involving similar triangles.

Introduction

Geometric shapes, also called figures, are an important part of the study of geometry. The **triangle** is one of the basic shapes in geometry. It is the simplest shape within a classification of shapes called **polygons**. All triangles have three sides and three angles, but they come in many different shapes and sizes. Within the group of all triangles, the characteristics of a triangle's sides and angles are used to classify it even further. Triangles have some important characteristics, and understanding these characteristics allows you to apply the ideas in real-world problems.

Classifying and Naming Triangles

Objective 1

A polygon is a closed plane figure with three or more straight sides. Polygons each have a special name based on the number of sides they have. For example, the polygon with three sides is called a triangle because "tri" is a prefix that means "three." Its name also indicates that this polygon has three angles. The prefix "poly" means many.

The table below shows and describes three classifications of triangles. Notice how the types of angles in the triangle are used to classify the triangle.

Name of Triangle	Picture of Triangle	Description
Acute Triangle		A triangle with 3 acute angles (3 angles measuring between 0° and 90°).
Obtuse Triangle		A triangle with 1 obtuse angle (1 angle measuring between 90° and 180°).
Right Triangle		A triangle containing one right angle (1 angle that measures 90°). Note that the right angle is shown with a corner mark and does not need to be labeled 90°.

The sum of the measures of the three interior angles of a triangle is always 180°. This fact can be applied to find the measure of the third angle of a triangle, if you are given the other two. Consider the examples below.

Example
Problem   **A triangle has two angles that measure 35° and 75°. Find the measure of the third angle.**
$35° + 75° + x = 180°$ The sum of the three interior angles of a triangle is 180°. $110° + x = 180°$ Find the value of x. $x = 180° - 110°$ $x = 70°$
Answer   The third angle of the triangle measures 70°.

Example
Problem   **One of the angles in a right triangle measures 57°. Find the measurement of the third angle.**
$57° + 90° + x = 180°$ The sum of the three angles of a triangle is 180°. One of the angles has a measure of 90° as it is a right triangle. $147° + x = 180°$ Simplify. $x = 180° - 147°$ Find the value of x. $x = 33°$
Answer   The third angle of the right triangle measures 33°.

There is an established convention for naming triangles. The labels of the vertices of the triangle, which are generally capital letters, are used to name a triangle.

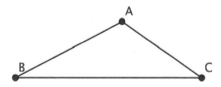

You can call this triangle ABC or $\triangle ABC$ since A, B, and C are vertices of the triangle. When naming the triangle, you can begin with any vertex. Then keep the letters in order as you go around the polygon. The triangle above could be named in a variety of ways: $\triangle ABC$, or $\triangle CBA$. The sides of the triangle are line segments *AB, AC,* and *CB.*

Just as triangles can be classified as acute, obtuse, or right based on their angles, they can also be classified by the length of their sides. Sides of equal length are called **congruent** sides. While we designate a segment joining points *A* and *B* by the notation $\overline{AB}$, we designate the length of a segment joining points *A* and *B* by the notation *AB* without a segment bar over it. The length *AB* is a number, and the segment $\overline{AB}$ is the collection of points that make up the segment.

Mathematicians show congruency by putting a hash mark symbol through the middle of sides of equal length. If the hash mark is the same on one or more sides, then those sides are congruent. If the sides have different hash marks, they are *not* congruent. The table below shows the classification of triangles by their side lengths.

Name of Triangle	Picture of Triangle	Description
Equilateral Triangle		A triangle whose three sides have the same length. These sides of equal length are called congruent sides.
Isosceles Triangle		A triangle with exactly two congruent sides.
Scalene Triangle		A triangle in which all three sides are a different length.

To describe a triangle even more specifically, you can use information about both its sides and its angles. Consider this example.

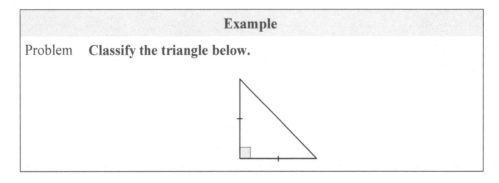

Example

Problem   **Classify the triangle below.**

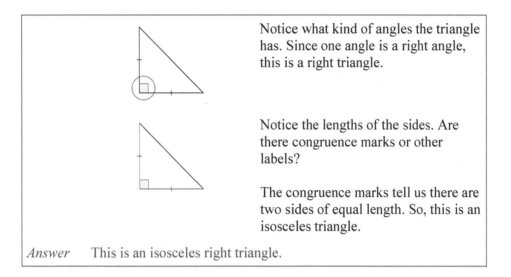

Notice what kind of angles the triangle has. Since one angle is a right angle, this is a right triangle.

Notice the lengths of the sides. Are there congruence marks or other labels?

The congruence marks tell us there are two sides of equal length. So, this is an isosceles triangle.

Answer   This is an isosceles right triangle.

Try it Now 1

Classify the given triangle.

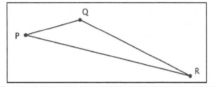

Identifying Congruent and Similar Triangles   Objective 2

Two triangles are congruent if they are exactly the same size and shape. In congruent triangles, the measures of **corresponding angles** and the lengths of **corresponding sides** are equal. Consider the two triangles shown below:

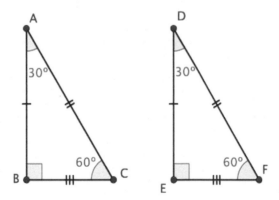

Since both $\angle B$ and $\angle E$ are right angles, these triangles are right triangles. Let's call these two triangles $\triangle ABC$ and $\triangle DEF$. These triangles are congruent if every pair of corresponding sides has equal lengths and every pair of corresponding angles has the same measure.

The corresponding sides are opposite the corresponding angles.

$\leftrightarrow$ means
"corresponds to"

$\angle B \leftrightarrow \angle E$

$\angle A \leftrightarrow \angle D$

$\angle C \leftrightarrow \angle F$

$\overline{AB} \leftrightarrow \overline{DE}$

$\overline{AC} \leftrightarrow \overline{DF}$

$\overline{BC} \leftrightarrow \overline{EF}$

$\triangle ABC$ and $\triangle DEF$ are congruent triangles as the corresponding sides and corresponding angles are equal.

Let's take a look at another pair of triangles. Below are the triangles $\triangle ABC$ and $\triangle RST$.

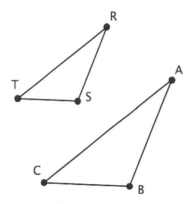

These two triangles are surely not congruent because $\triangle RST$ is clearly smaller in size than $\triangle ABC$. But, even though they are not the same size, they do resemble one another. They are the same shape. The corresponding angles of these triangles look like they might have the same exact measurement, and if they did they would be congruent angles and we would call the triangles similar triangles.

Congruent angles are marked with hash marks, just as congruent sides are.

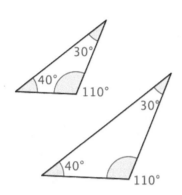

Image showing angle measurements of both triangles.

Image showing triangles ABC and RST using hash marks to show angle congruency.

We can also show congruent angles by using multiple bands within the angle, rather than multiple hash marks on one band. Below is an image using multiple bands within the angle.

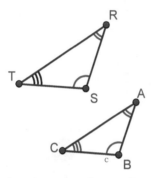

Image showing triangles ABC and RST
using bands to show angle congruency.

If the corresponding angles of two triangles have the same measurements they are called **similar triangles**. This name makes sense because they have the same shape, but not necessarily the same size. When a pair of triangles is similar, the corresponding sides are proportional to one another. That means that there is a consistent scale factor that can be used to compare the corresponding sides. In the previous example, the side lengths of the larger triangle are all 1.4 times the length of the smaller. So, similar triangles are proportional to one another.

Just because two triangles *look* similar does not mean they *are* similar triangles in the mathematical sense of the word. Checking that the corresponding angles have equal measure is one way of being sure the triangles are similar.

Corresponding Sides of Similar Triangles

Objective 3

There is another method for determining similarity of triangles that involves comparing the ratios of the lengths of the corresponding sides.

If the ratios of the pairs of corresponding sides are equal, the triangles are similar.

Consider the two triangles below.

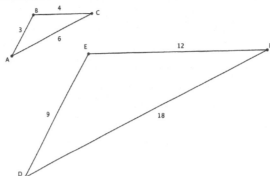

$\triangle ABC$ is *not* congruent to $\triangle DEF$ because the side lengths of $\triangle DEF$ are longer than those of $\triangle ABC$. So, are these triangles similar? If they are, the corresponding sides should be proportional.

Since these triangles are oriented in the same way, you can pair the left, right, and bottom sides: $\overline{AB}$ and $\overline{DE}$, $\overline{BC}$ and $\overline{EF}$, $\overline{AC}$ and $\overline{DE}$. (You might call these the two shortest sides, the two longest sides, and the two leftover sides and arrived at the same ratios). Now we will look at the ratios of their lengths.

$$\frac{AB}{DE} = \frac{BC}{EF} = \frac{AC}{DF}$$

Substituting the side length values into the proportion, you see that it is true:

$$\frac{3}{9} = \frac{4}{12} = \frac{6}{18}$$

If the corresponding sides are proportional, then the triangles are similar. Triangles *ABC* and *DEF* are similar, but not congruent.

Let's use this idea of proportional corresponding sides to determine whether two more triangles are similar.

Example
Problem   **Determine if the triangles below are similar by seeing if their corresponding sides are proportional.** 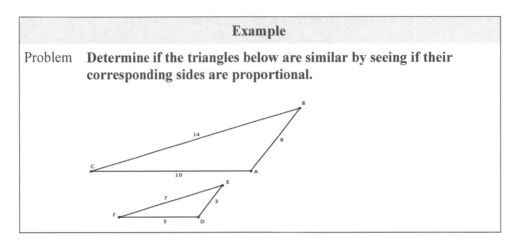

$\overline{CA} \leftrightarrow \overline{FD}$ $\overline{AB} \leftrightarrow \overline{DE}$ $\overline{BC} \leftrightarrow \overline{EF}$	First determine the corresponding sides, which are opposite corresponding angles.
$\dfrac{CA}{FD} = \dfrac{AB}{DE} = \dfrac{BC}{EF}$	Write the corresponding side lengths as ratios.
$\dfrac{10}{5} = \dfrac{6}{3} = \dfrac{14}{7}$ $2 = 2 = 2$	Substitute the side lengths into the ratios, and determine if the ratios of the corresponding sides are equivalent. They are, so the triangles are similar.
Answer	$\triangle ABC$ and $\triangle DEF$ are similar.

The mathematical symbol ~ means "is similar to". So, you can write $\triangle ABC$ is similar to $\triangle DEF$ as $\triangle ABC \sim \triangle DEF$.

Try it Now 2

Determine whether the two triangles are similar, congruent, or neither.

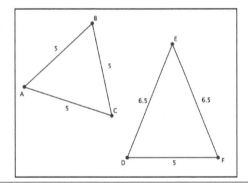

Finding Missing Measurements in Similar Triangles

Objective 4

You can find the missing measurements in a triangle if you know some measurements of a similar triangle. Let's look at an example.

Example
Problem $\triangle ABC$ and $\triangle XYZ$ are similar triangles. **What is the length of side BC?**

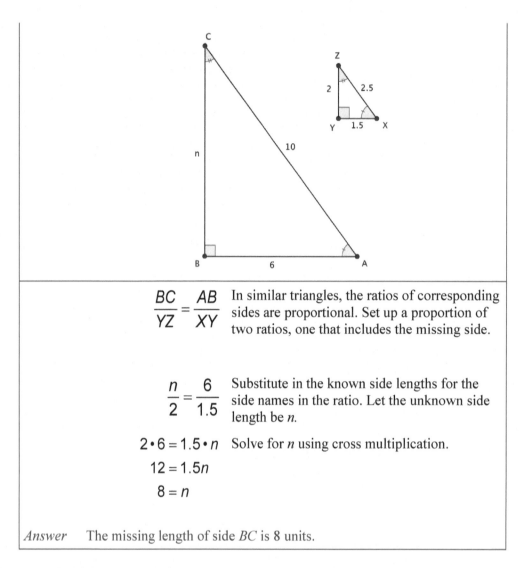

$$\frac{BC}{YZ} = \frac{AB}{XY}$$

In similar triangles, the ratios of corresponding sides are proportional. Set up a proportion of two ratios, one that includes the missing side.

$$\frac{n}{2} = \frac{6}{1.5}$$

Substitute in the known side lengths for the side names in the ratio. Let the unknown side length be n.

$2 \cdot 6 = 1.5 \cdot n$ Solve for n using cross multiplication.

$12 = 1.5n$

$8 = n$

Answer   The missing length of side *BC* is 8 units.

This process is fairly straightforward—but be careful that your ratios represent corresponding sides, recalling that corresponding sides are opposite corresponding angles.

Solving Application Problems Involving Similar Triangles

Objective 5

Applying knowledge of triangles, similarity, and congruence can be very useful for solving problems in real life. Just as you can solve for missing lengths of a triangle drawn on a page, you can use triangles to find unknown distances between locations or objects.

Let's consider the example of two trees and their shadows. Suppose the sun is shining down on two trees, one that is 6 feet tall and the other whose height is unknown. By measuring the length of each shadow on the ground, you can use triangle similarity to find the unknown height of the second tree.

First, let's figure out where the triangles are in this situation. The trees themselves create one pair of corresponding sides. The shadows cast on the ground are another pair of corresponding sides. The third side of these imaginary similar triangles runs from the top of each tree to the tip of its shadow on the ground. This is the hypotenuse of the triangle.

If you know that the trees and their shadows form similar triangles, you can set up a proportion to find the height of the tree.

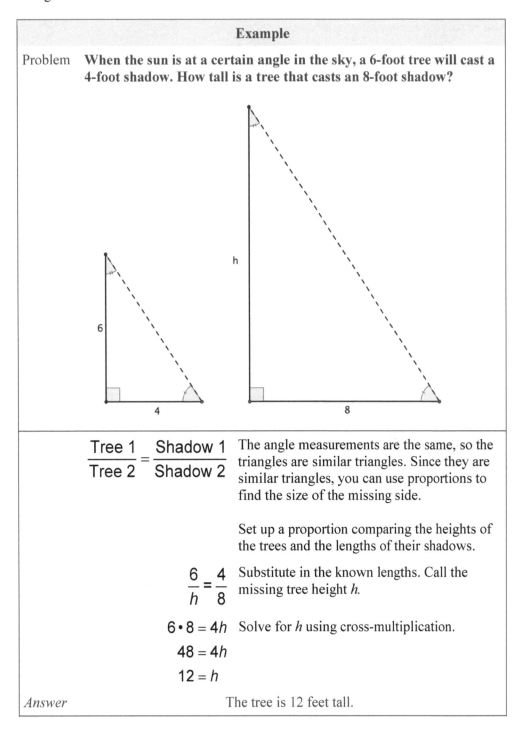

Example	
Problem	**When the sun is at a certain angle in the sky, a 6-foot tree will cast a 4-foot shadow. How tall is a tree that casts an 8-foot shadow?**

$$\frac{\text{Tree 1}}{\text{Tree 2}} = \frac{\text{Shadow 1}}{\text{Shadow 2}}$$

The angle measurements are the same, so the triangles are similar triangles. Since they are similar triangles, you can use proportions to find the size of the missing side.

Set up a proportion comparing the heights of the trees and the lengths of their shadows.

$$\frac{6}{h} = \frac{4}{8}$$

Substitute in the known lengths. Call the missing tree height *h*.

$$6 \cdot 8 = 4h$$

Solve for *h* using cross-multiplication.

$$48 = 4h$$

$$12 = h$$

Answer   The tree is 12 feet tall.

Summary

Triangles are one of the basic shapes in the real world. Triangles can be classified by the characteristics of their angles and sides, and triangles can be compared based on these characteristics. The sum of the measures of the interior angles of any triangle is 180°. Congruent triangles are triangles of the same size and shape. They have corresponding sides of equal length and corresponding angles of the same measurement. Similar triangles have the same shape, but not necessarily the same size. The lengths of their sides are proportional. Knowledge of triangles can be a helpful in solving real-world problems.

Try it Now Answers

1. Obtuse scalene; this triangle has vertices P, Q, and R, one angle (angle Q) that is between 90° and 180°, and sides of three different lengths.

2. $\triangle ABC$ and $\triangle DEF$ are neither similar nor congruent; the corresponding angle measures are not known to be equal as shown by the absence of congruence marks on the angles. Also, the ratios of the corresponding sides are not equal: $\dfrac{6.5}{5} = \dfrac{6.5}{5} \neq \dfrac{5}{5}$.

6.1.3 Pythagorean Theorem

Learning Objective(s)
1 Use the Pythagorean Theorem to find the unknown side of a right triangle.
2 Solve application problems involving the Pythagorean Theorem.

Introduction

A long time ago, a Greek mathematician named **Pythagoras** discovered an interesting property about **right triangles**: the sum of the squares of the lengths of each of the triangle's **legs** is the same as the square of the length of the triangle's **hypotenuse**. This property—which has many applications in science, art, engineering, and architecture—is now called the **Pythagorean Theorem**.

Let's take a look at how this theorem can help you learn more about the construction of triangles. And the best part—you don't even have to speak Greek to apply Pythagoras' discovery.

The Pythagorean Theorem

Pythagoras studied right triangles, and the relationships between the legs and the hypotenuse of a right triangle, before deriving his theory.

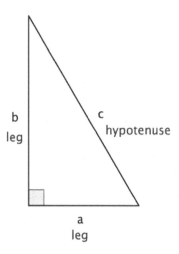

The Pythagorean Theorem
If a and b are the lengths of the legs of a right triangle and c is the length of the hypotenuse, then the sum of the squares of the lengths of the legs is equal to the square of the length of the hypotenuse.

This relationship is represented by the formula: $a^2 + b^2 = c^2$

In the box above, you may have noticed the word "square," as well as the small 2s to the top right of the letters in $a^2 + b^2 = c^2$. To **square** a number means to multiply it by itself. So, for example, to square the number 5 you multiply 5 • 5, and to square the number 12, you multiply 12 • 12. Some common squares are shown in the table below.

Number	Number Times Itself	Square
1	$1^2 = 1 \cdot 1$	1
2	$2^2 = 2 \cdot 2$	4
3	$3^2 = 3 \cdot 3$	9
4	$4^2 = 4 \cdot 4$	16
5	$5^2 = 5 \cdot 5$	25
10	$10^2 = 10 \cdot 10$	100

When you see the equation $a^2 + b^2 = c^2$, you can think of this as "the length of side a times itself, plus the length of side b times itself is the same as the length of side c times itself."

Let's try out all of the Pythagorean Theorem with an actual right triangle.

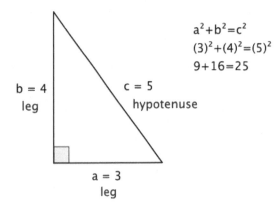

$$a^2 + b^2 = c^2$$
$$(3)^2 + (4)^2 = (5)^2$$
$$9 + 16 = 25$$

This theorem holds true for this right triangle—the sum of the squares of the lengths of both legs is the same as the square of the length of the hypotenuse. And, in fact, it holds true for all right triangles.

The Pythagorean Theorem can also be represented in terms of area. In any right triangle, the area of the square drawn from the hypotenuse is equal to the sum of the areas of the squares that are drawn from the two legs. You can see this illustrated below in the same 3-4-5 right triangle.

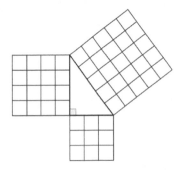

Note that the Pythagorean Theorem only works with *right* triangles.

Finding the Length of the Hypotenuse   Objective 1

You can use the Pythagorean Theorem to find the length of the hypotenuse of a right triangle if you know the length of the triangle's other two sides, called the legs. Put another way, if you know the lengths of *a* and *b*, you can find *c*.

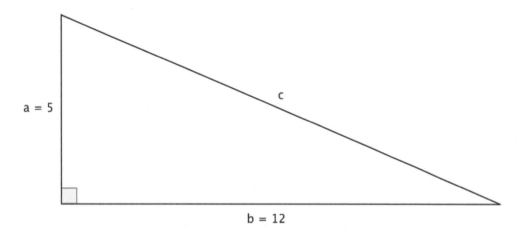

In the triangle above, you are given measures for legs *a* and *b*: 5 and 12, respectively. You can use the Pythagorean Theorem to find a value for the length of *c*, the hypotenuse.

$a^2 + b^2 = c^2$	The Pythagorean Theorem.
$(5)^2 + (12)^2 = c^2$	Substitute known values for *a* and *b*.
$25 + 144 = c^2$	Evaluate.
$169 = c^2$	Simplify. To find the value of *c*, think about a number that, when multiplied by itself, equals 169. Does 10 work? How about 11? 12? 13? (You can use a calculator to multiply if the numbers are unfamiliar.)
$13 = c$	The square root of 169 is 13.

Using the formula, you find that the length of *c*, the hypotenuse, is 13.

In this case, you did not know the value of *c*—you were given the square of the length of the hypotenuse, and had to figure it out from there. When you are given an equation like $169 = c^2$ and are asked to find the value of *c*, this is called finding the **square root** of a number. (Notice you found a number, *c*, whose square was 169.)

Finding a square root takes some practice, but it also takes knowledge of multiplication, division, and a little bit of trial and error. Look at the table below.

Number x	Number y which, when multiplied by itself, equals number x	Square root y
1	$1 \cdot 1$	1
4	$2 \cdot 2$	2
9	$3 \cdot 3$	3
16	$4 \cdot 4$	4
25	$5 \cdot 5$	5
100	$10 \cdot 10$	10

It is a good habit to become familiar with the squares of the numbers from 0–10, as these arise frequently in mathematics. If you can remember those square numbers—or if you can use a calculator to find them—then finding many common square roots will be just a matter of recall.

Try it Now 1

For which of these triangles is $(3)^2 + (3)^2 = r^2$?

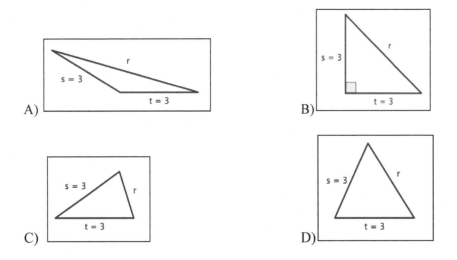

A)   B)

C)   D)

Finding the Length of a Leg

You can use the same formula to find the length of a right triangle's leg if you are given measurements for the lengths of the hypotenuse and the other leg. Consider the example below.

Example
Problem   **Find the length of side a in the triangle below. Use a calculator to estimate the square root to one decimal place.**

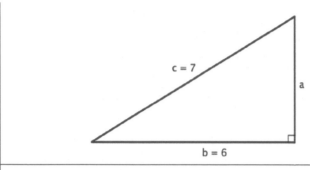

$a = ?$ In this right triangle, you are given
$b = 6$ the measurements for the
$c = 7$ hypotenuse, c, and one leg, b. The hypotenuse is always opposite the right angle and it is always the longest side of the triangle.

$a^2 + b^2 = c^2$ To find the length of leg a, substitute the known values into the
$a^2 + 6^2 = 7^2$ Pythagorean Theorem.

$a^2 + 36 = 49$ Solve for a^2. Think: what number, when added to 36, gives you 49?

$a^2 = 13$

$a \approx 3.6$ Use a calculator to find the square root of 13. The calculator gives an answer of 3.6055…, which you can round to 3.6. (Since you are approximating, you use the symbol $\approx$.)

Answer $a \approx 3.6$

Try it Now 2

Which of the following correctly uses the Pythagorean Theorem to find the missing side, x?

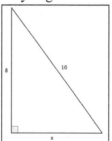

A) $8^2 + 10^2 = x^2$
B) $x + 8 = 10$
C) $x^2 + 8^2 = 10^2$
D) $x^2 + 10^2 = 8^2$

Using the Pythagorean Theorem to Solve Real-world Problems

The Pythagorean Theorem is perhaps one of the most useful formulas you will learn in mathematics because there are so many applications of it in real world settings. Architects and engineers use this formula extensively when building ramps, bridges, and buildings. Look at the following examples.

Example

Problem	The owners of a house want to convert a stairway leading from the ground to their back porch into a ramp. The porch is 3 feet off the ground, and due to building regulations, the ramp must start 12 feet away from the base of the porch. How long will the ramp be?
	Use a calculator to find the square root, and round the answer to the nearest tenth.

To solve a problem like this one, it often makes sense to draw a simple diagram showing where the legs and hypotenuse of the triangle lie.

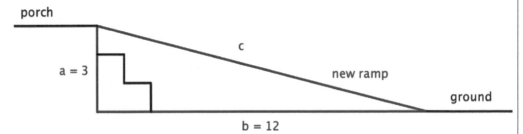

$a = 3$ Identify the legs and the hypotenuse of
$b = 12$ the triangle. You know that the triangle is
$c = ?$ a *right* triangle since the ground and the
          raised portion of the porch are
          perpendicular—this means you can use
          the Pythagorean Theorem to solve this
          problem. Identify a, b, and c.

$$a^2 + b^2 = c^2$$ Use the Pythagorean Theorem to find the
$$3^2 + 12^2 = c^2$$ length of c.
$$9 + 144 = c^2$$
$$153 = c^2$$

$$12.4 = c$$ Use a calculator to find c. The square root
               of 153 is 12.369…, so you can round that
               to 12.4.

| *Answer* | The ramp will be 12.4 feet long. |

Example	
Problem	**A sailboat has a large sail in the shape of a right triangle. The longest edge of the sail measures 17 yards, and the bottom edge of the sail is 8 yards. How tall is the sail?**
	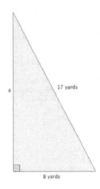 Draw an image to help you visualize the problem. In a right triangle, the hypotenuse will always be the longest side, so here it must be 17 yards. The problem also tells you that the bottom edge of the triangle is 8 yards.
	$a^2 + b^2 = c^2$ Setup the Pythagorean Theorem. $$a^2 + 8^2 = 17^2$$ $$a^2 + 64 = 289$$ $$a^2 = 225$$ $a = 15$ $15 \cdot 15 = 225$, so $a = 15$.
Answer	The height of the sail is 15 yards.

Summary

The Pythagorean Theorem states that in any right triangle, the sum of the squares of the lengths of the triangle's legs is the same as the square of the length of the triangle's hypotenuse. This theorem is represented by the formula $a^2 + b^2 = c^2$. Put simply, if you know the lengths of two sides of a right triangle, you can apply the Pythagorean Theorem to find the length of the third side. Remember, this theorem only works for right triangles.

Try it Now Answers

1. B) ![small triangle figure]; this is a right triangle; when you sum the squares of the lengths of the sides, you get the square of the length of the hypotenuse.

2. C) $x^2 + 8^2 = 10^2$; in this triangle, the hypotenuse has length 10, and the legs have length 8 and x. Substituting into the Pythagorean Theorem you have: $x^2 + 8^2 = 10^2$; this equation is the same as $x^2 + 64 = 100$, or $x^2 = 36$. What number, times itself, equals 36? That would make $x = 6$.

6.2   Perimeter, Circumference, and Area

6.2.1 *Quadrilaterals*

Learning Objective(s)
1 Identify properties, including angle measurements, of quadrilaterals.

Introduction

Quadrilaterals are a special type of polygon. As with triangles and other polygons, quadrilaterals have special properties and can be classified by characteristics of their angles and sides. Understanding the properties of different quadrilaterals can help you in solving problems that involve this type of polygon.

Defining a Quadrilateral

Picking apart the name "quadrilateral" helps you understand what it refers to. The prefix "quad-" means "four," and "lateral" is derived from the Latin word for "side." So a quadrilateral is a four-sided polygon.

Since it is a **polygon**, you know that it is a two-dimensional figure made up of straight sides. A quadrilateral also has four angles formed by its four sides. Below are some examples of quadrilaterals. Notice that each figure has four straight sides and four angles.

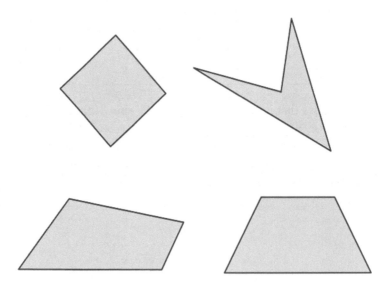

Objective 1

Interior Angles of a Quadrilateral

The sum of the interior angles of any quadrilateral is 360°. Consider the two examples below.

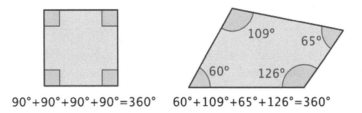

$$90°+90°+90°+90°=360° \qquad 60°+109°+65°+126°=360°$$

You could draw many quadrilaterals such as these and carefully measure the four angles. You would find that for every quadrilateral, the sum of the interior angles will always be 360°.

You can also use your knowledge of triangles as a way to understand why the sum of the interior angles of any quadrilateral is 360°. Any quadrilateral can be divided into two triangles as shown in the images below.

In the first image, the quadrilaterals have each been divided into two triangles. The angle measurements of one triangle are shown for each.

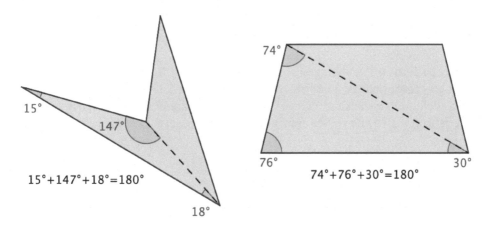

These measurements add up to 180°. Now look at the measurements for the other triangles—they also add up to 180°!

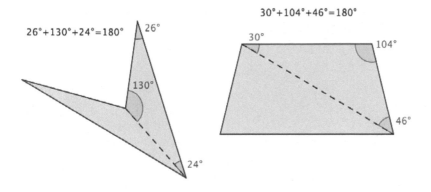

Since the sum of the interior angles of any triangle is 180° and there are two triangles in a quadrilateral, the sum of the angles for each quadrilateral is 360°.

Specific Types of Quadrilaterals

Let's start by examining the group of quadrilaterals that have two pairs of parallel sides. These quadrilaterals are called **parallelograms**. They take a variety of shapes, but one classic example is shown below.

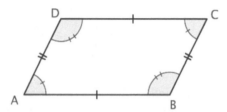

Imagine extending the pairs of opposite sides. They would never intersect because they are parallel. Notice, also, that the opposite angles of a parallelogram are congruent, as are the opposite sides. (Remember that "congruent" means "the same size.") The geometric symbol for congruent is $\cong$, so you can write $\angle A \cong \angle C$ and $\angle B \cong \angle D$. The parallel sides are also the same length: $\overline{AB} \cong \overline{DC}$ and $\overline{BC} \cong \overline{AD}$. These relationships are true for all parallelograms.

There are two special cases of parallelograms that will be familiar to you from your earliest experiences with geometric shapes. The first special case is called a **rectangle**. By definition, a rectangle is a parallelogram because its pairs of opposite sides are parallel. A rectangle also has the special characteristic that all of its angles are right angles; all four of its angles are congruent.

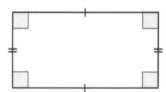

The other special case of a parallelogram is a special type of rectangle, a **square**. A square is one of the most basic geometric shapes. It is a special case of a parallelogram that has four congruent sides and four right angles.

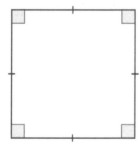

A square is also a rectangle because it has two sets of parallel sides and four right angles. A square is also a parallelogram because its opposite sides are parallel. So, a square can be classified in any of these three ways, with "parallelogram" being the least specific description and "square," the most descriptive.

Another quadrilateral that you might see is called a **rhombus**. All four sides of a rhombus are congruent. Its properties include that each pair of opposite sides is parallel, also making it a parallelogram.

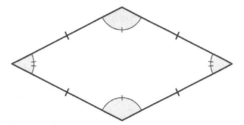

In summary, all squares are rectangles, but not all rectangles are squares. All rectangles are parallelograms, but not all parallelograms are rectangles. And *all* of these shapes are quadrilaterals.

The diagram below illustrates the relationship between the different types of quadrilaterals.

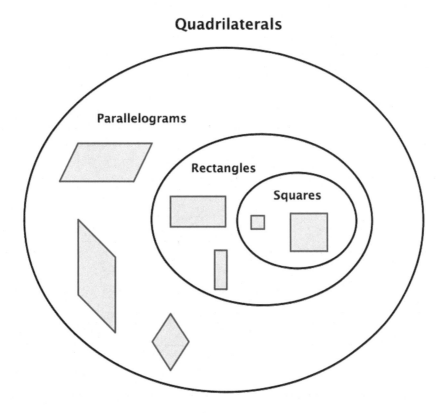

You can use the properties of parallelograms to solve problems. Consider the example that follows.

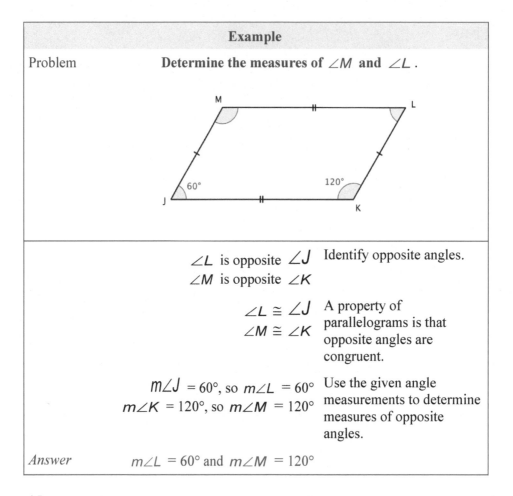

	Example	
Problem	**Determine the measures of $\angle M$ and $\angle L$.**	

$\angle L$ is opposite $\angle J$ $\angle M$ is opposite $\angle K$	Identify opposite angles.
$\angle L \cong \angle J$ $\angle M \cong \angle K$	A property of parallelograms is that opposite angles are congruent.
$m\angle J = 60°$, so $m\angle L = 60°$ $m\angle K = 120°$, so $m\angle M = 120°$	Use the given angle measurements to determine measures of opposite angles.
Answer	$m\angle L = 60°$ and $m\angle M = 120°$

Trapezoids

There is another special type of quadrilateral. This quadrilateral has the property of having *only one* pair of opposite sides that are parallel. Here is one example of a **trapezoid**.

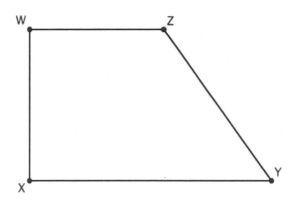

Notice that $\overline{XY} \parallel \overline{WZ}$, and that $\overline{WX}$ and $\overline{ZY}$ are not parallel. You can easily imagine that if you extended sides $\overline{WX}$ and $\overline{ZY}$, they would intersect above the figure.

If the non-parallel sides of a trapezoid are congruent, the trapezoid is called an **isosceles trapezoid**. Like the similarly named triangle that has two sides of equal length, the isosceles trapezoid has a pair of opposite sides of equal length. The other pair of opposite sides is parallel. Below is an example of an isosceles trapezoid.

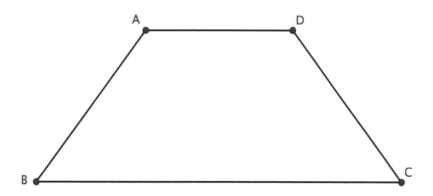

In this trapezoid *ABCD*, $\overline{BC} \parallel \overline{AD}$ and $\overline{AB} \cong \overline{CD}$.

Try it Now 1

Which of the following statements is true?

A) Some trapezoids are parallelograms.
B) All trapezoids are quadrilaterals.
C) All rectangles are squares.
D) A shape cannot be a parallelogram and a quadrilateral.

You can use the properties of quadrilaterals to solve problems involving trapezoids. Consider the example below.

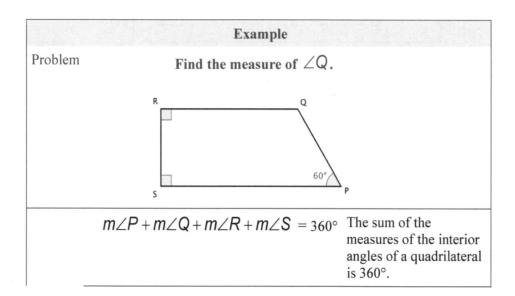

	Example
Problem	**Find the measure of $\angle Q$.**
	$m\angle P + m\angle Q + m\angle R + m\angle S = 360°$ The sum of the measures of the interior angles of a quadrilateral is 360°.

$$m\angle R = 90°$$ The square symbol

$$m\angle S = 90°$$ indicates a right angle.

$$60° + m\angle Q + 90° + 90° = 360°$$ Since three of the four angle measures are given, you can find the fourth angle measurement.

$$m\angle Q + 240° = 360°$$ Calculate the

$$m\angle Q = 120°$$ measurement of $\angle Q$.

From the image, you can see that it is an obtuse angle, so its measure must be greater than 90°.

Answer $$m\angle Q = 120°$$

The table below summarizes the special types of quadrilaterals and some of their properties.

Name of Quadrilateral	Quadrilateral	Description
Parallelogram		2 pairs of parallel sides. Opposite sides and opposite angles are congruent.
Rectangle		2 pairs of parallel sides. 4 right angles (90°). Opposite sides are parallel and congruent. All angles are congruent.

Square	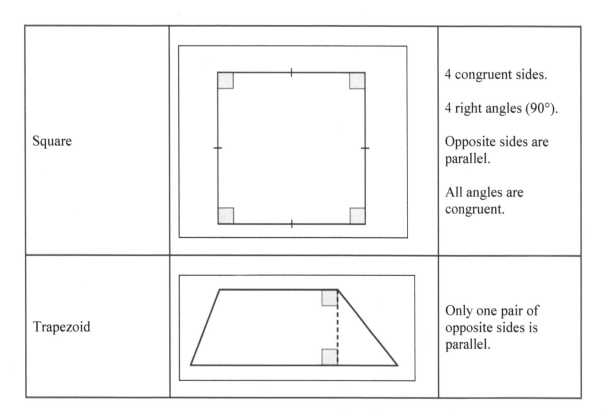	4 congruent sides. 4 right angles (90°). Opposite sides are parallel. All angles are congruent.
Trapezoid		Only one pair of opposite sides is parallel.

Summary

A quadrilateral is a mathematical name for a four-sided polygon. Parallelograms, squares, rectangles, and trapezoids are all examples of quadrilaterals. These quadrilaterals earn their distinction based on their properties, including the number of pairs of parallel sides they have and their angle and side measurements.

Try it Now Answers

1. B) All trapezoids are quadrilaterals; trapezoids are four-sided polygons, so they are all quadrilaterals.

6.2.2 Perimeter and Area

Learning Objective(s)
1 Find the perimeter of a polygon.
2 Find the area of a polygon.
3 Find the area and perimeter of non-standard polygons.

Introduction

Perimeter and **area** are two important and fundamental mathematical topics. They help you to quantify physical space and also provide a foundation for more advanced mathematics found in algebra, trigonometry, and calculus. Perimeter is a measurement of the distance around a shape and area gives us an idea of how much surface the shape covers.

Knowledge of area and perimeter is applied practically by people on a daily basis, such as architects, engineers, and graphic designers, and is math that is very much needed by people in general. Understanding how much space you have and learning how to fit shapes together exactly will help you when you paint a room, buy a home, remodel a kitchen, or build a deck.

Perimeter

Objective 1

The perimeter of a two-dimensional shape is the distance around the shape. You can think of wrapping a string around a triangle. The length of this string would be the perimeter of the triangle. Or walking around the outside of a park, you walk the distance of the park's perimeter. Some people find it useful to think "peRIMeter" because the edge of an object is its rim and peRIMeter has the word "rim" in it.

If the shape is a **polygon**, then you can add up all the lengths of the sides to find the perimeter. Be careful to make sure that all the lengths are measured in the same units. You measure perimeter in linear units, which is one dimensional. Examples of units of measure for length are inches, centimeters, or feet.

Example		
Problem	**Find the perimeter of the given figure. All measurements indicated are inches.**	
	$P = 5 + 3 + 6 + 2 + 3 + 3$	Since all the sides are measured in inches, just add the lengths of all six sides to get the perimeter.
Answer	$P = 22$ inches	Remember to include units.

This means that a tightly wrapped string running the entire distance around the polygon would measure 22 inches long.

	Example
Problem	**Find the perimeter of a triangle with sides measuring 6 cm, 8 cm, and 12 cm.**
	$P = 6 + 8 + 12$ Since all the sides are measured in centimeters, just add the lengths of all three sides to get the perimeter.
Answer	$P = 26$ centimeters

Sometimes, you need to use what you know about a polygon in order to find the perimeter. Let's look at the rectangle in the next example.

	Example
Problem	**A rectangle has a length of 8 centimeters and a width of 3 centimeters. Find the perimeter.**

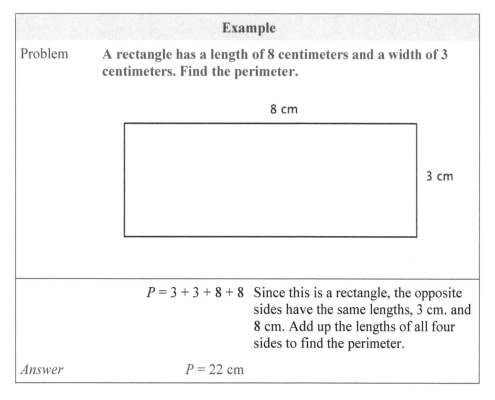

	$P = 3 + 3 + 8 + 8$ Since this is a rectangle, the opposite sides have the same lengths, 3 cm. and 8 cm. Add up the lengths of all four sides to find the perimeter.
Answer	$P = 22$ cm

Notice that the perimeter of a rectangle always has two pairs of equal length sides. In the above example, you could have also written $P = 2(3) + 2(8) = 6 + 16 = 22$ cm. The formula for the perimeter of a rectangle is often written as $P = 2l + 2w$, where l is the length of the rectangle and w is the width of the rectangle.

Area of Parallelograms
Objective 2

The area of a two-dimensional figure describes the amount of surface the shape covers. You measure area in square units of a fixed size. Examples of square units of measure are square inches, square centimeters, or square miles. When finding the area of a polygon, you count how many squares of a certain size will cover the region inside the polygon.

Let's look at a 4 x 4 square.

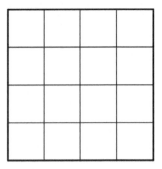

You can count that there are 16 squares, so the area is 16 square units. Counting out 16 squares doesn't take too long, but what about finding the area if this is a larger square or the units are smaller? It could take a long time to count.

Fortunately, you can use multiplication. Since there are 4 rows of 4 squares, you can multiply 4 • 4 to get 16 squares! And this can be generalized to a formula for finding the area of a square with any length, *s*: *Area = s • s = s²*.

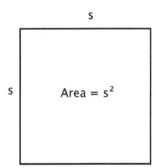

You can write "in²" for square inches and "ft²" for square feet.

To help you find the area of the many different categories of polygons, mathematicians have developed formulas. These formulas help you find the measurement more quickly than by simply counting. The formulas you are going to look at are all developed from the understanding that you are counting the number of square units *inside* the polygon. Let's look at a rectangle.

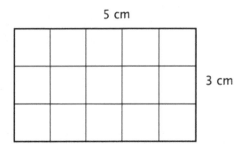

You can count the squares individually, but it is much easier to multiply 3 times 5 to find the number more quickly. And, more generally, the area of any rectangle can be found by multiplying *length* times *width*.

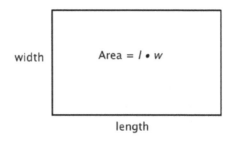

width | Area = *l* • *w*

length

Example		
Problem	A rectangle has a length of 8 centimeters and a width of 3 centimeters. Find the area.	
	8 cm 3 cm	
	$A = l \cdot w$	Start with the formula for the area of a rectangle, which multiplies the length times the width.
	$A = 8 \cdot 3$	Substitute 8 for the length and 3 for the width.
Answer	$A = 24$ cm^2	Be sure to include the units, in this case square cm.

It would take 24 squares, each measuring 1 cm on a side, to cover this rectangle.

The formula for the area of any parallelogram (remember, a rectangle is a type of parallelogram) is the same as that of a rectangle: *Area = l • w*. Notice in a rectangle, the length and the width are perpendicular. This should also be true for all parallelograms. *Base* (*b*) for the length (of the base), and *height* (*h*) for the width of the line perpendicular to the base is often used. So, the formula for a parallelogram is generally written, $A = b \cdot h$.

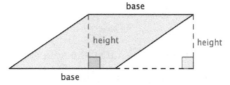

base

height | height

base

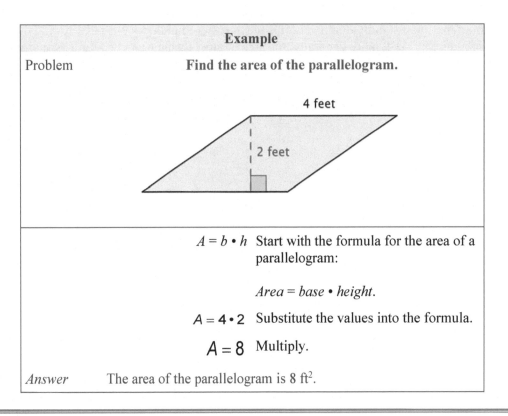

Example

| Problem | **Find the area of the parallelogram.** |

$A = b \cdot h$ Start with the formula for the area of a parallelogram:

$Area = base \cdot height.$

$A = 4 \cdot 2$ Substitute the values into the formula.

$A = 8$ Multiply.

Answer   The area of the parallelogram is 8 ft^2.

Try it Now 1

Find the area of a parallelogram with a height of 12 feet and a base of 9 feet.

Area of Triangles and Trapezoids

Objective 3

The formula for the area of a triangle can be explained by looking at a right triangle. Look at the image below—a rectangle with the same height and base as the original triangle. The area of the triangle is one half of the rectangle!

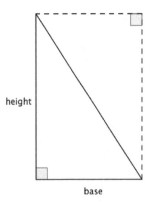

Since the area of two congruent triangles is the same as the area of a rectangle, you can come up with the formula $Area = \frac{1}{2}bh$ to find the area of a triangle.

When you use the formula for a triangle to find its area, it is important to identify a base and its corresponding height, which is perpendicular to the base.

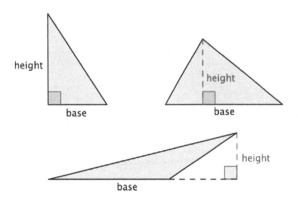

	Example
Problem	**A triangle has a height of 4 inches and a base of 10 inches. Find the area.**

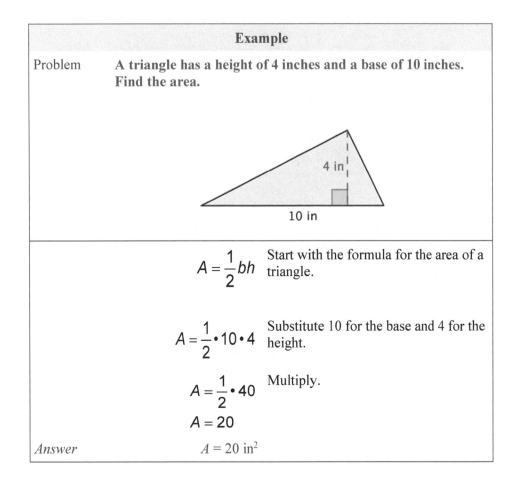

$A = \dfrac{1}{2}bh$	Start with the formula for the area of a triangle.
$A = \dfrac{1}{2} \cdot 10 \cdot 4$	Substitute 10 for the base and 4 for the height.
$A = \dfrac{1}{2} \cdot 40$	Multiply.
$A = 20$	
Answer	$A = 20 \text{ in}^2$

Now let's look at the trapezoid. To find the area of a trapezoid, take the average length of the two parallel bases and multiply that length by the height: $A = \dfrac{(b_1 + b_2)}{2}h$.

An example is provided below. Notice that the height of a trapezoid will always be perpendicular to the bases (just like when you find the height of a parallelogram).

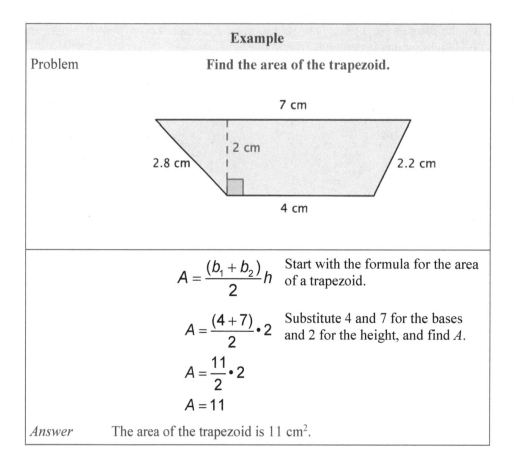

Example	
Problem	**Find the area of the trapezoid.**

$$A = \frac{(b_1 + b_2)}{2}h$$ Start with the formula for the area of a trapezoid.

$$A = \frac{(4+7)}{2} \cdot 2$$ Substitute 4 and 7 for the bases and 2 for the height, and find A.

$$A = \frac{11}{2} \cdot 2$$

$$A = 11$$

Answer The area of the trapezoid is 11 cm².

Area Formulas

Use the following formulas to find the areas of different shapes.

square: $A = s^2$

rectangle: $A = l \cdot w$

parallelogram: $A = b \cdot h$

triangle: $A = \frac{1}{2}b \cdot h$

trapezoid: $A = \frac{(b_1 + b_2)}{2}h$

Working with Perimeter and Area

Often you need to find the area or perimeter of a shape that is not a standard polygon. Artists and architects, for example, usually deal with complex shapes. However, even complex shapes can be thought of as being composed of smaller, less complicated shapes, like rectangles, trapezoids, and triangles.

To find the perimeter of non-standard shapes, you still find the distance around the shape by adding together the length of each side.

Finding the area of non-standard shapes is a bit different. You need to create regions *within* the shape for which you can find the area, and add these areas together. Have a look at how this is done below.

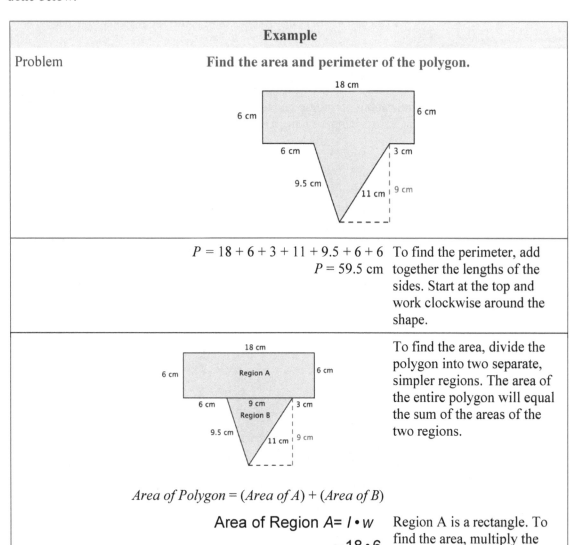

	Example
Problem	**Find the area and perimeter of the polygon.**

$P = 18 + 6 + 3 + 11 + 9.5 + 6 + 6$
$P = 59.5$ cm

To find the perimeter, add together the lengths of the sides. Start at the top and work clockwise around the shape.

Area of Polygon = (Area of A) + (Area of B)

Area of Region *A* = *l • w*
$= 18 • 6$
$= 108$

To find the area, divide the polygon into two separate, simpler regions. The area of the entire polygon will equal the sum of the areas of the two regions.

Region A is a rectangle. To find the area, multiply the length (18) by the width (6).

The area of Region A is 108 cm^2.

Area of Region B $= \frac{1}{2}b \cdot h$

$$= \frac{1}{2} \cdot 9 \cdot 9$$

$$= \frac{1}{2} \cdot 81$$

$$= 40.5$$

Region B is a triangle. To find the area, use the formula $\frac{1}{2}bh$, where the base is 9 and the height is 9.

The area of Region B is 40.5 cm².

108 cm² + 40.5 cm² = 148.5 cm².   Add the regions together.

Answer
Perimeter = 59.5 cm
Area = 148.5 cm²

You also can use what you know about perimeter and area to help solve problems about situations like buying fencing or paint, or determining how big a rug is needed in the living room. Here's a fencing example.

Example

Problem   **Rosie is planting a garden with the dimensions shown below. She wants to put a thin, even layer of mulch over the entire surface of the garden. The mulch costs $3 a square foot. How much money will she have to spend on mulch?**

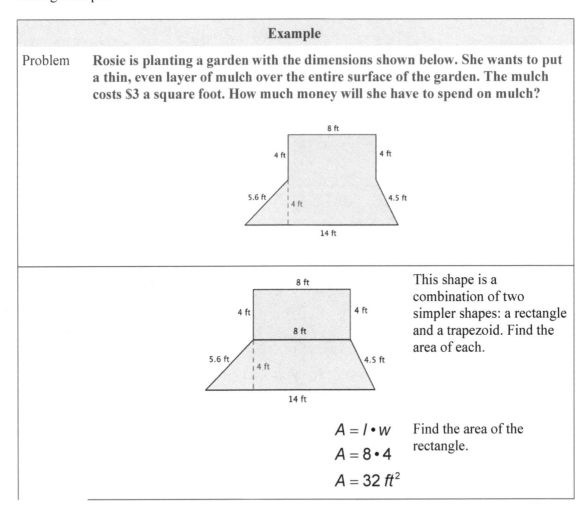

$A = l \cdot w$ Find the area of the rectangle.

$A = 8 \cdot 4$

$A = 32\ ft^2$

This shape is a combination of two simpler shapes: a rectangle and a trapezoid. Find the area of each.

$$A = \frac{(b_1 + b_2)}{2} h$$

Find the area of the trapezoid.

$$A = \frac{(14 + 8)}{2} \cdot 4$$

$$A = \frac{22}{2} \cdot 4$$

$$A = 11 \cdot 4$$

$$A = 44 \ ft^2$$

$32 \ ft^2 + 44 \ ft^2 = 76 \ ft^2$ Add the measurements.

$76 \ ft^2 \cdot \$3 = \$228$ Multiply by \$3 to find out how much Rosie will have to spend.

Answer   Rosie will spend \$228 to cover her garden with mulch.

Try it Now 2

Find the area of the given shape.

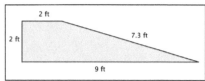

Summary

The perimeter of a two-dimensional shape is the distance around the shape. It is found by adding up all the sides (as long as they are all the same unit). The area of a two-dimensional shape is found by counting the number of squares that cover the shape. Many formulas have been developed to quickly find the area of standard polygons, like triangles and parallelograms.

Try it Now Answers

1. 108 ft²; the height of the parallelogram is 12 and the base of the parallelogram is 9; the area is 12 times 9, or 108 ft².

2. 11 ft²; this shape is a trapezoid, so you can use the formula $A = \frac{(b_1 + b_2)}{2} h$ to find the area:

$A = \frac{(2 + 9)}{2} \cdot 2$.

6.2.3 Circles

Learning Objective(s)
1 Identify properties of circles.
2 Find the circumference of a circle.
3 Find the area of a circle.
4 Find the area and perimeter of composite geometric figures.

Introduction

Circles are a common shape. You see them all over—wheels on a car, Frisbees passing through the air, compact discs delivering data. These are all circles.

A circle is a two-dimensional figure just like polygons and quadrilaterals. However, circles are measured differently than these other shapes—you even have to use some different terms to describe them. Let's take a look at this interesting shape.

Properties of Circles

Objective 1

A circle represents a set of points, all of which are the same distance away from a fixed, middle point. This fixed point is called the center. The distance from the center of the circle to any point on the circle is called the **radius**.

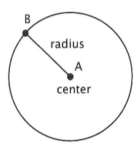

When two radii (the plural of radius) are put together to form a line segment across the circle, you have a **diameter**. The diameter of a circle passes through the center of the circle and has its endpoints on the circle itself.

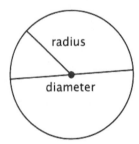

The diameter of any circle is two times the length of that circle's radius. It can be represented by the expression $2r$, or "two times the radius." So, if you know a circle's radius, you can multiply it by 2 to find the diameter; this also means that if you know a circle's diameter, you can divide by 2 to find the radius.

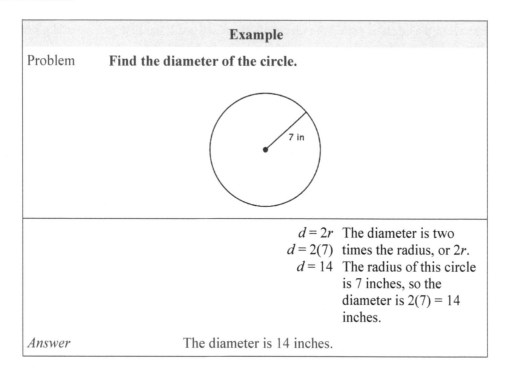

Example	
Problem	**Find the diameter of the circle.**

$d = 2r$ The diameter is two
$d = 2(7)$ times the radius, or $2r$.
$d = 14$ The radius of this circle is 7 inches, so the diameter is $2(7) = 14$ inches.

Answer   The diameter is 14 inches.

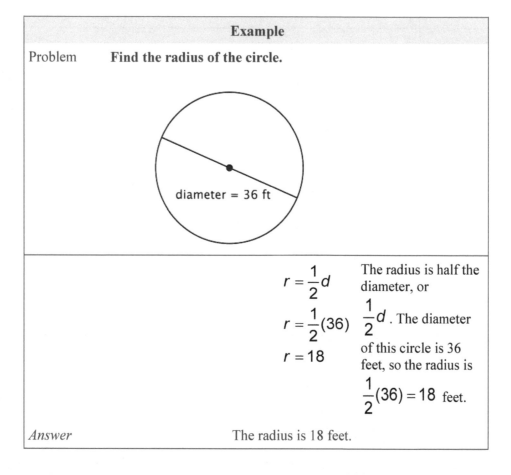

Example	
Problem	**Find the radius of the circle.**

$r = \dfrac{1}{2}d$ The radius is half the diameter, or

$r = \dfrac{1}{2}(36)$ $\dfrac{1}{2}d$. The diameter

$r = 18$ of this circle is 36 feet, so the radius is $\dfrac{1}{2}(36) = 18$ feet.

Answer   The radius is 18 feet.

Circumference

The distance around a circle is called the **circumference**. (Recall, the distance around a polygon is the perimeter.)

One interesting property about circles is that the ratio of a circle's circumference and its diameter is the same for all circles. No matter the size of the circle, the ratio of the circumference and diameter will be the same.

Some actual measurements of different items are provided below. The measurements are accurate to the nearest millimeter or quarter inch (depending on the unit of measurement used). Look at the ratio of the circumference to the diameter for each one—although the items are different, the ratio for each is approximately the same.

Item	Circumference (C) (rounded to nearest hundredth)	Diameter (d)	Ratio $\dfrac{C}{d}$
Cup	253 mm	79 mm	$\dfrac{253}{79} = 3.2025...$
Quarter	84 mm	27 mm	$\dfrac{84}{27} = 3.1111...$
Bowl	37.25 in	11.75 in	$\dfrac{37.25}{11.75} = 3.1702...$

The circumference and the diameter are approximate measurements, since there is no precise way to measure these dimensions exactly. If you were able to measure them more precisely, however, you would find that the ratio $\dfrac{C}{d}$ would move towards 3.14 for each of the items given. The mathematical name for the ratio $\dfrac{C}{d}$ is **pi**, and is represented by the Greek letter π.

π is a non-terminating, non-repeating decimal, so it is impossible to write it out completely? The first 10 digits of π are 3.141592653; it is often rounded to 3.14 or estimated as the fraction $\dfrac{22}{7}$.

Note that both 3.14 and $\dfrac{22}{7}$ are *approximations* of π, and are used in calculations where it is not important to be precise.

Since you know that the ratio of circumference to diameter (or π) is consistent for all circles, you can use this number to find the circumference of a circle if you know its diameter.

$$\frac{C}{d} = \pi \text{, so } C = \pi d$$

Also, since $d = 2r$, then $C = \pi d = \pi (2r) = 2\pi r$.

Circumference of a Circle

To find the circumference (C) of a circle, use one of the following formulas:

If you know the diameter (d) of a circle: $C = \pi d$

If you know the radius (r) of a circle: $C = 2\pi r$

Example	
Problem	**Find the circumference of the circle.**

diameter = 9 in

$C = \pi d$ $C = \pi \cdot 9$ $C \approx 3.14 \cdot 9$ $C \approx 28.26$	To calculate the circumference given a diameter of 9 inches, use the formula $C = \pi d$. Use 3.14 as an approximation for π. Since you are using an approximation for π, you cannot give an exact measurement of the circumference. Instead, you use the symbol $\approx$ to indicate "approximately equal to."
Answer	The circumference is 9π or approximately 28.26 inches.

Example	
Problem	**Find the circumference of a circle with a radius of 2.5 yards.**
	$C = 2\pi r$ $C = 2\pi \cdot 2.5$ $C = \pi \cdot 5$ $C \approx 3.14 \cdot 5$ $C \approx 15.7$ To calculate the circumference of a circle given a radius of 2.5 yards, use the formula $C = 2\pi r$. Use 3.14 as an approximation for π.
Answer	The circumference is 5π or approximately 15.7 yards.

Try it Now 1

A circle has a radius of 8 inches. What is its circumference, rounded to the nearest inch?

Area

Pi, π, is an important number in geometry. You have already used it to calculate the circumference of a circle. You use π when you are figuring out the *area* of a circle, too.

Area of a Circle

To find the area (A) of a circle, use the formula: $A = \pi r^2$

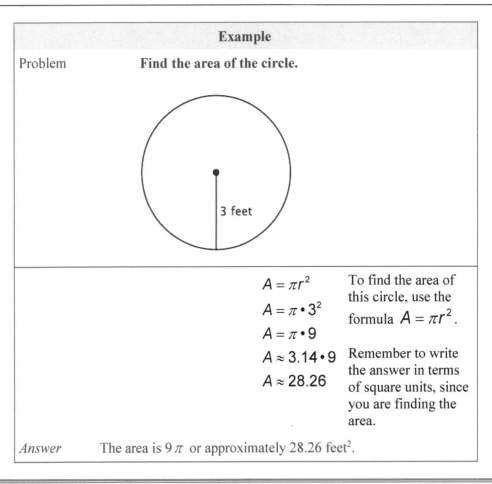

	Example
Problem	**Find the area of the circle.**

3 feet

$A = \pi r^2$ To find the area of this circle, use the formula $A = \pi r^2$.

$A = \pi \cdot 3^2$

$A = \pi \cdot 9$

$A \approx 3.14 \cdot 9$ Remember to write the answer in terms of square units, since you are finding the area.

$A \approx 28.26$

Answer   The area is 9π or approximately 28.26 feet^2.

Try it Now 2

A button has a diameter of 20 millimeters. What is the area of the button? Use 3.14 as an approximation of π.

Composite Figures

Now that you know how to calculate the circumference and area of a circle, you can use this knowledge to find the perimeter and area of composite figures. The trick to figuring out these types of problems is to identify shapes (and parts of shapes) within the composite figure, calculate their individual dimensions, and then add them together.

For example, look at the image below. Is it possible to find the perimeter?

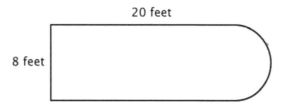

The first step is to identify simpler figures within this composite figure. You can break it down into a rectangle and a semicircle, as shown below.

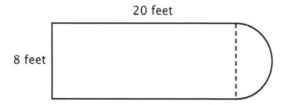

You know how to find the perimeter of a rectangle, and you know how to find the circumference of a circle. Here, the perimeter of the three solid sides of the rectangle is $8 + 20 + 20 = 48$ feet. (Note that only three sides of the rectangle will add into the perimeter of the composite figure because the other side is not at an edge; it is covered by the semicircle!)

To find the circumference of the semicircle, use the formula $C = \pi d$ with a diameter of 8 feet, then take half of the result. The circumference of the semicircle is 4π, or approximately 12.56 feet, so the total perimeter is about 60.56 feet.

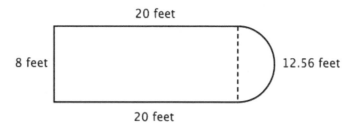

Example	
Problem	**Find the perimeter (to the nearest hundredth) of the composite figure, made up of a semi-circle and a triangle.**

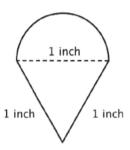

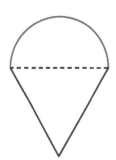

Identify smaller shapes within the composite figure. This figure contains a semicircle and a triangle.

Diameter $(d) = 1$

$C = \pi d$

$C = \pi(1)$

$C = \pi$

Find the circumference of the circle. Then divide by 2 to find the circumference of the semi-circle.

Circumference of semicircle $= \dfrac{1}{2}\pi$ or approximately 1.57 inches

$1 + 1 + \dfrac{1}{2}\pi \approx 3.57$ inches

Find the total perimeter by adding the circumference of the semicircle and the lengths of the two legs. Since our measurement of the semi-circle's circumference is approximate, the perimeter will be an approximation also.

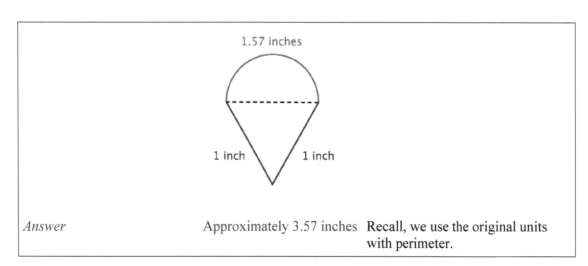

1.57 inches

1 inch   1 inch

Answer   Approximately 3.57 inches   Recall, we use the original units
                                                        with perimeter.

Example

Problem   **Find the area of the composite figure, made up of three-quarters of a circle**
            and a square, to the nearest hundredth.

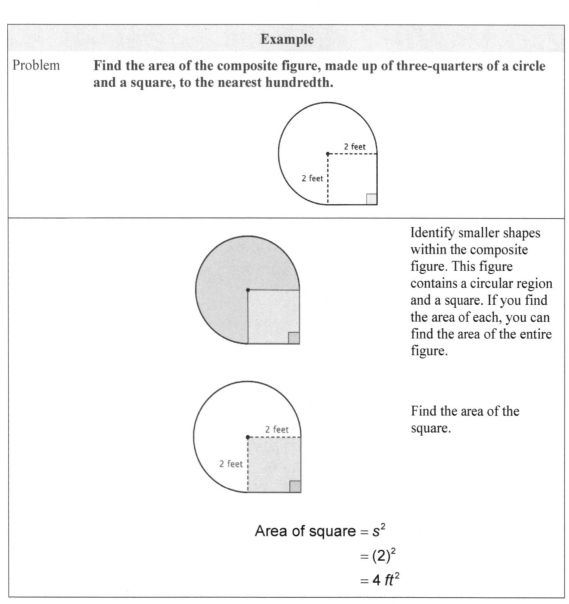

2 feet

2 feet

Identify smaller shapes
within the composite
figure. This figure
contains a circular region
and a square. If you find
the area of each, you can
find the area of the entire
figure.

Find the area of the
square.

2 feet

2 feet

$$\text{Area of square} = s^2$$
$$= (2)^2$$
$$= 4 \ ft^2$$

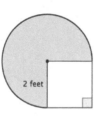

Find the area of the circular region. The radius is 2 feet.

Note that the region is $\dfrac{3}{4}$ of a whole circle, so you need to multiply the area of the circle by $\dfrac{3}{4}$. Use 3.14 as an approximation for π.

$$\text{Area of full circle} = \pi r^2$$
$$= \pi(2)^2$$
$$= 4\pi \; ft^2$$

$$\text{Area of region} = \frac{3}{4} \cdot 4\pi$$
$$= 3\pi$$
$$\approx 3 \cdot 3.14 \; ft^2$$

This is approximately 9.42 feet^2.

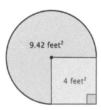

Add the two regions together. Since your measurement of the circular's area is approximate, the area of the figure will be an approximation also.

$$4 \text{ feet}^2 + 3\pi \text{ feet}^2 = \text{approximately } 13.42 \text{ feet}^2$$

Answer   The area is approximately 13.42 feet^2.

Recall, we use square units for area.

Try it Now 3

What is the area (to the nearest hundredth) of the figure shown below? (Both rounded regions are semi-circles.)

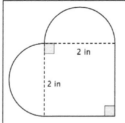

Summary

Circles are an important geometric shape. The distance around a circle is called the circumference, and the interior space of a circle is called the area. Calculating the circumference and area of a circle requires a number called pi (π), which is a non-terminating, non-repeating decimal. Pi is often approximated by the values 3.14 and $\dfrac{22}{7}$. You can find the perimeter or area of composite shapes—including shapes that contain circular sections—by applying the circumference and area formulas where appropriate.

Try it Now Answers

1. 50 inches; if the radius is 8 inches, the correct formula for circumference when the radius is given is $C = 2\pi r$ The correct answer is 50 inches.
2. 314 mm^2; the diameter is 20 mm, so the radius must be 10 mm. Then, using the formula $A = \pi r^2$, you find $A = \pi \cdot 10^2 = \pi \cdot 100 \approx 314$ mm^2.
3. 7.14 in^2; imagine the two semi-circles being put together to create one circle. The radius of the circle is 1 inch; this means the area of the circle is $\pi r^2 = \pi \cdot 1^2 = \pi$. The area of the square is $2 \cdot 2$ = 4. Adding those together yields 7.14 in^2.

6.3   Volume of Geometric Solids

Learning Objective(s)
1 Identify geometric solids.
2 Find the volume of geometric solids.
3 Find the volume of a composite geometric solid.

Introduction

Living in a two-dimensional world would be pretty boring. Thankfully, all of the physical objects that you see and use every day—computers, phones, cars, shoes—exist in three dimensions. They all have length, width, and height. (Even very thin objects like a piece of paper are three-dimensional. The thickness of a piece of paper may be a fraction of a millimeter, but it does exist.)

In the world of geometry, it is common to see three-dimensional figures. In mathematics, a flat side of a three-dimensional figure is called a **face**. **Polyhedrons** are shapes that have four or more faces, each one being a polygon. These include cubes, prisms, and pyramids. Sometimes you may even see single figures that are composites of two of these figures. Let's take a look at some common polyhedrons.

Identifying Solids

Objective 1

The first set of solids contains rectangular bases. Have a look at the table below, which shows each figure in both solid and transparent form.

Name	Definition	Solid Form	Transparent Form
Cube	A six-sided polyhedron that has congruent squares as faces.		
Rectangular prism	A polyhedron that has three pairs of congruent, rectangular, parallel faces.		
Pyramid	A polyhedron with a polygonal base and a collection of triangular faces that meet at a point.		

Notice the different names that are used for these figures. A **cube** is different than a square, although they are sometimes confused with each other—a cube has three dimensions, while a square only has two. Likewise, you would describe a shoebox as a **rectangular prism** (not simply a rectangle), and the ancient **pyramids** of Egypt as…well, as pyramids (not triangles).

In this next set of solids, each figure has a circular base.

Name	Definition	Solid Form	Transparent Form
Cylinder	A solid figure with a pair of circular, parallel bases and a round, smooth face between them.		
Cone	A solid figure with a single circular base and a round, smooth face that diminishes to a single point.		

Take a moment to compare a pyramid and a **cone**. Notice that a pyramid has a rectangular base and flat, triangular faces; a cone has a circular base and a smooth, rounded body.

Finally, let's look at a shape that is unique: a **sphere**.

Name	Definition	Solid Form	Transparent Form
Sphere	A solid, round figure where every point on the surface is the same distance from the center.		

There are many spherical objects all around you—soccer balls, tennis balls, and baseballs being three common items. While they may not be perfectly spherical, they are generally referred to as spheres.

Example

| Problem | **A three-dimensional figure has the following properties:**
    •   **It has a rectangular base.**
    •   **It has four triangular faces.**
What kind of a solid is it? |

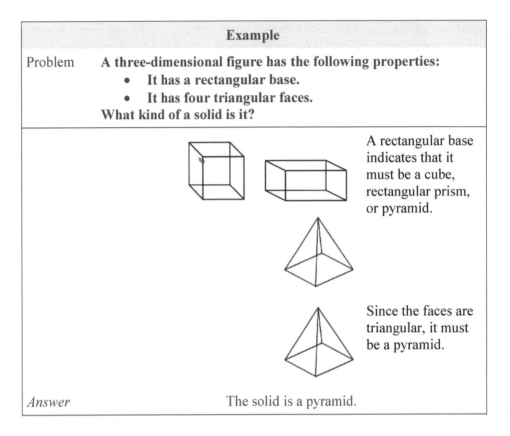

A rectangular base indicates that it must be a cube, rectangular prism, or pyramid.

Since the faces are triangular, it must be a pyramid.

| *Answer* | The solid is a pyramid. |

Volume

Recall that perimeter measures one dimension (length), and area measures two dimensions (length and width). To measure the amount of space a three-dimensional figure takes up, you use another measurement called **volume**.

To visualize what "volume" measures, look back at the transparent image of the rectangular prism mentioned earlier (or just think of an empty shoebox). Imagine stacking identical cubes inside that box so that there are no gaps between any of the cubes. Imagine filling up the entire box in this manner. If you counted the number of cubes that fit inside that rectangular prism, you would have its volume.

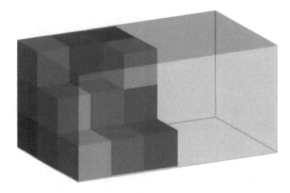

Volume is measured in cubic units. The shoebox illustrated above may be measured in cubic inches (usually represented as in^3 or $inches^3$), while the Great Pyramid of Egypt would be more appropriately measured in cubic meters (m^3 or $meters^3$).

To find the volume of a geometric solid, you could create a transparent version of the solid, create a bunch of 1x1x1 cubes, and then stack them carefully inside. However, that would take a long time! A much easier way to find the volume is to become familiar with some geometric formulas, and to use those instead.

Let's go through the geometric solids once more and list the volume formula for each.

As you look through the list below, you may notice that some of the volume formulas look similar to their area formulas. To find the volume of a rectangular prism, you find the area of the base and then multiply that by the height.

Name	Transparent Form	Volume Formula
Cube		$V = a \cdot a \cdot a = a^3$ a = the length of one side
Rectangular prism		$V = l \cdot w \cdot h$ l = length w = width h = height
Pyramid		$V = \dfrac{l \cdot w \cdot h}{3}$ l = length w = width h = height

Remember that all cubes are rectangular prisms, so the formula for finding the volume of a cube is the area of the base of the cube times the height.

Now let's look at solids that have a circular base.

Name	Transparent Form	Volume Formula
Cylinder		$V = \pi \bullet r^2 \bullet h$ r = radius h = height
Cone		$V = \dfrac{\pi \bullet r^2 \bullet h}{3}$ r = radius h = height

Here you see the number π again.

The volume of a **cylinder** is the area of its base, πr^2, times its height, h.

Compare the formula for the volume of a cone ($V = \dfrac{\pi \bullet r^2 \bullet h}{3}$) with the formula for the volume of

a pyramid ($V = \dfrac{l \bullet w \bullet h}{3}$). The numerator of the cone formula is the volume formula for a

cylinder, and the numerator of the pyramid formula is the volume formula for a rectangular prism.
Then divide each by 3 to find the volume of the cone and the pyramid. Looking for patterns and
similarities in the formulas can help you remember which formula refers to a given solid.

Finally, the formula for a sphere is provided below. Notice that the radius is cubed, not squared

and that the quantity πr^3 is multiplied by $\dfrac{4}{3}$.

Name	Wireframe Form	Volume Formula
Sphere		$V = \dfrac{4}{3}\pi r^3$ r = radius

Applying the Formulas

You know how to identify the solids, and you also know the volume formulas for these solids. To calculate the actual volume of a given shape, all you need to do is substitute the solid's dimensions into the formula and calculate.

In the examples below, notice that cubic units (meters³, inches³, feet³) are used.

Example
Problem   **Find the volume of a cube with side lengths of 6 meters.**

$$V = a \bullet a \bullet a = a^3 \quad \text{Identify the proper formula to use.}$$

$$a = \text{side length}$$

$$V = 6 \bullet 6 \bullet 6 = 6^3 \quad \text{Substitute } a = 6 \text{ into the formula.}$$

$$6 \bullet 6 \bullet 6 = 216 \quad \text{Calculate the volume.}$$

Answer   Volume = 216 meters³   Recall, we use cubic units with volume.

Example
Problem   **Find the volume of the shape shown below.**

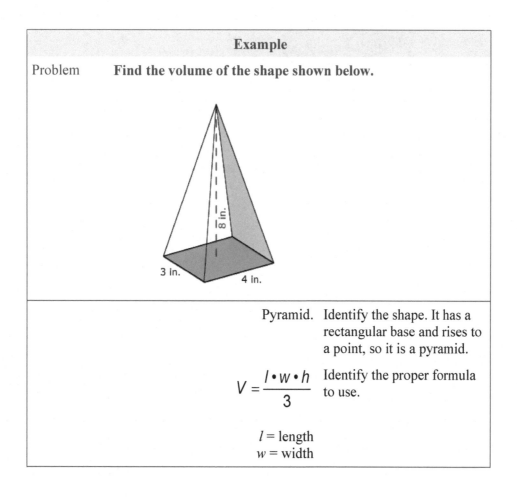

Pyramid.   Identify the shape. It has a rectangular base and rises to a point, so it is a pyramid.

$$V = \frac{l \bullet w \bullet h}{3} \quad \text{Identify the proper formula to use.}$$

$$l = \text{length}$$
$$w = \text{width}$$

h = height

4 = length   Use the image to identify the
3 = width   dimensions. Then substitute
8 = height $l = 4$, $w = 3$, and $h = 8$ into
              the formula.

$$V = \frac{4 \cdot 3 \cdot 8}{3}$$

$$V = \frac{96}{3}$$ Calculate the volume.

$$= 32$$

Answer   The volume of the pyramid is 32 inches³.

Example

Problem **Find the volume of the shape shown below.**

Use 3.14 for π, and round the answer to the nearest hundredth.

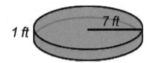

Cylinder.   Identify the shape. It has a
             circular base and has
             uniform thickness (or
             height), so it is a cylinder.

$$V = \pi \cdot r^2 \cdot h$$ Identify the proper formula
                                 to use.

$$V = \pi \cdot 7^2 \cdot 1$$ Use the image to identify the
                                 dimensions. Then substitute
                                 $r = 7$ and $h = 1$ into the
                                 formula.

$$V = \pi \cdot 49 \cdot 1$$ Calculate the volume, using
$$= 49\pi$$ 3.14 as an approximation for
$$\approx 153.86$$ π.

Answer   The volume is 49π or approximately 153.86 feet³.

Try it Now 1

Find the volume of a rectangular prism that is 8 inches long, 3 inches wide, and 10 inches tall.

Composite Solids

Composite geometric solids are made from two or more geometric solids. You can find the volume of these solids as well, as long as you are able to figure out the individual solids that make up the composite shape.

Look at the image of a capsule below. Each end is a half-sphere. You can find the volume of the solid by taking it apart. What solids can you break this shape into?

You can break it into a cylinder and two half-spheres.

Two half-spheres form a whole one, so if you know the volume formulas for a cylinder and a sphere, you can find the volume of this capsule.

Example
Problem   **If the radius of the spherical ends is 6 inches, find the volume of the solid below. Use 3.14 for π. Round your final answer to the nearest whole number.**

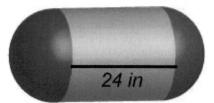

24 in

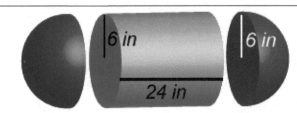

Identify the composite solids. This capsule can be thought of as a cylinder with a half-sphere on each end.

Volume of a cylinder: $\pi \cdot r^2 \cdot h$

Volume of a sphere: $\dfrac{4}{3} \pi r^3$

Identify the proper formulas to use.

Volume of a cylinder: $\pi \cdot 6^2 \cdot 24$

Volume of a sphere: $\dfrac{4}{3} \pi \cdot 6^3$

Substitute the dimensions into the formulas.

The height of a cylinder refers to the section between the two circular bases. This dimension is given as 24 inches, so $h = 24$.

The radius of the sphere is 6 inches. You can use $r = 6$ in both formulas.

$$V = \pi \cdot 36 \cdot 24$$

Volume of the cylinder: $= 864 \cdot \pi$

$$\approx 2712.96$$

Calculate the volume of the cylinder and the sphere.

$$V = \dfrac{4}{3} \pi \cdot 216$$

Volume of the sphere: $= 288 \cdot \pi$

$$\approx 904.32$$

Volume of capsule: Add the volumes.

$$2712.96 + 904.32 \approx 3617.28$$

Answer   The volume of the capsule is $1,152\,\pi$ or approximately 3617 inches³.

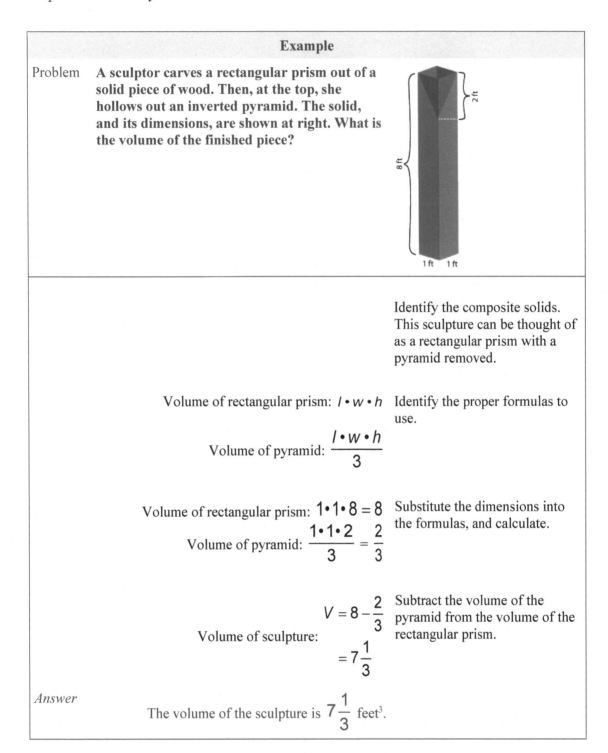

	Example
Problem	**A sculptor carves a rectangular prism out of a solid piece of wood. Then, at the top, she hollows out an inverted pyramid. The solid, and its dimensions, are shown at right. What is the volume of the finished piece?**

	Identify the composite solids. This sculpture can be thought of as a rectangular prism with a pyramid removed.
Volume of rectangular prism: $l \cdot w \cdot h$ Volume of pyramid: $\dfrac{l \cdot w \cdot h}{3}$	Identify the proper formulas to use.
Volume of rectangular prism: $1 \cdot 1 \cdot 8 = 8$ Volume of pyramid: $\dfrac{1 \cdot 1 \cdot 2}{3} = \dfrac{2}{3}$	Substitute the dimensions into the formulas, and calculate.
Volume of sculpture: $V = 8 - \dfrac{2}{3}$ $= 7\dfrac{1}{3}$	Subtract the volume of the pyramid from the volume of the rectangular prism.
Answer	The volume of the sculpture is $7\dfrac{1}{3}$ feet^3.

Try it Now 2

A machine takes a solid cylinder with a height of 9 mm and a diameter of 7 mm, and bores a hole all the way through it. The hole that it creates has a diameter of 3 mm. Which of the following expressions would correctly find the volume of the solid?

A) $(\pi \cdot 7^2 \cdot 9) - (\pi \cdot 3^2 \cdot 9)$

B) $(\pi \cdot 3.5^2 \cdot 9) - (\pi \cdot 1.5^2 \cdot 9)$

C) $(\pi \cdot 7^2 \cdot 9) + (\pi \cdot 3^2 \cdot 9)$

D) $(\pi \cdot 3.5^2 \cdot 9) + (\pi \cdot 1.5^2 \cdot 9)$

Summary

Three-dimensional solids have length, width, and height. You use a measurement called volume to figure out the amount of space that these solids take up. To find the volume of a specific geometric solid, you can use a volume formula that is specific to that solid. Sometimes, you will encounter composite geometric solids. These are solids that combine two or more basic solids. To find the volume of these, identify the simpler solids that make up the composite figure, find the volumes of those solids, and combine them as needed.

Try it Now Answers

1. 240 inches^3; to find the volume of the rectangular prism, use the formula $V = l \cdot w \cdot h$, and then substitute in the values for the length, width, and height. 8 inches $\cdot$ 3 inches $\cdot$ 10 inches = 240 inches^3.

2. B) $(\pi \cdot 3.5^2 \cdot 9) - (\pi \cdot 1.5^2 \cdot 9)$; you find the volume of the entire cylinder by multiplying $\pi \cdot 3.5^2 \cdot 9$, then subtract the empty cylinder in the middle, which is found by multiplying $\pi \cdot 1.5^2 \cdot 9$.

Glossary

acute triangle	An angle measuring less than 90°.
angle	A figure formed by the joining of two rays with a common endpoint.
area	The amount of space inside a two-dimensional shape, measured in square units.
circumference	The distance around a circle, calculated by the formula $C = \pi d$.
complementary angles	Two angles whose measurements add up to 90°.
cone	A solid figure with a single circular base and a round, smooth face that diminishes to a single point.
congruent	Having the same size and shape.
corresponding angles	Angles of separate figures that are in the same position within each figure.
corresponding sides	Sides of separate figures that are opposite corresponding angles.
cube	A six-sided polyhedron that has congruent squares as faces.
cylinder	A solid figure with a pair of circular, parallel bases and a round, smooth face between them.
diameter	The length across a circle, passing through the center of the circle. A diameter is equal to the length of two radii.
equilateral triangle	A triangle with 3 equal sides. Equilateral triangles also have three angles that measure the same.
face	The flat surface of a solid figure.
hypotenuse	The side opposite the right angle in any right triangle. The hypotenuse is the longest side of any right triangle.
isosceles trapezoid	A trapezoid with one pair of parallel sides and another pair of opposite sides that are congruent.
isosceles triangle	A triangle with 2 equal sides and 2 equal angles.
leg, legs	In a right triangle, one of the two sides creating a right angle.
line	A line is a one-dimensional figure, which extends without end in two directions.

line segment	A finite section of a line between any two points that lie on the line.
obtuse angle, obtuse angles	An angle measuring more than 90° and less than 180°.
obtuse triangle	A triangle with one angle that measures between 90° and 180°.
parallel lines	Two or more lines that lie in the same plane but which never intersect.
parallelogram, parallelograms	A quadrilateral with two pairs of parallel sides.
perimeter	The distance around a two-dimensional shape.
perpendicular lines	Two lines that lie in the same plane and intersect at a 90° angle.
pi	The ratio of a circle's circumference to its diameter. Pi is denoted by the Greek letter π. It is often approximated as 3.14 or $\dfrac{22}{7}$.
plane	In geometry, a two-dimensional surface that continues infinitely. Any three individual points that don't lie on the same line will lie on exactly one plane.
point	A zero-dimensional object that defines a specific location on a plane. It is represented by a small dot.
polygon, polygons	A closed plane figure with three or more straight sides.
polyhedron, polyhedrons	A solid whose faces are polygons.
pyramid, pyramids	A polyhedron with a polygonal base and a collection of triangular faces that meet at a point.
Pythagorean Theorem	The formula that relates the lengths of the sides of any right triangle: $a^2 + b^2 = c^2$, where c is the hypotenuse, and a and b are the legs of the right triangle.
quadrilateral, quadrilaterals	A four-sided polygon.
radius	The distance from the center of a circle to any point on the circle.
ray	A half-line that begins at one point and goes on forever in one direction.
rectangle	A quadrilateral with two pairs of parallel sides and four right angles.

rectangular prism	A polyhedron that has three pairs of congruent, rectangular, parallel faces.
rhombus	A quadrilateral with four congruent sides.
right angle	An angle measuring exactly 90°.
right triangle, right triangles	A triangle containing a right angle.
scalene triangle	A triangle in which all three sides are a different length.
similar	Having the same shape but not necessarily the same size.
sphere	A solid, round figure where every point on the surface is the same distance from the center.
square	A quadrilateral whose sides are all congruent and which has four right angles.
straight angle	An angle measuring exactly 180°.
supplementary angles	Two angles whose measurements add up to 180°.
trapezoid	A quadrilateral with one pair of parallel sides.
triangle	A polygon with three sides.
vertex	A turning point in a graph. Also the endpoint of the two rays that form an angle.
volume	A measurement of how much it takes to fill up a three-dimensional figure. Volume is measured in cubic units.

Exercises

1. Classify the angle below as acute, obtuse, or right.

2. Classify the angle below as acute, obtuse, or right.

3. Classify the angle shown as Acute, Obtuse, or Right

4. Use the picture below to answer the following questions. *Note,* ∠*COD is a right angle.*

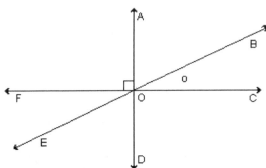

   a) Which angle is supplementary to ∠BOC?
   b) Which angle is complementary to ∠BOC?
   c) What is the measure of ∠EOF?
   d) What is the measure of ∠AOE?
   e) What is the measure of ∠BOF?

5. Find the unknown angle measure.

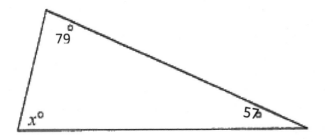

6. Find the unknown angle measure.

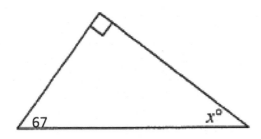

7. Find the unknown angle measure.

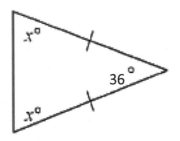

8. Find the unknown angle measure.

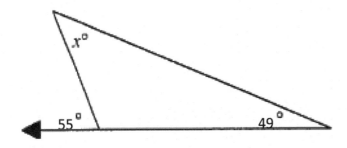

9. Find the unknown angle measures.

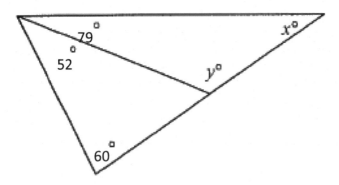

10. Find the unknown angle measure.

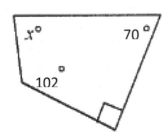

11. Find the length of the hypotenuse of the given right triangle pictured below. *Round to two decimal places.*

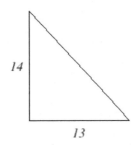

12. Find the length of the leg x. *Enter the exact value, not a decimal approximation.*

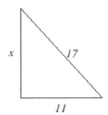

13. Find the perimeter of the figure pictured below.

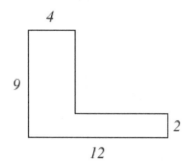

14. Find the perimeter of the rectangle pictured below.

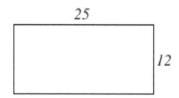

15. Find the perimeter of the parallelogram shown below.

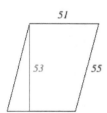

16. Find the circumference of the circle pictured below. *Round your answer to the nearest hundredth.*

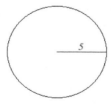

17. Find the circumference of the circle pictured below. *Round your answer to the nearest hundredth.*

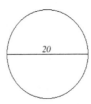

18. Find the area of the rectangle pictured below.

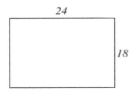

19. Find the area of the figure pictured below and state the correct units.

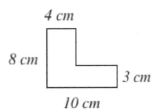

20. Find the area of the parallelogram shown below.

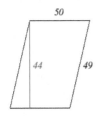

21. The area of a triangle can be found using the formula: Area $= \frac{1}{2} \cdot$ base $\cdot$ height

    Find the area of the triangle pictured below, where the measurements are given in meters (m).

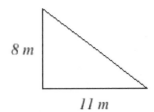

22. Find the area of the circle pictured below. *Round your answer to the nearest hundredth.*

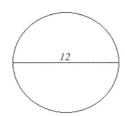

23. Find the area of the shaded area. *Round your answer to the nearest tenth.*

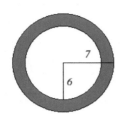

24. Match the formula for each volume to the figure to which it applies.

Figure		Volume
_____ Right Circular Cylinder	A.	$V = \pi r^2 h$
_____ Rectangular Solid	B.	$V = \frac{4}{3}\pi r^3$
_____ Sphere	C.	$V = l$

25. The volume of a cylinder with height h and radius r can be found using the formula $V = \pi r^2 h$.

    Sketch a cylinder with radius 7 feet and height 4 feet, then find the volume and select the correct units. *Round your answer to the nearest tenth.*

26. The volume of a cone with height h and radius r can be found using the formula $V = \frac{1}{3}\pi r^2 h$.

    Sketch a cone with radius 9 feet and height 3 feet, then find the volume and select the correct units. *Round your answer to the nearest tenth.*

27. A sports ball has a diameter of 26 cm. Find the volume of the ball and select the correct units. *Round your answer to 2 decimal places.*

Chapter 7: Finance

We have to work with money every day. While balancing your checkbook or calculating monthly expenditures requires only arithmetic, when we start saving, planning for retirement, or need a loan, we need further mathematics for more sophisticated ways to financially plan.

Table of Contents

7.1   Simple Interest

Discussing interest starts with the **principal**, or amount your account starts with. This could be a starting investment, or the starting amount of a loan. Interest, in its most simple form, is calculated as a percent of the principal. For example, if you borrowed $100 from a friend and agree to repay it with 5% interest, then the amount of interest you would pay would just be 5% of 100: $100(0.05) = $5. The total amount you would repay would be $105, the original principal plus the interest.

Simple One-time Interest

$$I = P_0 r$$

$$A = P_0 + I = P_0 + P_0 r = P_0(1 + r)$$

I is the interest
A is the end amount: principal plus interest
P_0 is the principal (starting amount)
r is the interest rate (in decimal form. Example: 5% = 0.05)

Example 1

A friend asks to borrow $300 and agrees to repay it in 30 days with 3% interest. How much interest will you earn?

$P_0 = \$300$ the principal
$r = 0.03$ 3% rate
$I = \$300(0.03) = \$9.$ You will earn $9 interest.

One-time simple interest is only common for extremely short-term loans. For longer term loans, it is common for interest to be paid on a daily, monthly, quarterly, or annual basis. In that case, interest would be earned regularly. For example, bonds are essentially a loan made to the bond issuer (a company or government) by you, the bond holder. In return for the loan, the issuer agrees to pay interest, often annually. Bonds have a maturity date, at which time the issuer pays back the original bond value.

Example 2

Suppose your city is building a new park, and issues bonds to raise the money to build it. You obtain a $1,000 bond that pays 5% interest annually that matures in 5 years. How much interest will you earn?

Each year, you would earn 5% interest: $1000(0.05) = $50 in interest. So, over the course of five years, you would earn a total of $250 in interest. When the bond matures, you would receive back the $1,000 you originally paid, leaving you with a total of $1,250.

We can generalize this idea of simple interest over time.

Simple Interest over Time

$$I = P_0 rt$$

$$A = P_0 + I = P_0 + P_0 rt = P_0(1 + rt)$$

I is the interest
A is the end amount: principal plus interest
P_0 is the principal (starting amount)
r is the interest rate in decimal form
t is time

The units of measurement (years, months, etc.) for the time should match the time period for the interest rate.

APR – Annual Percentage Rate

Interest rates are usually given as an **annual percentage rate (APR)** – the total interest that will be paid in the year. If the interest is paid in smaller time increments, the APR will be divided up.

For example, a 6% APR paid monthly would be divided into twelve 0.5% payments.
A 4% annual rate paid quarterly would be divided into four 1% payments.

Example 3

Treasury Notes (T-notes) are bonds issued by the federal government to cover its expenses. Suppose you obtain a $1,000 T-note with a 4% annual rate, paid semi-annually, with a maturity in 4 years. How much interest will you earn?

Since interest is being paid semi-annually (twice a year), the 4% interest will be divided into two 2% payments.

P_0 = $1,000   the principal
r = 0.02   2% rate per half-year
t = 8   4 years = 8 half-years

I = $1000(0.02)(8) = $160

You will earn $160 interest total over the four years.

Try it Now 1

A loan company charges $30 interest for a one month loan of $500. Find the annual interest rate they are charging.

7.2 Compound Interest

With simple interest, we were assuming that we pocketed the interest when we received it. In a standard bank account, any interest we earn is automatically added to our balance, and we earn interest on that interest in future years. This reinvestment of interest is called **compounding**.

Suppose that we deposit $1,000 in a bank account offering 3% interest, compounded monthly. How will our money grow?

The 3% interest is an annual percentage rate (APR) – the total interest to be paid during the year. Since interest is being paid monthly, each month, we will earn $\frac{3\%}{12} = 0.25\%$ per month.

In the first month,
$P_0 = \$1000$
$r = 0.0025\ (0.25\%)$
$I = \$1000\ (0.0025) = \$2.50$
$A = \$1000 + \$2.50 = \$1,002.50$

In the first month, we will earn $2.50 in interest, raising our account balance to $1,002.50. In the second month,

$P_0 = \$1,002.50$
$I = \$1002.50\ (0.0025) = \$2.51\ (\text{rounded})$
$A = \$1002.50 + \$2.51 = \$1005.01$

Notice that in the second month we earned more interest than we did in the first month. This is because we earned interest not only on the original $1,000 we deposited, but we also earned interest on the $2.50 of interest we earned the first month. This is the key advantage that **compounding** of interest gives us.

Calculating out a few more months:

Month	Starting balance	Interest earned	Ending Balance
1	1000.00	2.50	1002.50
2	1002.50	2.51	1005.01
3	1005.01	2.51	1007.52
4	1007.52	2.52	1010.04
5	1010.04	2.53	1012.57
6	1012.57	2.53	1015.10
7	1015.10	2.54	1017.64
8	1017.64	2.54	1020.18
9	1020.18	2.55	1022.73
10	1022.73	2.56	1025.29
11	1025.29	2.56	1027.85
12	1027.85	2.57	1030.42

To find an equation to represent this, if P_m represents the amount of money after m months, then we could write the recursive equation:

$P_0 = \$1000$
$P_m = (1+0.0025)P_{m-1}$

You probably recognize this as the recursive form of exponential growth. If not, we could go through the steps to build an explicit equation for the growth:
$P_0 = \$1000$
$P_1 = 1.0025P_0 = 1.0025\,(1000)$
$P_2 = 1.0025P_1 = 1.0025\,(1.0025\,(1000)) = 1.0025^{\,2}(1000)$
$P_3 = 1.0025P_2 = 1.0025\,(1.0025^2(1000)) = 1.0025^3(1000)$
$P_4 = 1.0025P_3 = 1.0025\,(1.0025^3(1000)) = 1.0025^4(1000)$

Observing a pattern, we could conclude

$P_m = (1.0025)^m(\$1000)$

Notice that the \$1000 in the equation was P_0, the starting amount. We found 1.0025 by adding one to the growth rate divided by 12, since we were compounding 12 times per year. Generalizing our result, we could write

$$P_m = P_0\left(1+\frac{r}{k}\right)^m$$

In this formula:
m is the number of compounding periods (months in our example)
r is the annual interest rate
k is the number of compounds per year.

While this formula works fine, it is more common to use a formula that involves the number of years, rather than the number of compounding periods. If N is the number of years, then $m=Nk$. Making this change gives us the standard formula for compound interest.

Compound Interest

$$P_N = P_0\left(1+\frac{r}{k}\right)^{Nk}$$

P_N is the balance in the account after N years.
P_0 is the starting balance of the account (also called initial deposit, or principal)
r is the annual interest rate in decimal form
k is the number of compounding periods in one year.

If the compounding is done annually (once a year), $k = 1$.
If the compounding is done quarterly, $k = 4$.
If the compounding is done monthly, $k = 12$.
If the compounding is done daily, $k = 365$.

The most important thing to remember about using this formula is that it assumes that we put money in the account <u>once</u> and let it sit there earning interest.

Example 4

A certificate of deposit (CD) is a savings instrument that many banks offer. It usually gives a higher interest rate, but you cannot access your investment for a specified length of time. Suppose you deposit $3000 in a CD paying 6% interest, compounded monthly. How much will you have in the account after 20 years?

In this example,

$P_0 = \$3000$ the initial deposit
$r = 0.06$ 6% annual rate
$k = 12$ 12 months in 1 year
$N = 20$ since we're looking for how much we'll have after 20 years

So $P_{20} = 3000\left(1 + \dfrac{0.06}{12}\right)^{20\times12} = \$9930.61$ (round your answer to the nearest penny)

Let us compare the amount of money earned from compounding against the amount you would earn from simple interest.

Years	Simple Interest ($15 per month)	6% compounded monthly = 0.5% each month.
5	$3900	$4046.55
10	$4800	$5458.19
15	$5700	$7362.28
20	$6600	$9930.61
25	$7500	$13394.91
30	$8400	$18067.73
35	$9300	$24370.65

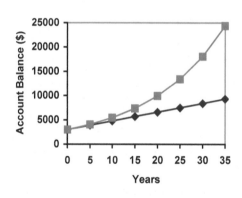

As you can see, over a long period of time, compounding makes a large difference in the account balance. You may recognize this as the difference between linear growth and exponential growth.

Evaluating exponents on the calculator

When we need to calculate something like 5^3 it is easy enough to just multiply $5 \cdot 5 \cdot 5 = 125$. But when we need to calculate something like 1.005^{240}, it would be very tedious to calculate this by multiplying 1.005 by itself 240 times! So to make things easier, we can harness the power of our scientific calculators.

Most scientific calculators have a button for exponents. It is typically either labeled like:

$\boxed{\wedge}$, $\boxed{y^x}$, or $\boxed{x^y}$.

To evaluate 1.005^{240} we'd type 1.005 $\boxed{\wedge}$ 240, or 1.005 $\boxed{y^x}$ 240. Try it out - you should get something around 3.3102044758.

Example 5

You know that you will need $40,000 for your child's education in 18 years. If your account earns 4% compounded quarterly, how much would you need to deposit now to reach your goal?

In this example, we're looking for P_0.

$r = 0.04$	4%
$k = 4$	4 quarters in 1 year
$N = 18$	Since we know the balance in 18 years
$P_{18} = \$40,000$	The amount we have in 18 years

In this case, we're going to have to set up the equation, and solve for P_0.

$$40000 = P_0 \left(1 + \frac{0.04}{4}\right)^{18 \times 4}$$

$$40000 = P_0 (2.0471)$$

$$P_0 = \frac{40000}{2.0471} = \$19539.84$$

So, you would need to deposit $19,539.84 now to have $40,000 in 18 years.

> **Rounding**
> It is important to be very careful about rounding when calculating things with exponents. In general, you want to keep as many decimals during calculations as you can. Be sure to **keep at least 3 significant digits** (numbers after any leading zeros). Rounding 0.00012345 to 0.000123 will usually give you a "close enough" answer, but keeping more digits is always better.

Example 6

The reason we shouldn't "over-round" is displayed by this example. Suppose you were investing $1,000 at 5% interest compounded monthly for 30 years.

$P_0 = \$1000$	the initial deposit
$r = 0.05$	5%
$k = 12$	12 months in 1 year
$N = 30$	since we're looking for the amount after 30 years

If we first compute r/k, we find $0.05/12 = 0.00416666666667$

Here is the effect of rounding this to different values:

r/k rounded to:	Gives P_{30} to be:	Error
0.004	$4208.59	$259.15
0.0042	$4521.45	$53.71
0.00417	$4473.09	$5.35
0.004167	$4468.28	$0.54
0.0041667	$4467.80	$0.06
no rounding	$4467.74	

If you're working in a bank, of course you wouldn't round at all. For our purposes, the answer we got by rounding to 0.00417, three significant digits, is close enough - $5 off of $4,500 isn't too bad. Certainly, keeping that fourth decimal place wouldn't have hurt.

Using your calculator

In many cases, you can avoid rounding completely by how you enter things in your calculator. For example, in the example above, we needed to calculate

$$P_{30} = 1000\left(1 + \frac{0.05}{12}\right)^{12 \times 30}$$

We can quickly calculate $12 \times 30 = 360$, giving $P_{30} = 1000\left(1 + \frac{0.05}{12}\right)^{360}$.

Now we can use the calculator.

Type this	Calculator shows
0.05 $\div$ 12 $=$	0.00416666666667
$+$ 1 $=$	1.00416666666667
y^x 360 $=$	4.46774431400613
$\times$ 1000 $=$	4467.74431400613

The previous steps were assuming you have a "one operation at a time" calculator; a more advanced calculator will often allow you to type in the entire expression to be evaluated. If you have a calculator like this, you will probably just need to enter:

1000 $\times$ (1 $+$ 0.05 $\div$ 12) y^x 360 $=$

7.3   Annuities

For most of us, we aren't able to put a large sum of money in the bank today. Instead, we save for the future by depositing a smaller amount of money from each paycheck into the bank. This idea is called a **savings annuity**. Most retirement plans like 401k plans or IRA plans are examples of savings annuities.

An annuity can be described recursively in a fairly simple way. Recall that basic compound interest follows from the relationship

$$P_m = \left(1 + \frac{r}{k}\right)P_{m-1}$$

For a savings annuity, we simply need to add a deposit, d, to the account with each compounding period:

$$P_m = \left(1 + \frac{r}{k}\right)P_{m-1} + d$$

Taking this equation from recursive form to explicit form is a bit trickier than with compound interest. It will be easiest to see by working with an example rather than working in general.

Suppose we will deposit $100 each month into an account paying 6% interest. We assume that the account is compounded with the same frequency as we make deposits unless stated otherwise. In this example:
$r = 0.06$ (6%)
$k = 12$ (12 compounds/deposits per year)
$d = \$100$ (our deposit per month)

Writing out the recursive equation gives

$$P_m = \left(1 + \frac{0.06}{12}\right)P_{m-1} + 100 = (1.005)P_{m-1} + 100$$

Assuming we start with an empty account, we can begin using this relationship:

$P_0 = 0$
$P_1 = (1.005)P_0 + 100 = 100$
$P_2 = (1.005)P_1 + 100 = (1.005)(100) + 100 = 100(1.005) + 100$
$P_3 = (1.005)P_2 + 100 = (1.005)(100(1.005) + 100) + 100 = 100(1.005)^2 + 100(1.005) + 100$

Continuing this pattern, after m deposits, we'd have saved

$$P_m = 100(1.005)^{m-1} + 100(1.005)^{m-2} + \cdots + 100(1.005) + 100$$

In other words, after m months, the first deposit will have earned compound interest for m-1 months. The second deposit will have earned interest for m-2 months. Last months deposit would have earned only one month worth of interest. The most recent deposit will have earned no interest yet.

This equation leaves a lot to be desired, though – it doesn't make calculating the ending balance any easier! To simplify things, multiply both sides of the equation by 1.005:

$$1.005 P_m = 1.005 \left(100(1.005)^{m-1} + 100(1.005)^{m-2} + \cdots + 100(1.005) + 100 \right)$$

Distributing on the right side of the equation gives

$$1.005 P_m = 100(1.005)^m + 100(1.005)^{m-1} + \cdots + 100(1.005)^2 + 100(1.005)$$

Now we'll line this up with like terms from our original equation, and subtract each side

$$
\begin{aligned}
1.005 P_m &= 100(1.005)^m + 100(1.005)^{m-1} + \cdots + 100(1.005) \\
P_m &= 100(1.005)^{m-1} + \cdots + 100(1.005) + 100
\end{aligned}
$$

Almost all the terms cancel on the right-hand side when we subtract, leaving

$$1.005 P_m - P_m = 100(1.005)^m - 100$$

Solving for P_m

$$0.005 P_m = 100\left((1.005)^m - 1 \right)$$

$$P_m = \frac{100\left((1.005)^m - 1 \right)}{0.005}$$

Replacing m months with $12N$, where N is measured in years, gives

$$P_N = \frac{100\left((1.005)^{12N} - 1 \right)}{0.005}$$

Recall 0.005 was r/k and 100 was the deposit d. 12 was k, the number of deposit each year. Generalizing this result, we get the saving annuity formula.

Annuity Formula

$$P_N = \frac{d\left(\left(1+\dfrac{r}{k}\right)^{Nk} - 1\right)}{\left(\dfrac{r}{k}\right)}$$

P_N is the balance in the account after N years.
d is the regular deposit (the amount you deposit each year, each month, etc.)
r is the annual interest rate in decimal form.
k is the number of compounding periods in one year.

If the compounding frequency is not explicitly stated, assume there are the same number of compounds in a year as there are deposits made in a year.

If the compounding frequency isn't stated then use these as rule of thumb:

- If you make your deposits every month, use monthly compounding, $k = 12$.
- If you make your deposits every year, use yearly compounding, $k = 1$.
- If you make your deposits every quarter, use quarterly compounding, $k = 4$.

and so on.

When do you use this
Annuities assume that you put money in the account <u>on a regular schedule (every month, year, quarter, etc.)</u> and let it sit there earning interest.

Compound interest assumes that you put money in the account <u>once</u> and let it sit there earning interest.

Compound interest: <u>One</u> deposit
Annuity: <u>Many</u> deposits.

Example 7

A traditional individual retirement account (IRA) is a special type of retirement account in which the money you invest is exempt from income taxes until you withdraw it. If you deposit $100 each month into an IRA earning 6% interest, how much will you have in the account after 20 years?

In this example,
$d = \$100$ the monthly deposit
$r = 0.06$ 6% annual rate
$k = 12$ since we're doing monthly deposits, we'll compound monthly
$N = 20$ we want the amount after 20 years

Putting this into the equation, we get

$$P_{20} = \frac{100\left(\left(1+\dfrac{0.06}{12}\right)^{20(12)}-1\right)}{\left(\dfrac{0.06}{12}\right)}$$

$$P_{20} = \frac{100\left((1.005)^{240}-1\right)}{(0.005)}$$

$$P_{20} = \frac{100(3.310-1)}{(0.005)}$$

$$P_{20} = \frac{100(2.310)}{(0.005)} = \$46200$$

The account will grow to $46,200 after 20 years.

Notice that you deposited into the account a total of $24,000 ($100 a month for 240 months). The difference between what you end up with and how much you put in is the <u>interest earned</u>. In this case, it is $46,200 - $24,000 = $22,200.

Example 8

You want to have $200,000 in your account when you retire in 30 years. Your retirement account earns 8% interest. How much do you need to deposit each month to meet your retirement goal?

In this example, we're looking for d.

$r = 0.08$	8% annual rate
$k = 12$	since we're depositing monthly
$N = 30$	30 years
$P_{30} = \$200,000$	The amount we want to have in 30 years

In this case, we're going to have to set up the equation, and solve for d.

$$200,000 = \frac{d\left(\left(1+\dfrac{0.08}{12}\right)^{30(12)}-1\right)}{\left(\dfrac{0.08}{12}\right)}$$

$$200,000 = \frac{d\left((1.00667)^{360}-1\right)}{(0.00667)}$$

$$200,000 = d(1491.57)$$

$$d = \frac{200,000}{1491.57} = \$134.09$$

So, you would need to deposit $134.09 each month to have $200,000 in 30 years if your account earns 8% interest

In general, if we need to obtain the amount of the deposits, we can just rewrite the annuity formula as

$$d = \frac{P_N \cdot \frac{r}{k}}{\left(1 + \frac{r}{k}\right)^{Nk} - 1}$$

Try it Now 2

A more conservative investment account pays 3% interest. If you deposit $5 a day into this account, how much will you have after 10 years? How much is from interest?

7.4 Payout Annuities

In the last section, you learned about annuities. In an annuity, you start with nothing, put money into an account on a regular basis, and end up with money in your account.

In this section, we will learn about a variation called a **payout annuity**. With a payout annuity, you start with money in the account, and pull money out of the account on a regular basis. Any remaining money in the account earns interest. After a fixed amount of time, the account will end up empty.

Payout annuities are typically used after retirement. Perhaps you have saved $500,000 for retirement, and want to take money out of the account each month to live on. You want the money to last you 20 years. This is a payout annuity. The formula is derived in a similar way as we did for savings annuities. The details are omitted here.

Payout Annuity Formula

$$P_0 = \frac{d\left(1-\left(1+\frac{r}{k}\right)^{-Nk}\right)}{\left(\frac{r}{k}\right)}$$

P_0 is the balance in the account at the beginning (starting amount, or principal).
d is the regular withdrawal (the amount you take out each year, each month, etc.)
r is the annual interest rate (in decimal form. Example: 5% = 0.05)
k is the number of compounding periods in one year.
N is the number of years we plan to take withdrawals

Like with annuities, the compounding frequency is not always explicitly given, but is determined by how often you take the withdrawals.

When do you use this
Payout annuities assume that you take money from the account on a regular schedule (every month, year, quarter, etc.) and let the rest sit there earning interest.

Compound interest: One deposit
Annuity: Many deposits
Payout Annuity: Many withdrawals

Example 9

After retiring, you want to be able to take $1000 every month for a total of 20 years from your retirement account. The account earns 6% interest. How much will you need in your account when you retire?

In this example,
$d = \$1000$ the monthly withdrawal

$r = 0.06$	6% annual rate
$k = 12$	since we're doing monthly withdrawals, we'll compound monthly
$N = 20$	since were taking withdrawals for 20 years

We're looking for P_0; how much money needs to be in the account at the beginning. Putting this into the equation, we get

$$P_0 = \frac{1000\left(1-\left(1+\dfrac{0.06}{12}\right)^{-20(12)}\right)}{\left(\dfrac{0.06}{12}\right)}$$

$$P_0 = \frac{1000\times\left(1-(1.005)^{-240}\right)}{(0.005)}$$

$$P_0 = \frac{1000\times(1-0.302)}{(0.005)} = \$139,600$$

You will need to have $139,600 in your account when you retire.

Notice that you withdrew a total of $240,000 ($1000 a month for 240 months). The difference between what you pulled out and what you started with is the <u>interest earned</u>. In this case, it is $240,000 - $139,600 = $100,400 in interest.

Evaluating negative exponents on your calculator

With these problems, you need to raise numbers to negative powers. Most calculators have a separate button for negating a number that is different than the subtraction button. Some calculators label this $\boxed{(-)}$, some with $\boxed{+/-}$. The button is often near the = key or the decimal point.

If your calculator displays operations on it (typically a calculator with multiline display), to calculate 1.005^{-240} you'd type something like 1.005 $\boxed{\wedge}$ $\boxed{(-)}$ 240

If your calculator only shows one value at a time, then usually you hit the (-) key after a number to negate it, so you'd enter 1.005 $\boxed{y^x}$ 240 $\boxed{(-)}$ =

Give it a try - you should get $1.005^{-240} = 0.302096$

Example 10

You know you will have $500,000 in your account when you retire. You want to be able to take monthly withdrawals from the account for a total of 30 years. Your retirement account earns 8% interest. How much will you be able to withdraw each month?

In this example, we're looking for d.

$r = 0.08$	8% annual rate
$k = 12$	since we're withdrawing monthly

$$N = 30 \qquad \text{30 years}$$
$$P_0 = \$500,000 \qquad \text{we are beginning with } \$500,000$$

In this case, we're going to have to set up the equation, and solve for d.

$$500,000 = \frac{d\left(1-\left(1+\dfrac{0.08}{12}\right)^{-30(12)}\right)}{\left(\dfrac{0.08}{12}\right)}$$

$$500,000 = \frac{d\left(1-\left(1.00667\right)^{-360}\right)}{\left(0.00667\right)}$$

$$500,000 = d(136.232)$$

$$d = \frac{500,000}{136.232} = \$3670.21$$

You would be able to withdraw $3,670.21 each month for 30 years.

In general, if we wanted to obtain the amount of each withdrawal, we can rewrite the payout annuity formula as

$$d = \frac{P_0 \cdot \dfrac{r}{k}}{\left(1 - \left(1 + \dfrac{r}{k}\right)^{-Nk}\right)}$$

Try it Now 3

A donor gives $100,000 to a university, and specifies that it is to be used to give annual scholarships for the next 20 years. If the university can earn 4% interest, how much can they give in scholarships each year?

7.5   Loans

In the last section, you learned about payout annuities.

In this section, you will learn about conventional loans (also called amortized loans or installment loans). Examples include auto loans and home mortgages. These techniques do not apply to payday loans, add-on loans, or other loan types where the interest is calculated up front.

One great thing about loans is that they use exactly the same formula as a payout annuity. To see why, imagine that you had $10,000 invested at a bank, and started taking out payments while earning interest as part of a payout annuity, and after 5 years your balance was zero. Flip that around, and imagine that you are acting as the bank, and a car lender is acting as you. The car lender invests $10,000 in you. Since you're acting as the bank, you pay interest. The car lender takes payments until the balance is zero.

Loans Formula

$$P_0 = \frac{d\left(1 - \left(1 + \frac{r}{k}\right)^{-Nk}\right)}{\left(\frac{r}{k}\right)}$$

P_0 is the balance in the account at the beginning (the principal, or amount of the loan).
d is your loan payment (your monthly payment, annual payment, etc)
r is the annual interest rate in decimal form.
k is the number of compounding periods in one year.
N is the length of the loan, in years

Like before, the compounding frequency is not always explicitly given, but is determined by how often you make payments.

When do you use this
The loan formula assumes that you make loan payments <u>on a regular schedule (every month, year, quarter, etc.)</u> and are paying interest on the loan.

Compound interest: <u>One</u> deposit
Annuity: <u>Many</u> deposits
Payout Annuity: <u>Many withdrawals</u>
Loans: <u>Many payments</u>

Example 11

You can afford $200 per month as a car payment. If you can get an auto loan at 3% interest for 60 months (5 years), how expensive of a car can you afford? In other words, what amount loan can you pay off with $200 per month?

In this example,

$d = \$200$	the monthly loan payment
$r = 0.03$	3% annual rate
$k = 12$	since we're doing monthly payments, we'll compound monthly
$N = 5$	since we're making monthly payments for 5 years

We're looking for P_0, the starting amount of the loan.

$$P_0 = \frac{200\left(1-\left(1+\frac{0.03}{12}\right)^{-5(12)}\right)}{\left(\frac{0.03}{12}\right)}$$

$$P_0 = \frac{200\left(1-(1.0025)^{-60}\right)}{(0.0025)}$$

$$P_0 = \frac{200(1-0.861)}{(0.0025)} = \$11,120$$

You can afford a $11,120 loan.

You will pay a total of $12,000 ($200 per month for 60 months) to the loan company. The difference between the amount you pay and the amount of the loan is the <u>interest paid</u>. In this case, you're paying $12,000-$11,120 = $880 interest total.

Example 12

You want to take out a $140,000 mortgage (home loan). The interest rate on the loan is 6%, and the loan is for 30 years. How much will your monthly payments be?

In this example, we're looking for d.

$r = 0.06$	6% annual rate
$k = 12$	since we're paying monthly
$N = 30$	30 years
$P_0 = \$140,000$	the starting loan amount

In this case, we're going to have to set up the equation, and solve for d.

$$140,000 = \frac{d\left(1 - \left(1 + \dfrac{0.06}{12}\right)^{-30(12)}\right)}{\left(\dfrac{0.06}{12}\right)}$$

$$140,000 = \frac{d\left(1 - (1.005)^{-360}\right)}{(0.005)}$$

$$140,000 = d(166.792)$$

$$d = \frac{140,000}{166.792} = \$839.37$$

You will make payments of $839.37 per month for 30 years.

You're paying a total of $302,173.20 to the loan company $839.37 per month for 360 months. You are paying a total of $302,173.20 - $140,000 = $162,173.20 in interest over the life of the loan.

In general, if we wanted to obtain the amount of each payment, we can rewrite the loan formula as

$$d = \frac{P_0 \cdot \dfrac{r}{k}}{\left(1 - \left(1 + \dfrac{r}{k}\right)^{-Nk}\right)}$$

Try it Now 4

Janine bought $3,000 of new furniture on credit. Because her credit score isn't very good, the store is charging her a fairly high interest rate on the loan: 16%. If she agreed to pay off the furniture over 2 years, how much will she have to pay each month?

7.6   Remaining Loan Balance

With loans, it is often desirable to determine what the remaining loan balance will be after some number of years. For example, if you purchase a home and plan to sell it in five years, you might want to know how much of the loan balance you will have paid off and how much you have to pay from the sale.

To determine the remaining loan balance after some number of years, we first need to know the loan payments, if we don't already know them. Remember that only a portion of your loan payments go towards the loan balance; a portion is going to go towards interest. For example, if your payments were $1,000 a month, after a year you will *not* have paid off $12,000 of the loan balance.

To determine the remaining loan balance, we can think "how much of the loan will these loan payments be able to pay off in the remaining time on the loan?"

Example 13

If a mortgage at a 6% interest rate has payments of $1,000 a month, how much will the loan balance be 10 years from the end the loan?

To determine this, we are looking for the amount of the loan that can be paid off by $1,000 a month payments in 10 years. In other words, we're looking for P_0 when

$d = \$1,000$	the monthly loan payment
$r = 0.06$	6% annual rate
$k = 12$	since we're doing monthly payments, we'll compound monthly
$N = 10$	since we're making monthly payments for 10 more years

$$P_0 = \frac{1000\left(1-\left(1+\frac{0.06}{12}\right)^{-10(12)}\right)}{\left(\frac{0.06}{12}\right)}$$

$$P_0 = \frac{1000\left(1-(1.005)^{-120}\right)}{(0.005)}$$

$$P_0 = \frac{1000(1-0.5496)}{(0.005)} = \$90,073.45$$

The loan balance with 10 years remaining on the loan will be $90,073.45.

Often times answering remaining balance questions requires two steps:

1) Calculating the monthly payments on the loan
2) Calculating the remaining loan balance based on the *remaining time* on the loan

Example 14

A couple purchases a home with a $180,000 mortgage at 4% for 30 years with monthly payments. What will the remaining balance on their mortgage be after 5 years?

First, we will calculate their monthly payments, i.e., we're looking for d.

$r = 0.04$ 4% annual rate
$k = 12$ since they're paying monthly
$N = 30$ 30 years
$P_0 = \$180,000$ the starting loan amount

We set up the equation and solve for d.

$$180,000 = \frac{d\left(1-\left(1+\dfrac{0.04}{12}\right)^{-30(12)}\right)}{\left(\dfrac{0.04}{12}\right)}$$

$$180,000 = \frac{d\left(1-(1.00333)^{-360}\right)}{(0.00333)}$$

$$180,000 = d(209.562)$$

$$d = \frac{180,000}{209.562} = \$858.93$$

Now that we know the monthly payments, we can determine the remaining balance. We want the remaining balance after 5 years, when 25 years will be remaining on the loan. So, we calculate the loan balance that will be paid off with the monthly payments over those 25 years.

$d = \$858.93$ the monthly loan payment we calculated above
$r = 0.04$ 4% annual rate
$k = 12$ since they're doing monthly payments
$N = 25$ since they'd be making monthly payments for 25 more years

$$P_0 = \frac{858.93\left(1-\left(1+\dfrac{0.04}{12}\right)^{-25(12)}\right)}{\left(\dfrac{0.04}{12}\right)}$$

$$P_0 = \frac{858.93\left(1-(1.00333)^{-300}\right)}{(0.00333)}$$

$$P_0 = \frac{858.93(1-0.369)}{(0.00333)} = \$155,793.91$$

The loan balance after 5 years, with 25 years remaining on the loan, will be $155,793.91

Over that 5 years, the couple has paid off $180,000 - $155,793.91 = $24,206.09 of the loan balance. They have paid a total of $858.93 a month for 5 years (60 months), for a total of $51,535.80, so $51,535.80 - $24,206.09 = $27,329.71 of what they have paid so far has been interest.

7.7   Solving for time

Often, we are interested in how long it will take to accumulate money or how long we'd need to extend a loan to bring payments down to a reasonable level.

Note: This section assumes you've covered solving exponential equations using logarithms, either in prior classes or in the growth models chapter.

Example 15
___

If you invest \$2,000 at 6% compounded monthly, how long will it take the account to double in value?

This is a compound interest problem, since we are depositing money once and allowing it to grow. In this problem,

P_0 = \$2000   the initial deposit
r = 0.06   6% annual rate
k = 12   12 months in 1 year

So, our general equation is $P_N = 2000\left(1+\dfrac{0.06}{12}\right)^{N\times12}$. We also know that we want our ending

amount to be double of \$2,000, which is \$4,000, so we're looking for N so that P_N = 4,000. To solve this, we set our equation for P_N equal to 4,000.

$$4000 = 2000\left(1+\frac{0.06}{12}\right)^{N\times12}$$ Divide both sides by 2,000

$$2 = (1.005)^{12N}$$ To solve for the exponent, take the log of both sides

$$\log(2) = \log\left((1.005)^{12N}\right)$$ Use the exponent property of logs on the right side

$$\log(2) = 12N\log(1.005)$$ Now we can divide both sides by 12log(1.005)

$$\frac{\log(2)}{12\log(1.005)} = N$$

Approximating this to three decimal places, we obtain N = 11.581. Thus, it will take about 11.581 years for the account to double in value.

Note that your answer may come out slightly differently if you had evaluated the logs to decimals and rounded during your calculations, but your answer should be close. For example if you rounded log(2) to 0.301 and log(1.005) to 0.00217, then your final answer would have been about 11.577 years.

Example 16

If you invest $100 each month into an account earning 3% compounded monthly, how long will it take the account to grow to $10,000?

This is a savings annuity problem since we are making regular deposits into the account.

$d = \$100$	the monthly deposit
$r = 0.03$	3% annual rate
$k = 12$	since we're doing monthly deposits, we'll compound monthly

We don't know N, but we want P_N to be $10,000.

Putting this into the equation, we get

$$10,000 = \frac{100\left(\left(1+\dfrac{0.03}{12}\right)^{N(12)}-1\right)}{\left(\dfrac{0.03}{12}\right)}$$ Simplifying the fractions a bit

$$10,000 = \frac{100\left((1.0025)^{12N}-1\right)}{0.0025}$$

We want to isolate the exponential term, 1.0025^{12N}, so multiply both sides by 0.0025

$25 = 100\left((1.0025)^{12N}-1\right)$ Divide both sides by 100

$0.25 = (1.0025)^{12N}-1$ Add 1 to both sides

$1.25 = (1.0025)^{12N}$ Now take the log of both sides

$\log(1.25) = \log\left((1.0025)^{12N}\right)$ Use the exponent property of logs

$\log(1.25) = 12N\log(1.0025)$ Divide by 12log(1.0025)

$\dfrac{\log(1.25)}{12\log(1.0025)} = N$

Approximating to three decimal places, we get $N = 7.447$ years. Thus, it will take about 7.447 years to grow the account to $10,000.

Try it Now 6

Joel is considering putting a $1,000 laptop purchase on his credit card, which has an interest rate of 12% compounded monthly. How long will it take him to pay off the purchase if he makes payments of $30 a month?

Try it Now Answers

1.

I = $30 of interest

P_0 = $500 principal

r = unknown

t = 1 month

Using $I = P_0 r t$, we get $30 = 500 \cdot r \cdot 1$. Solving, we get $r = 0.06$, or 6%. Since the time was monthly, this is the monthly interest. The annual rate would be 12 times this: 72% interest.

2.

d = $5   the daily deposit

r = 0.03   3% annual rate

k = 365   since we're doing daily deposits, we'll compound daily

N = 10   we want the amount after 10 years

$$P_{10} = \frac{5\left(\left(1+\dfrac{0.03}{365}\right)^{365 \times 10} - 1\right)}{\dfrac{0.03}{365}} = \$21{,}282.07$$

We would have deposited a total of $5 \cdot 365 \cdot 10 =$ $18,250, so $3,032.07 is from interest

3.

d = unknown

r = 0.04   4% annual rate

k = 1   since we're doing annual scholarships

N = 20   20 years

P_0 = 100,000   we're starting with $100,000

$$100{,}000 = \frac{d\left(1-\left(1+\dfrac{0.04}{1}\right)^{-20 \times 1}\right)}{\dfrac{0.04}{1}}$$

Solving for d gives $7,358.18 each year that they can give in scholarships.

It is worth noting that usually donors instead specify that only interest is to be used for scholarship, which makes the original donation last indefinitely. If this donor had specified that, $100,000(0.04) =$ $4,000 a year would have been available.

4.

d = unknown

r = 0.16   16% annual rate

k = 12   since we're making monthly payments

N = 2   2 years to repay

P_0 = 3,000   we're starting with a $3,000 loan

Try it Now Answers continued

$$3,000 = \dfrac{d\left(1-\left(1+\dfrac{0.16}{12}\right)^{-2\times12}\right)}{\dfrac{0.16}{12}}$$

Solving for *d* gives $146.89 as monthly payments.

In total, she will pay $3,525.36 to the store, meaning she will pay $525.36 in interest over the two years.

5.
    a. This is a payout annuity problem. She can pull out $1833.60 a quarter.
    b. This is a savings annuity problem. He will have saved up $7,524.11/
    c. This is compound interest problem. She would need to deposit $22,386.46.
    d. This is a loans problem. She can buy $4,609.33 of new equipment.
    e. This is a savings annuity problem. You would need to save $200.46 each month

6.
$d = \$30$ The monthly payments
$r = 0.12$ 12% annual rate
$k = 12$ since we're making monthly payments
$P_0 = 1,000$ we're starting with a $1,000 loan
We are solving for *N*, the time to pay off the loan

$$1,000 = \dfrac{30\left(1-\left(1+\dfrac{0.12}{12}\right)^{-N(12)}\right)}{\dfrac{0.12}{12}}$$

Solving for *N* gives 3.396. It will take about 3.4 years to pay off the purchase.

Exercises

Skills

1. A friend lends you $200 for a week, which you agree to repay with 5% one-time interest. How much will you have to repay?

2. Suppose you obtain a $3,000 T-note with a 3% annual rate, paid quarterly, with maturity in 5 years. How much interest will you earn?

3. A T-bill is a type of bond that is sold at a discount over the face value. For example, suppose you buy a 13-week T-bill with a face value of $10,000 for $9,800. This means that in 13 weeks, the government will give you the face value, earning you $200. What annual interest rate have you earned?

4. Suppose you are looking to buy a $5,000 face value 26-week T-bill. If you want to earn at least 1% annual interest, what is the most you should pay for the T-bill?

5. You deposit $300 in an account earning 5% interest compounded annually. How much will you have in the account in 10 years?

6. How much will $1,000 deposited in an account earning 7% interest compounded annually be worth in 20 years?

7. You deposit $2,000 in an account earning 3% interest compounded monthly.
   a. How much will you have in the account in 20 years?
   b. How much interest will you earn?

8. You deposit $10,000 in an account earning 4% interest compounded monthly.
   a. How much will you have in the account in 25 years?
   b. How much interest will you earn?

9. How much would you need to deposit in an account now in order to have $6,000 in the account in 8 years? Assume the account earns 6% interest compounded monthly.

10. How much would you need to deposit in an account now in order to have $20,000 in the account in 4 years? Assume the account earns 5% interest.

11. You deposit $200 each month into an account earning 3% interest compounded monthly.
    a. How much will you have in the account in 30 years?
    b. How much total money will you put into the account?
    c. How much total interest will you earn?

12. You deposit $1,000 each year into an account earning 8% compounded annually.
    a. How much will you have in the account in 10 years?
    b. How much total money will you put into the account?
    c. How much total interest will you earn?

13. Jose has determined he needs to have $800,000 for retirement in 30 years. His account earns 6% interest.
    a. How much would you need to deposit in the account each month?
    b. How much total money will you put into the account?
    c. How much total interest will you earn?

14. You wish to have $3,000 in 2 years to buy a fancy new stereo system. How much should you deposit each quarter into an account paying 8% compounded quarterly?

15. You want to be able to withdraw $30,000 each year for 25 years. Your account earns 8% interest.
    a. How much do you need in your account at the beginning?
    b. How much total money will you pull out of the account?
    c. How much of that money is interest?

16. How much money will I need to have at retirement so I can withdraw $60,000 a year for 20 years from an account earning 8% compounded annually?
    a. How much do you need in your account at the beginning?
    b. How much total money will you pull out of the account?
    c. How much of that money is interest?

17. You have $500,000 saved for retirement. Your account earns 6% interest. How much will you be able to pull out each month, if you want to be able to take withdrawals for 20 years?

18. Loren already knows that he will have $500,000 when he retires. If he sets up a payout annuity for 30 years in an account paying 10% interest, how much could the annuity provide each month?

19. You can afford a $700 per month mortgage payment. You've found a 30-year loan at 5% interest.
    a. How big of a loan can you afford?
    b. How much total money will you pay the loan company?
    c. How much of that money is interest?

20. Marie can afford a $250 per month car payment. She's found a 5-year loan at 7% interest.
    a. How expensive of a car can she afford?
    b. How much total money will she pay the loan company?
    c. How much of that money is interest?

21. You want to buy a $25,000 car. The company is offering a 2% interest rate for 48 months (4 years). What will your monthly payments be?

22. You decide finance a $12,000 car at 3% compounded monthly for 4 years. What will your monthly payments be? How much interest will you pay over the life of the loan?

23. You want to buy a $200,000 home. You plan to pay 10% as a down payment, and take out a 30-year loan for the rest.
    a. How much is the loan amount going to be?
    b. What will your monthly payments be if the interest rate is 5%?
    c. What will your monthly payments be if the interest rate is 6%?

24. Lynn bought a $300,000 house, paying 10% down, and financing the rest at 6% interest for 30 years.
    a. Find her monthly payments.
    b. How much interest will she pay over the life of the loan?

25. Emile bought a car for $24,000 three years ago. The loan had a 5-year term at 3% interest rate. How much does he still owe on the car?

26. A friend bought a house 15 years ago, taking out a $120,000 mortgage at 6% for 30 years. How much does she still owe on the mortgage?

27. Pat deposits $6,000 into an account earning 4% compounded monthly. How long will it take the account to grow to $10,000?

28. Kay is saving $200 a month into an account earning 5% interest. How long will it take her to save $20,000?

29. James has $3,000 in credit card debt, which charges 14% interest. How long will it take to pay off the card if he makes the minimum payment of $60 a month?

30. Chris has saved $200,000 for retirement, and it is in an account earning 6% interest. If she withdraws $3,000 a month, how long will the money last?

Concepts

31. Suppose you invest $50 a month for 5 years into an account earning 8% compounded monthly. After 5 years, you leave the money, without making additional deposits, in the account for another 25 years. How much will you have in the end?

32. Suppose you put off making investments for the first 5 years, and instead made deposits of $50 a month for 25 years into an account earning 8% compounded monthly. How much will you have in the end?

33. Mike plans to make contributions to his retirement account for 15 years. After the last contribution, he will start withdrawing $10,000 a quarter for 10 years. Assuming Mike's account earns 8% compounded quarterly, how large must his quarterly contributions be during the first 15 years, in order to accomplish his goal?

34. Kendra wants to be able to make withdrawals of $60,000 a year for 30 years after retiring in 35 years. How much will she have to save each year up until retirement if her account earns 7% interest?

35. You have $2,000 to invest, and want it to grow to $3,000 in two years. What interest rate would you need to find to make this possible?

36. You have $5,000 to invest, and want it to grow to $20,000 in ten years. What interest rate would you need to find to make this possible?

37. You plan to save $600 a month for the next 30 years for retirement. What interest rate would you need to have $1,000,000 at retirement?

38. You really want to buy a used car for $11,000, but can only afford $200 a month. What interest rate would you need to find to be able to afford the car, assuming the loan is for 60 months?

Exploration

39. Pay day loans are short term loans that you take out against future paychecks: The company advances you money against a future paycheck. Either visit a pay day loan company, or look one up online. Be forewarned that many companies do not make their fees obvious, so you might need to do some digging or look at several companies.
    a. Explain the general method by which the loan works.
    b. We will assume that we need to borrow $500 and that we will pay back the loan in 14 days. Determine the total amount that you would need to pay back and the effective loan rate. The effective loan rate is the percentage of the original loan amount that you pay back. It is not the same as the APR (annual rate) that is probably published.
    c. If you cannot pay back the loan after 14 days, you will need to get an extension for another 14 days. Determine the fees for an extension, determine the total amount you will be paying for the now 28-day loan, and compute the effective loan rate.

40. Suppose that 10 years ago you bought a home for $110,000, paying 10% as a down payment, and financing the rest at 9% interest for 30 years.
    a. Let's consider your existing mortgage:
        i. How much money did you pay as your down payment?
        ii. How much money was your mortgage (loan) for?
        iii. What is your current monthly payment?
        iv. How much total interest will you pay over the life of the loan?
    b. This year, you check your loan balance. Only part of your payments has been going to pay down the loan; the rest has been going towards interest. You see that you still have $88,536 left to pay on your loan. Your house is now valued at $150,000.
        i. How much of the loan have you paid off? (i.e., how much have you reduced the loan balance by? Keep in mind that interest is charged each month - it's not part of the loan balance.)
        ii. How much money have you paid to the loan company so far?
        iii. How much interest have you paid so far?

iv. How much equity do you have in your home (equity is value minus remaining debt)

c. Since interest rates have dropped, you consider refinancing your mortgage at a lower 6% rate.

i. If you took out a new 30-year mortgage at 6% for your remaining loan balance, what would your new monthly payments be?

ii. How much interest will you pay over the life of the new loan?

d. Notice that if you refinance, you are going to be making payments on your home for another 30 years. In addition to the 10 years you've already been paying, that's a total of 40 years.

i. How much will you save each month because of the lower monthly payment?

ii. How much total interest will you be paying (you need to consider the amount from 2c and 3b)

iii. Does it make sense to refinance? (there isn't a correct answer to this question. Just give your opinion and your reason)

Chapter 8: Statistics, Collecting Data

Like most people, you probably feel that it is important to "take control of your life," but what does this mean? Partly, it means being able to properly evaluate data and claims that bombard you every day. If we cannot distinguish good reasoning from faulty reasoning, then we are vulnerable to manipulation and decisions that are not in our best interest. Statistics provides tools that we need in order to react intelligently to information we hear or read. In this sense, statistics is one of the most important things that we can study.

To be more specific, here are some claims that we have heard on several occasions. (We are *not* saying that each one of these claims is true!)

- 4 out of 5 dentists recommend Dentyne.
- Almost 85% of lung cancers in men and 45% in women are tobacco-related.
- Condoms are effective 94% of the time.
- Native Americans are significantly more likely to be hit crossing the streets than are people of other ethnicities.
- People tend to be more persuasive when they look others directly in the eye and speak loudly and quickly.
- Women make 75 cents to every dollar a man makes when they work the same job.
- A surprising new study shows that eating egg whites can increase one's life span.
- People predict that it is very unlikely there will ever be another baseball player with a batting average over 400.
- There is an 80% chance that in a room full of 30 people that at least two people will share the same birthday.
- 79.48% of all statistics are made up on the spot.

All of these claims are statistical in character. We suspect that some of them sound familiar; if not, we bet that you have heard other claims like them. Notice how diverse the examples are; they come from psychology, health, law, sports, business, etc. Indeed, data and data-interpretation show up in discourse from virtually every facet of contemporary life.

Statistics are often presented in an effort to add credibility to an argument or advice. We can see this by paying attention to television advertisements. Many of the numbers thrown about in this way do not represent careful statistical analysis. They can be misleading, and push consumers into decisions that might find cause to regret. For these reasons, learning about statistics is a long step towards taking "control of your life." (It is not, of course, the only step needed for this purpose.) These next two chapters will help you learn statistical essentials. It will make you into an intelligent consumer of statistical claims.

You can take the first step right away. To be an intelligent consumer of statistics, your first reflex must be to question the statistics that you encounter. The British Prime Minister Benjamin Disraeli famously said, "There are three kinds of lies -- lies, damned lies, and statistics." This quote reminds us why it is so important to understand statistics. So, let us invite you to reform your statistical habits from now on. No longer will you blindly accept numbers or findings. Instead, you will begin to think about the numbers, their sources, and most importantly, the procedures used to generate them.

We have put the emphasis on defending ourselves against fraudulent claims wrapped up as statistics. Just as important as detecting the deceptive use of statistics is the appreciation of the proper use of statistics. You must also learn to recognize statistical evidence that supports a stated conclusion. When a research team is testing a new treatment for a disease, statistics allows them to conclude based on a relatively small trial that there is good evidence their drug is effective. Statistics allowed prosecutors in the 1950's and 60's to demonstrate racial bias existed in jury panels. Statistics are all around you, sometimes used well, sometimes not. We must learn how to distinguish the two cases.

Table of Contents

8.1   Populations and samples

Before we begin gathering and analyzing data we need to characterize the **population** we are studying. If we want to study the amount of money spent on textbooks by a typical first-year college student, our population might be all first-year students at your college. Or it might be:

- All first-year community college students in the state of Washington.
- All first-year students at public colleges and universities in the state of Washington.
- All first-year students at all colleges and universities in the state of Washington.
- All first-year students at all colleges and universities in the entire United States.
- And so on.

Population
The **population** of a study is the group the collected data is intended to describe.

Sometimes the intended population is called the **target population**, since if we design our study badly, the collected data might not actually be representative of the intended population.

Why is it important to specify the population? We might get different answers to our question as we vary the population we are studying. First-year students at the University of Washington might take slightly more diverse courses than those at your college, and some of these courses may require less popular textbooks that cost more; or, on the other hand, the University Bookstore might have a larger pool of used textbooks, reducing the cost of these books to the students. Whichever the case (and it is likely that some combination of these and other factors are in play), the data we gather from your college will probably not be the same as that from the University of Washington. Particularly when conveying our results to others, we want to be clear about the population we are describing with our data.

Example 1

A newspaper website contains a poll asking people their opinion on a recent news article. What is the population?

While the target (intended) population may have been all people, the real population of the survey is readers of the website.

If we were able to gather data on every member of our population, say the average (we will define "average" more carefully in a subsequent section) amount of money spent on textbooks by each first-year student at your college during the 2009-2010 academic year, the resulting number would be called a **parameter**.

Parameter
A **parameter** is a value (average, percentage, etc.) calculated using all the data from a population

---

We seldom see parameters, however, since surveying an entire population is usually very time-consuming and expensive, unless the population is very small or we already have the data collected.

Census
A survey of an entire population is called a **census**.

You are probably familiar with two common censuses: the official government Census that attempts to count the population of the U.S. every ten years, and voting, which asks the opinion of all eligible voters in a district. The first of these demonstrates one additional problem with a census: the difficulty in finding and getting participation from everyone in a large population, which can bias, or skew, the results.

There are occasionally times when a census is appropriate, usually when the population is fairly small. For example, if the manager of Starbucks wanted to know the average number of hours her employees worked last week, she should be able to pull up payroll records or ask each employee directly.

Since surveying an entire population is often impractical, we usually select a **sample** to study;

> **Sample**
> A **sample** is a smaller subset of the entire population, ideally one that is fairly representative of the whole population.

We discuss sampling methods in greater detail in a later section. For now, let us assume that samples are chosen in an appropriate manner. If we survey a sample, say 100 first-year students at your college, and find the average amount of money spent by these students on textbooks, the resulting number is called a **statistic**.

> **Statistic**
> A **statistic** is a value (average, percentage, etc.) calculated using the data from a sample.

Example 2

A researcher wanted to know how citizens of Tacoma felt about a voter initiative. To study this, she goes to the Tacoma Mall and randomly selects 500 shoppers and asks them their opinion. 60% indicate they are supportive of the initiative. What is the sample and population? Is the 60% value a parameter or a statistic?

The sample is the 500 shoppers questioned. The population is less clear. While the intended population of this survey was Tacoma citizens, the effective population was mall shoppers. There is no reason to assume that the 500 shoppers questioned would be representative of all Tacoma citizens.

The 60% value was based on the sample, so it is a statistic.

Try it Now 1

To determine the average length of trout in a lake, researchers catch 20 fish and measure them. What is the sample and population in this study?

Try it Now 2

A college reports that the average age of their students is 28 years old. Is this a statistic or a parameter?

8.2   Categorizing data

Once we have gathered data, we might wish to classify it. Roughly speaking, data can be classified as categorical data or quantitative data.

Quantitative and categorical data
Categorical (qualitative) data are pieces of information that allow us to classify the objects under investigation into various categories.

Quantitative data are responses that are numerical in nature and with which we can perform meaningful arithmetic calculations.

Example 3

We might conduct a survey to determine the name of the favorite movie that each person in a math class saw in a movie theater.

When we conduct such a survey, the responses would look like: *Finding Nemo, The Hulk,* or *Terminator 3: Rise of the Machines.* We might count the number of people who give each answer, but the answers themselves do not have any numerical values: we cannot perform computations with an answer like "*Finding Nemo.*" This would be categorical data.

Example 4

A survey could ask the number of movies you have seen in a movie theater in the past 12 months (0, 1, 2, 3, 4, ...).

This would be quantitative data.

Other examples of quantitative data would be the running time of the movie you saw most recently (104 minutes, 137 minutes, 104 minutes, ...) or the amount of money you paid for a movie ticket the last time you went to a movie theater ($5.50, $7.75, $9, ...).

Sometimes, determining whether or not data is categorical or quantitative can be a bit trickier.

Example 5

Suppose we gather respondents' ZIP codes in a survey to track their geographical location.

ZIP codes are numbers, but we can't do any meaningful mathematical calculations with them (it doesn't make sense to say that 98036 is "twice" 49018 — that's like saying that Lynnwood, WA is "twice" Battle Creek, MI, which doesn't make sense at all), so ZIP codes are really categorical data.

Example 6

A survey about the movie you most recently attended includes the question "How would you rate the movie you just saw?" with these possible answers:

1 - it was awful
2 - it was just OK
3 - I liked it
4 - it was great
5 - best movie ever!

Again, there are numbers associated with the responses, but we can't really do any calculations with them: a movie that rates a 4 is not necessarily twice as good as a movie that rates a 2, whatever that means; if two people see the movie and one of them thinks it stinks and the other thinks it's the best ever it doesn't necessarily make sense to say that "on average they liked it."

As we study movie-going habits and preferences, we shouldn't forget to specify the population under consideration. If we survey 3-7 year olds the runaway favorite might be *Finding Nemo*, and 13-17 year olds might prefer *Terminator 3,* and 33-37 year olds might prefer...well, *Finding Nemo*.

Try it Now 3

Classify each measurement as categorical or quantitative
a. Eye color of a group of people
b. Daily high temperature of a city over several weeks
c. Annual income

8.3   Sampling methods

As we mentioned in a previous section, the first thing we should do before conducting a survey is to identify the population that we want to study. Suppose we are hired by a politician to determine the amount of support he has among the electorate should he decide to run for another term. What population should we study? Every person in the district? Not every person is eligible to vote, and regardless of how strongly someone likes or dislikes the candidate, they don't have much to do with him being re-elected if they are not able to vote.

What about eligible voters in the district? That might be better, but if someone is eligible to vote but does not register by the deadline, they won't have any say in the election either. What about registered voters? Many people are registered but choose not to vote. What about "likely voters?"

This is the criteria used in much political polling, but it is sometimes difficult to define a "likely voter." Is it someone who voted in the last election? In the last general election? In the last presidential election? Should we consider someone who just turned 18 a "likely voter?" They weren't eligible to vote in the past, so how do we judge the likelihood that they will vote in the next election?

In November 1998, former professional wrestler Jesse "The Body" Ventura was elected governor of Minnesota. Up until right before the election, most polls showed he had little chance of winning. There were several contributing factors to the polls not reflecting the actual intent of the electorate:

- Ventura was running on a third-party ticket and most polling methods are better suited to a two-candidate race.
- Many respondents to polls may have been embarrassed to tell pollsters that they were planning to vote for a professional wrestler.
- The mere fact that the polls showed Ventura had little chance of winning might have prompted some people to vote for him in protest to send a message to the major-party candidates.

But one of the major contributing factors was that Ventura recruited a substantial amount of support from young people, particularly college students, who had never voted before and who registered specifically to vote in the gubernatorial election. The polls did not deem these young people likely voters (since in most cases young people have a lower rate of voter registration and a turnout rate for elections) and so the polling samples were subject to **sampling bias**: they omitted a portion of the electorate that was weighted in favor of the winning candidate.

Sampling bias
A sampling method is biased if every member of the population doesn't have equal likelihood of being in the sample.

So even identifying the population can be a difficult job, but once we have identified the population, how do we choose an appropriate sample? Remember, although we would prefer to survey all members of the population, this is usually impractical unless the population is very small, so we choose a sample. There are many ways to sample a population, but there is one goal we need to keep in mind: we would like the sample to be *representative of the population.*

Returning to our hypothetical job as a political pollster, we would not anticipate very accurate results if we drew all of our samples from among the customers at a Starbucks, nor would we expect that a sample drawn entirely from the membership list of the local Elks club would provide a useful picture of district-wide support for our candidate.

One way to ensure that the sample has a reasonable chance of mirroring the population is to employ *randomness.* The most basic random method is simple random sampling.

> **Simple random sample**
> A **random sample** is where each member of the population has an equal probability of being chosen.
>
> A **simple random sample** is where every member of the population and any group of members has an equal probability of being chosen.

Example 7

> If we could somehow identify all likely voters in the state, put each of their names on a piece of paper, toss the slips into a (very large) hat and draw 1,000 slips out of the hat, we would have a simple random sample.

In practice, computers are better suited for this sort of endeavor than millions of slips of paper and extremely large headgear.

It is always possible, however, that even a random sample might end up not being totally representative of the population. If we repeatedly take samples of 1,000 people from among the population of likely voters in the state of Washington, some of these samples might tend to have a slightly higher percentage of Democrats (or Republicans) than does the general population; some samples might include more older people and some samples might include more younger people; etc. In most cases, this **sampling variability** is not significant.

> **Sampling variability**
> The natural variation of samples is called **sampling variability**.
>
> This is unavoidable and expected in random sampling, and in most cases is not an issue.

To help account for variability, pollsters might instead use a **stratified sample**.

> **Stratified sampling**
> In **stratified sampling**, a population is divided into a number of subgroups (or strata). Random samples are then taken from each subgroup with sample sizes proportional to the size of the subgroup in the population.

Example 8

> Suppose in a particular state that previous data indicated that the electorate was comprised of 39% Democrats, 37% Republicans and 24% independents. In a sample of 1,000 people, they would then expect to get about 390 Democrats, 370 Republicans and 240 independents. To accomplish this, they could randomly select 390 people from among those voters known to be Democrats, 370 from those known to be Republicans, and 240 from those with no party affiliation.

Stratified sampling can also be used to select a sample with people in desired age groups, a specified mix ratio of males and females, etc. A variation on this technique is called **quota sampling**.

> **Quota sampling**
> **Quota sampling** is a variation on stratified sampling, wherein samples are collected in each subgroup until the desired quota is met.

Example 9

> Suppose the pollsters call people at random, but once they have met their quota of 390 Democrats, they only gather people who do not identify themselves as a Democrat.

You may have had the experience of being called by a telephone pollster who started by asking you your age, income, etc. and then thanked you for your time and hung up before asking any "real" questions. Most likely, they already had contacted enough people in your demographic group and were looking for people who were older or younger, richer or poorer, etc. Quota sampling is usually a bit easier than stratified sampling, but also does not ensure the same level of randomness.

Another sampling method is **cluster sampling**, in which the population is divided into groups, and one or more groups are randomly selected to be in the sample.

> **Cluster sampling**
> In **cluster sampling**, the population is divided into subgroups (clusters), and a set of subgroups are selected to be in the sample

Example 10

> For example, if the college wanted to survey students, since students are already divided into classes, they could randomly select 10 classes and give the survey to all the students just in those classes. The classes are the clusters, and we can randomly select clusters (classes in this case) to survey.

Other sampling methods include **systematic sampling**.

> **Systematic sampling**
> In **systematic sampling**, every n^{th} member of the population is selected to be in the sample.

Example 11

To select a sample using systematic sampling, a pollster calls every 100th name in the phone book.

Systematic sampling is not as random as a simple random sample (if your name is Albert Aardvark and your sister Alexis Aardvark is right after you in the phone book, there is no way you could both end up in the sample), but it can yield acceptable samples.

Perhaps the worst types of sampling methods are **convenience samples** and **voluntary response samples**.

> **Convenience sampling and voluntary response sampling**
> **Convenience sampling** is samples chosen by selecting whoever is convenient.
>
> **Voluntary response sampling** is allowing the sample to volunteer.

Example 12

A pollster stands on a street corner and interviews the first 100 people who agree to speak to him. This is a convenience sample.

Example 13

A website has a survey asking readers to give their opinion on a tax proposal. This is a self-selected sample, or voluntary response sample, in which respondents volunteer to participate.

Usually voluntary response samples are skewed towards people who have a particularly strong opinion about the subject of the survey or who just have way too much time on their hands and enjoy taking surveys.

Try it Now 4

In each case, indicate what sampling method was used.

a. Every 4th person in the class was selected
b. A sample was selected to contain 25 men and 35 women
c. Viewers of a new show are asked to vote on the show's website
d. A website randomly selects 50 of their customers to send a satisfaction survey to
e. To survey voters in a town, a polling company randomly selects 10 city blocks, and interviews everyone who lives on those blocks.

8.4   How to mess things up

There are number of ways that a study can be ruined before you even start collecting data. The first we have already explored – **sampling** or **selection bias**, which is when the sample is not representative of the population. One example of this is **voluntary response bias**, which is bias introduced by only collecting data from those who volunteer to participate. This is not the only potential source of bias.

Sources of bias
Sampling bias – when the sample is not representative of the population
Voluntary response bias – the sampling bias that often occurs when the sample is volunteers
Self-interest study – bias that can occur when the researchers have an interest in the outcome
Response bias – when the responder gives inaccurate responses for any reason
Perceived lack of anonymity – when the responder fears giving an honest answer might negatively affect them
Loaded questions – when the question wording influences the responses
Non-response bias – when people refusing to participate in the study can influence the validity of the outcome

---

Example 14

Consider a recent study which found that chewing gum may raise math grades in teenagers[1]. This study was conducted by the Wrigley Science Institute, a branch of the Wrigley chewing gum company. This is an example of a **self-interest study**; one in which the researches have a vested interest in the outcome of the study. While this does not necessarily ensure that the study was biased, it certainly suggests that we should subject the study to extra scrutiny.

Example 15

A survey asks people "when was the last time you visited your doctor?" This might suffer from **response bias**, since many people might not remember exactly when they last saw a doctor and give inaccurate responses.

Sources of response bias may be innocent, such as bad memory, or as intentional as pressuring by the pollster. Consider, for example, how many voting initiative petitions people sign without even reading them.

Example 16

A survey asks participants a question about their interactions with members of other races. Here, a **perceived lack of anonymity** could influence the outcome. The respondent might not want to be perceived as racist even if they are, and give an untruthful answer.

---

[1] Reuters. http://news.yahoo.com/s/nm/20090423/od_uk_nm/oukoe_uk_gum_learning. Retrieved 4/27/09

Example 17

> An employer puts out a survey asking their employees if they have a drug abuse problem and need treatment help. Here, answering truthfully might have consequences; responses might not be accurate if the employees do not feel their responses are anonymous or fear retribution from their employer.

Example 18

> A survey asks "do you support funding research of alternative energy sources to reduce our reliance on high-polluting fossil fuels?" This is an example of a **loaded** or **leading question** – questions whose wording leads the respondent towards an answer.

Loaded questions can occur intentionally by pollsters with an agenda, or accidentally through poor question wording. Also, a concern is **question order**, where the order of questions changes the results. A psychology researcher provides an example[2]:

> "My favorite finding is this: we did a study where we asked students, 'How satisfied are you with your life? How often do you have a date?' The two answers were not statistically related - you would conclude that there is no relationship between dating frequency and life satisfaction. But when we reversed the order and asked, 'How often do you have a date? How satisfied are you with your life?' the statistical relationship was a strong one. You would now conclude that there is nothing as important in a student's life as dating frequency."

Example 19

> A telephone poll asks the question "Do you often have time to relax and read a book?", and 50% of the people called refused to answer the survey. It is unlikely that the results will be representative of the entire population. This is an example of **non-response bias**, introduced by people refusing to participate in a study or dropping out of an experiment. When people refuse to participate, we can no longer be so certain that our sample is representative of the population.

Try it Now 5

In each situation, identify a potential source of bias.

a. A survey asks how many sexual partners a person has had in the last year
b. A radio station asks readers to phone in their choice in a daily poll.
c. A substitute teacher wants to know how students in the class did on their last test. The teacher asks the 10 students sitting in the front row to state their latest test score.
d. High school students are asked if they have consumed alcohol in the last two weeks.
e. The Beef Council releases a study stating that consuming red meat poses little cardiovascular risk.
f. A poll asks "Do you support a new transportation tax, or would you prefer to see our public transportation system fall apart?"

---

[2] Swartz, Norbert. http://www.umich.edu/~newsinfo/MT/01/Fal01/mt6f01.html. Retrieved 3/31/2009

8.5 Experiments

So far, we have primarily discussed **observational studies** – studies in which conclusions would be drawn from observations of a sample or the population. In some cases, these observations might be unsolicited, such as studying the percentage of cars that turn right at a red light even when there is a "no turn on red" sign. In other cases, the observations are solicited, like in a survey or a poll.

In contrast, it is common to use **experiments** when exploring how subjects react to an outside influence. In an experiment, some kind of **treatment** is applied to the subjects and the results are measured and recorded.

> **Observational studies and experiments**
> An **observational study** is a study based on observations or measurements
>
> An **experiment** is a study in which the effects of a **treatment** are measured

Here are some examples of experiments:

Example 20

a. A pharmaceutical company tests a new medicine for treating Alzheimer's disease by administering the drug to 50 elderly patients with recent diagnoses. The treatment here is the new drug.

b. A gym tests out a new weight loss program by enlisting 30 volunteers to try out the program. The treatment here is the new program.

c. You test a new kitchen cleaner by buying a bottle and cleaning your kitchen. The new cleaner is the treatment.

d. A psychology researcher explores the effect of music on temperament by measuring people's temperament while listening to different types of music. The music is the treatment.

Try it Now 6

Is each scenario describing an observational study or an experiment?

a. The weights of 30 randomly selected people are measured
b. Subjects are asked to do 20 jumping jacks, and then their heart rates are measured
c. Twenty coffee drinkers and twenty tea drinkers are given a concentration test

When conducting experiments, it is essential to isolate the treatment being tested.

Example 21

Suppose a middle school (junior high) finds that their students are not scoring well on the state's standardized math test. They decide to run an experiment to see if an alternate curriculum would improve scores. To run the test, they hire a math specialist to come in and teach a class using the new curriculum. To their delight, they see an improvement in test scores.

The difficulty with this scenario is that it is not clear whether the curriculum is responsible for the improvement, or whether the improvement is due to a math specialist teaching the class. This is called **confounding** – when it is not clear which factor or factors caused the observed effect. Confounding is the downfall of many experiments, though sometimes it is hidden.

> **Confounding**
> **Confounding** occurs when there are two potential variables that could have caused the outcome and it is not possible to determine which actually caused the result.

Example 22

> A drug company study about a weight loss pill might report that people lost an average of 8 pounds while using their new drug. However, in the fine print you find a statement saying that participants were encouraged to also diet and exercise. It is not clear in this case whether the weight loss is due to the pill, to diet and exercise, or a combination of both. In this case confounding has occurred.

Example 23

> Researchers conduct an experiment to determine whether students will perform better on an arithmetic test if they listen to music during the test. They first give the student a test without music, then give a similar test while the student listens to music. In this case, the student might perform better on the second test, regardless of the music, simply because it was the second test and they were warmed up.

There are a number of measures that can be introduced to help reduce the likelihood of confounding. The primary measure is to use a **control group**.

> **Control group**
> When using a control group, the participants are divided into two or more groups, typically a **control group** and a treatment group. The treatment group receives the treatment being tested; the control group does not receive the treatment.

Ideally, the groups are otherwise as similar as possible, isolating the treatment as the only potential source of difference between the groups. For this reason, the method of dividing groups is important. Some researchers attempt to ensure that the groups have similar characteristics (same number of females, same number of people over 50, etc.), but it is nearly impossible to control for every characteristic. Because of this, random assignment is very commonly used.

Example 24

> To determine if a two-day prep course would help high school students improve their scores on the SAT test, a group of students was randomly divided into two subgroups. The first group, the treatment group, was given a two-day prep course. The second group, the control group, was not given the prep course. Afterwards, both groups were given the SAT.

Example 25

> A company testing a new plant food grows two crops of plants in adjacent fields, the treatment group receiving the new plant food and the control group not. The crop yield would then be compared. By growing them at the same time in adjacent fields, they are controlling for weather and other confounding factors.

Sometimes not giving the control group anything does not completely control for confounding variables. For example, suppose a medicine study is testing a new headache pill by giving the treatment group the pill and the control group nothing. If the treatment group showed improvement, we would not know whether it was due to the medicine in the pill, or a response to have taken any pill. This is called a **placebo effect**.

> **Placebo effect**
> The **placebo effect** is when the effectiveness of a treatment is influenced by the patient's perception of how effective they think the treatment will be, so a result might be seen even if the treatment is ineffectual.

Example 26

> A study found that when doing painful dental tooth extractions, patients told they were receiving a strong painkiller while actually receiving a saltwater injection found as much pain relief as patients receiving a dose of morphine.[3]

To control the placebo effect, a **placebo**, or dummy treatment, is often given to the control group. This way, both groups are truly identical except for the specific treatment given.

> **Placebo and Placebo controlled experiments**
> A **placebo** is a dummy treatment given to control for the placebo effect.
>
> An experiment that gives the control group a placebo is called a **placebo controlled experiment**.

Example 27

> a. In a study for a new medicine that is dispensed in a pill form, a sugar pill could be used as a placebo.
> b. In a study on the effect of alcohol on memory, a non-alcoholic beer might be given to the control group as a placebo.
> c. In a study of a frozen meal diet plan, the treatment group would receive the diet food, and the control could be given standard frozen meals stripped of their original packaging.

In some cases, it is more appropriate to compare to a conventional treatment than a placebo. For example, in a cancer research study, it would not be ethical to deny any treatment to the control group or to give a placebo treatment. In this case, the currently acceptable medicine would be given to the second group, called a **comparison group** in this case. In our SAT

---

[3] Levine JD, Gordon NC, Smith R, Fields HL. (1981) Analgesic responses to morphine and placebo in individuals with postoperative pain. Pain. 10:379-89.

test example, the non-treatment group would most likely be encouraged to study on their own, rather than be asked to not study at all, to provide a meaningful comparison.

When using a placebo, it would defeat the purpose if the participant knew they were receiving the placebo.

Blind studies

A **blind study** is one in which the participant does not know whether or not they are receiving the treatment or a placebo.

A **double-blind study** is one in which those interacting with the participants don't know who is in the treatment group and who is in the control group.

Example 28

In a study about anti-depression medicine, you would not want the psychological evaluator to know whether the patient is in the treatment or control group either, as it might influence their evaluation, so the experiment should be conducted as a double-blind study.

It should be noted that not every experiment needs a control group.

Example 29

If a researcher is testing whether a new fabric can withstand fire, she simply needs to torch multiple samples of the fabric – there is no need for a control group.

Try it Now 7

To test a new lie detector, two groups of subjects are given the new test. One group is asked to answer all the questions truthfully, and the second group is asked to lie on one set of questions. The person administering the lie detector test does not know what group each subject is in.

Does this experiment have a control group? Is it blind, double-blind, or neither?

Try it Now Answers

1. The sample is the 20 fish. The population is all fish in the lake. The sample may be somewhat unrepresentative of the population since not all fish may be large enough to catch the bait.

2. This is a parameter, since the college would have access to data on all students (the population)

3. a. Categorical.   b. Quantitative   c. Quantitative

4. a. Systematic
   b. Stratified or Quota
   c. Voluntary response

d. Simple random

e. Cluster

5. a. Response bias – historically, men are likely to over-report, and women are likely to under-report to this question.

b. Voluntary response bias – the sample is self-selected

c. Sampling bias – the sample may not be representative of the whole class

d. Lack of anonymity

e. Self-interest study

f. Loaded question

6. a. Observational study

b. Experiment; the treatment is the jumping jacks

c. Experiment; the treatments are coffee and tea

7. The truth-telling group could be considered the control group, but really both groups are treatment groups here, since it is important for the lie detector to be able to correctly identify lies, and also not identify truth telling as lying. This study is blind, since the person running the test does not know what group each subject is in.

Exercises

Skills

1. A political scientist surveys 28 of the current 106 representatives in a state's congress. Of them, 14 said they were supporting a new education bill, 12 said there were not supporting the bill, and 2 were undecided.
    a. What is the population of this survey?
    b. What is the size of the population?
    c. What is the size of the sample?
    d. Give the sample statistic for the proportion of voters surveyed who said they were supporting the education bill.
    e. Based on this sample, we might expect how many of the representatives to support the education bill?

2. The city of Raleigh has 9500 registered voters. There are two candidates for city council in an upcoming election: Brown and Feliz. The day before the election, a telephone poll of 350 randomly selected registered voters was conducted. 112 said they'd vote for Brown, 207 said they'd vote for Feliz, and 31 were undecided.
    a. What is the population of this survey?
    b. What is the size of the population?
    c. What is the size of the sample?
    d. Give the sample statistic for the proportion of voters surveyed who said they'd vote for Brown.
    e. Based on this sample, we might expect how many of the 9500 voters to vote for Brown?

3. Identify the most relevant source of bias in this situation: A survey asks the following: Should the mall prohibit loud and annoying rock music in clothing stores catering to teenagers?

4. Identify the most relevant source of bias in this situation: To determine opinions on voter support for a downtown renovation project, a surveyor randomly questions people working in downtown businesses.

5. Identify the most relevant source of bias in this situation: A survey asks people to report their actual income and the income they reported on their IRS tax form.

6. Identify the most relevant source of bias in this situation: A survey randomly calls people from the phone book and asks them to answer a long series of questions.

7. Identify the most relevant source of bias in this situation: A survey asks the following: Should the death penalty be permitted if innocent people might die?

8. Identify the most relevant source of bias in this situation: A study seeks to investigate whether a new pain medication is safe to market to the public. They test by randomly selecting 300 men from a set of volunteers.

9. In a study, you ask the subjects their age in years. Is this data qualitative or quantitative?

10. In a study, you ask the subjects their gender. Is this data qualitative or quantitative?

11. Does this describe an observational study or an experiment: The temperature on randomly selected days throughout the year was measured.

12. Does this describe an observational study or an experiment? A group of students are told to listen to music while taking a test and their results are compared to a group not listening to music.

13. In a study, the sample is chosen by separating all cars by size, and selecting 10 of each size grouping. What is the sampling method?

14. In a study, the sample is chosen by writing everyone's name on a playing card, shuffling the deck, then choosing the top 20 cards. What is the sampling method?

15. A team of researchers is testing the effectiveness of a new HPV vaccine. They randomly divide the subjects into two groups. Group 1 receives new HPV vaccine, and Group 2 receives the existing HPV vaccine. The patients in the study do not know which group they are in.
    a. Which is the treatment group?
    b. Which is the control group (if there is one)?
    c. Is this study blind, double-blind, or neither?
    d. Is this best described as an experiment, a controlled experiment, or a placebo controlled experiment?

16. For the clinical trials of a weight loss drug containing *Garcinia cambogia* the subjects were randomly divided into two groups. The first received an inert pill along with an exercise and diet plan, while the second received the test medicine along with the same exercise and diet plan. The patients do not know which group they are in, nor do the fitness and nutrition advisors.
    a. Which is the treatment group?
    b. Which is the control group (if there is one)?
    c. Is this study blind, double-blind, or neither?
    d. Is this best described as an experiment, a controlled experiment, or a placebo controlled experiment?

Concepts

17. A teacher wishes to know whether the males in his/her class have more conservative attitudes than the females. A questionnaire is distributed assessing attitudes.
    a. Is this a sampling or a census?
    b. Is this an observational study or an experiment?
    c. Are there any possible sources of bias in this study?

18. A study is conducted to determine whether people learn better with spaced or massed practice. Subjects volunteer from an introductory psychology class. At the beginning of the semester 12 subjects volunteer and are assigned to the massed-practice group. At the end of the semester 12 subjects volunteer and are assigned to the spaced-practice condition.
    a. Is this a sampling or a census?
    b. Is this an observational study or an experiment?
    c. This study involves two kinds of non-random sampling: (1) Subjects are not randomly sampled from some specified population and (2) Subjects are not randomly assigned to groups. Which problem is more serious? What affect on the results does each have?

19. A farmer believes that playing Barry Manilow songs to his peas will increase their yield. Describe a controlled experiment the farmer could use to test his theory.

20. A sports psychologist believes that people are more likely to be extroverted as adults if they played team sports as children. Describe two possible studies to test this theory. Design one as an observational study and the other as an experiment. Which is more practical?

Exploration

21. Studies are often done by pharmaceutical companies to determine the effectiveness of a treatment program. Suppose that a new AIDS antibody drug is currently under study. It is given to patients once the AIDS symptoms have revealed themselves. Of interest is the average length of time, in months, patients live once starting the treatment. Two researchers each follow a different set of 50 AIDS patients from the start of treatment until their deaths.

    a. What is the population of this study?
    b. List two reasons why the data may differ.
    c. Can you tell if one researcher is correct and the other one is incorrect? Why?
    d. Would you expect the data to be identical? Why or why not?
    e. If the first researcher collected her data by randomly selecting 40 states, then selecting 1 person from each of those states. What sampling method is that?
    f. If the second researcher collected his data by choosing 40 patients he knew. What sampling method would that researcher have used? What concerns would you have about this data set, based upon the data collection method?

22. Find a newspaper or magazine article, or the online equivalent, describing the results of a recent study (the results of a poll are not sufficient). Give a summary of the study's findings, then analyze whether the article provided enough information to determine the validity of the conclusions. If not, produce a list of things that are missing from the article that would help you determine the validity of the study. Look for the things discussed in the text: population, sample, randomness, blind, control, placebos, etc.

Chapter 9: Statistics, Describing Data

Once we have collected data from surveys or experiments, we need to summarize and present the data in a way that will be meaningful to the reader. We will begin with graphical presentations of data then explore numerical summaries of data.

Table of Contents

9.1   Presenting Categorical Data Graphically

Categorical, or qualitative, data are pieces of information that allow us to classify the objects under investigation into various categories. We usually begin working with categorical data by summarizing the data into a **frequency table.**

> **Frequency Table**
> A frequency table is a table with two columns. One column lists the categories, and another for the frequencies with which the items in the categories occur (how many items fit into each category).

Example 1

An insurance company determines vehicle insurance premiums based on known risk factors. If a person is considered a higher risk, their premiums will be higher. One potential factor is the color of your car. The insurance company believes that people with some color cars are more likely to get in accidents. To research this, they examine police reports for recent total-loss collisions. The data is summarized in the frequency table below.

Color	Frequency
Blue	25
Green	52
Red	41
White	36
Black	39
Grey	23

Sometimes we need an even more intuitive way of displaying data. This is where charts and graphs come in. There are many, many ways of displaying data graphically, but we will concentrate on one very useful type of graph called a bar graph. In this section, we will work with bar graphs that display categorical data; the next section will be devoted to graphs that display quantitative data.

Bar graph
A **bar graph** is a graph that displays a bar for each category with the length of each bar indicating the frequency of that category.

Note, a bar graph is only used to display *categorical data*.

To construct a bar graph, we need to draw a vertical axis and a horizontal axis. The vertical direction will have a scale and measure the frequency of each category; the horizontal axis has no scale in this instance, but only the names of each category. The construction of a bar chart is most easily described by use of an example.

Example 2
_____

Using our car data from above, note the highest frequency is 52, so our vertical axis needs to go from 0 to 52, but we might as well use 0 to 55, so that we can put a hash mark every 5 units:

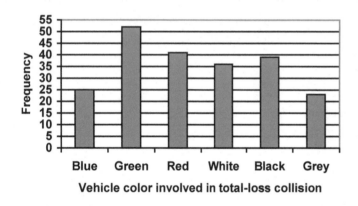

Notice that the height of each bar is determined by the frequency of the corresponding color. The horizontal gridlines are a nice touch, but not necessary. In practice, you will find it useful to draw bar graphs using graph paper, so the gridlines will already be in place, or using technology. Instead of gridlines, we might also list the frequencies at the top of each bar, like this:

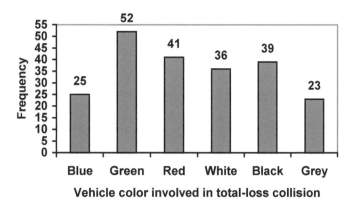

In this case, our chart might benefit from being reordered from largest to smallest frequency values. This arrangement can make it easier to compare similar values in the chart, even without gridlines. When we arrange the categories in decreasing frequency order like this, it is called a **Pareto chart**.

> **Pareto chart**
> A **Pareto chart** is a bar graph ordered from highest to lowest frequency

Example 3

Transforming our bar graph from earlier into a Pareto chart, we get:

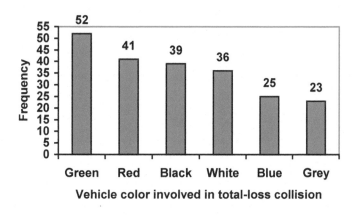

Example 4

In a survey[1], adults were asked whether they personally worried about a variety of environmental concerns. The numbers (out of 1012 surveyed) who indicated that they worried "a great deal" about some selected concerns are summarized below.

---

[1] Gallup Poll. Starch 5-8, 2009. http://www.pollingreport.com/enviro.htm

Environmental Issue	Frequency
Pollution of drinking water	597
Contamination of soil and water by toxic waste	526
Air pollution	455
Global warming	354

This data could be shown graphically in a bar graph:

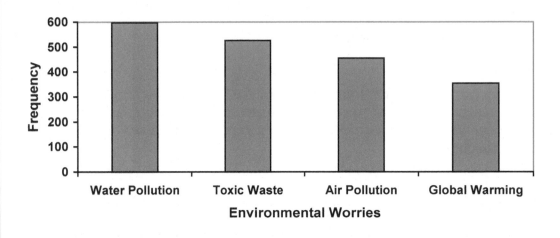

To show relative sizes, it is common to use a pie chart.

Pie Chart
A **pie chart** is a circle with wedges cut of varying sizes marked out like slices of pie or pizza. The relative sizes of the wedges correspond to the relative frequencies of the categories.

Example 5

For our vehicle color data, a pie chart might look like this:

Vehicle color involved in total-loss collisions

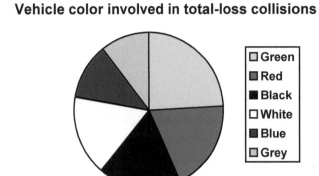

Pie charts can often benefit from including frequencies or relative frequencies (percents) in the chart next to the pie slices. Often, having the category names next to the pie slices also makes the chart clearer.

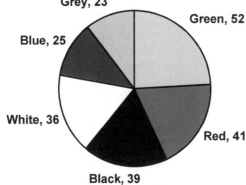

Vehicle color involved in total-loss collisions

Grey, 23 — Green, 52 — Blue, 25 — White, 36 — Red, 41 — Black, 39

Example 6

The pie chart to the right shows the percentage of voters supporting each candidate running for a local senate seat.

If there are 20,000 voters in the district, the pie chart shows that about 11% of those, about 2,200 voters, support Reeves.

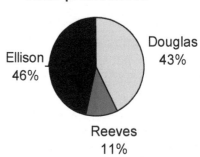

Voter preferences

Ellison 46% — Douglas 43% — Reeves 11%

Pie charts look nice, but are harder to draw by hand than bar charts since to draw them accurately we would need to compute the angle each wedge cuts out of the circle, then measure the angle with a protractor. Computers are much better suited to drawing pie charts. Common software programs like Microsoft Word or Excel, OpenOffice.org Write or Calc, or Google Docs are able to create bar graphs, pie charts, and other graph types. There are also numerous online tools that can create graphs[2].

Try it Now 1

Create a bar graph and a pie chart to illustrate the grades on a history exam below.
A: 12 students, B: 19 students, C: 14 students, D: 4 students, F: 5 students

---

[2] For example: http://nces.ed.gov/nceskids/createAgraph/ or http://docs.google.com

Don't get fancy with graphs! People sometimes add features to graphs that don't help to convey their information. For example, 3-dimensional bar charts like the one shown below are usually not as effective as their two-dimensional counterparts.

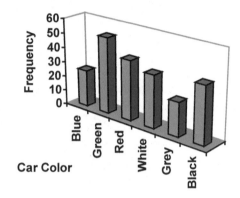

Here is another way that fanciness can lead to trouble. Instead of plain bars, it is tempting to substitute meaningful images. This type of graph is called a **pictogram**.

> **Pictogram**
> A **pictogram** is a statistical graphic in which the size of the picture is intended to represent the frequencies or size of the values being represented.

Example 7

A labor union might produce the graph to the right to show the difference between the average manager salary and the average worker salary.

Looking at the picture, it would be reasonable to guess that the manager salaries is 4 times as large as the worker salaries – the area of the bag looks about 4 times as large. However, the manager salaries are in fact only twice as large as worker salaries, which were reflected in the picture by making the manager bag twice as tall.

**Manager   Worker
Salaries   Salaries**

Another distortion in bar charts results from setting the baseline to a value other than zero. The baseline is the bottom of the vertical axis, representing the least number of cases that could have occurred in a category. Normally, this number should be zero.

Example 8

Compare the two graphs below showing support for same-sex marriage rights from a poll taken in December 2008[3]. The difference in the vertical scale on the first graph suggests a different story than the true differences in percentages; the second graph makes it look like twice as many people oppose marriage rights as support it.

---

[3]CNN/Opinion Research Corporation Poll. Dec 19-21, 2008, from http://www.pollingreport.com/civil.htm

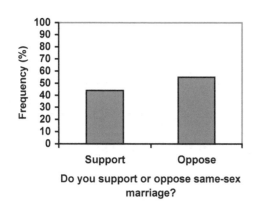

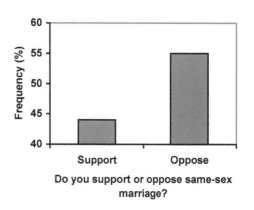

Try it Now 2

A poll was taken asking people if they agreed with the positions of the 4 candidates for a county office. Does the pie chart present a good representation of this data? Explain.

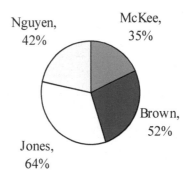

9.2   Presenting Quantitative Data Graphically

Quantitative, or numerical, data can also be summarized into frequency tables.

Example 9

A teacher records scores on a 20-point quiz for the 30 students in his class. The scores are

19 20 18 18 17 18 19 17 20 18 20 16 20 15 17 12 18 19 18 19 17 20 18 16 15 18 20 5 0 0

These scores could be summarized into a frequency table by grouping like values:

Score	Frequency
0	2
5	1
12	1
15	2
16	2
17	4
18	8
19	4
20	6

Using this table, it would be possible to create a standard bar chart from this summary, like we did for categorical data:

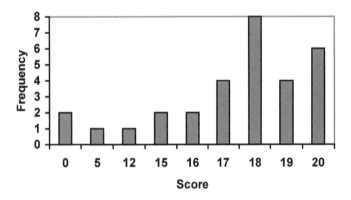

However, since the scores are numerical values, this chart doesn't really make sense; the first and second bars are five values apart, while the later bars are only one value apart. It would be more correct to treat the horizontal axis as a number line. This type of graph is called a **histogram**.

Histogram
A histogram is a graphical representation of *quantitative data*. The horizontal axis is a number line.

Example 10

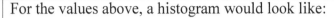

For the values above, a histogram would look like:

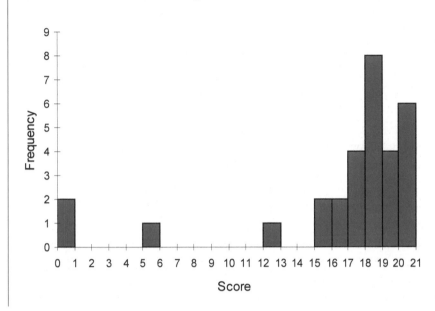

Notice that, in the histogram, a bar represents values on the horizontal axis from that on the left hand-side of the bar up to, but not including, the value on the right-hand side of the bar. Some people choose to have bars start at ½ values to avoid this ambiguity.

Unfortunately, not a lot of common software packages can correctly graph a histogram. About the best you can do in Excel or Word is a bar graph with no gap between the bars and spacing added to simulate a numerical horizontal axis.

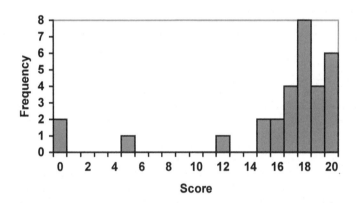

If we have a large number of widely varying data values, creating a frequency table that lists every possible value as a category would lead to an exceptionally long frequency table, and probably would not reveal any patterns. For this reason, it is common with quantitative data to group data into **class intervals**.

Class Intervals

Class intervals are groupings of the data. In general, we define class intervals so that

- Each interval is equal in size. For example, if the first class contains values from 120-129, the second class should include values from 130-139.
- Each interval has a **lower limit** and an **upper limit**, e.g., for interval 120-129, 120 is the lower limit and 129 is the upper limit.
- The **class width** is the difference between two consecutive lower limits.
- The **class width** is the same for every interval in the frequency table.
- We have somewhere between 5 and 20 classes, typically, depending upon the number of data we're working with.

Example 11

Suppose that we have collected weights from 100 male subjects as part of a nutrition study. For our weight data, we have values ranging from a low of 121 pounds to a high of 263 pounds, giving a total span of 263-121 = 142. We could create 7 intervals with a width of around 20, 14 intervals with a width of around 10, or somewhere in between. Often times, we have to experiment with a few possibilities to find something that represents the data well. Let us try using an interval width of 15. We could start at 121, or at 120 since it is a nice round number.

Interval	Frequency
120 - 134	4
135 – 149	14
150 – 164	16
165 – 179	28
180 – 194	12
195 – 209	8
210 – 224	7
225 – 239	6
240 – 254	2
255 - 269	3

Notice, the class width is 15 since 150-135 = 15, 165-150 = 15, and so on.

A histogram of this data would look like:

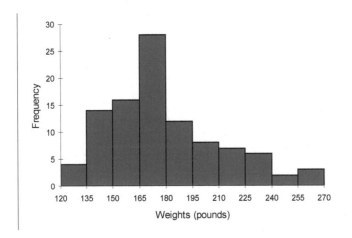

In many software packages, you can create a graph similar to a histogram by putting the class intervals as the labels on a bar chart.

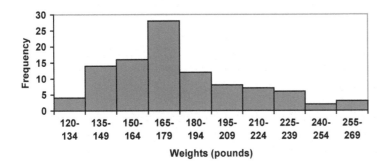

Other graph types such as pie charts are possible for quantitative data. The usefulness of different graph types will vary depending upon the number of intervals and the type of data being represented. For example, a pie chart of our weight data is difficult to read because of the quantity of intervals we used.

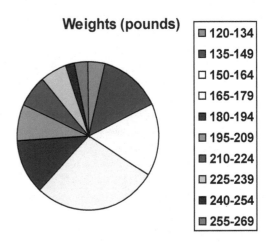

Try it Now 3

The total cost of textbooks for the term was collected from 36 students. Create a histogram for this data.

$140	$160	$160	$165	$180	$220	$235	$240	$250	$260	$280	$285
$285	$285	$290	$300	$300	$305	$310	$310	$315	$315	$320	$320
$330	$340	$345	$350	$355	$360	$360	$380	$395	$420	$460	$460

When collecting data to compare two groups, it is desirable to create a graph that compares quantities.

Example 12

The data below came from a task in which the goal is to move a computer mouse to a target on the screen as fast as possible. On 20 of the trials, the target was a small rectangle; on the other 20, the target was a large rectangle. Time to reach the target was recorded on each trial.

Interval (milliseconds)	Frequency small target	Frequency large target
300-399	0	0
400-499	1	5
500-599	3	10
600-699	6	5
700-799	5	0
800-899	4	0
900-999	0	0
1000-1099	1	0
1100-1199	0	0

One option to represent this data would be a comparative histogram or bar chart, in which bars for the small target group and large target group are placed next to each other.

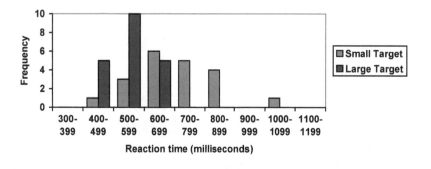

Frequency polygon

An alternative representation is a **frequency polygon**. A frequency polygon starts out like a histogram, but instead of drawing a bar, a point is placed in the **midpoint** of each interval at height equal to the frequency. The midpoint of an interval is

$$\frac{lower\ limit_2 - lower\ limit_1}{2}$$

Typically, the points are connected with straight lines to emphasize the distribution of the data.

Example 13

This graph makes it easier to see that reaction times were generally shorter for the larger target, and that the reaction times for the smaller target were more spread out.

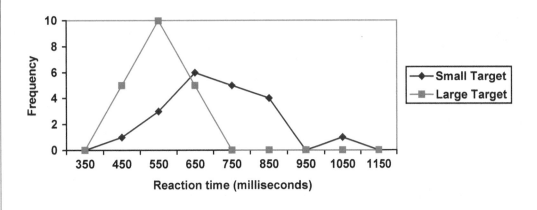

Numerical Summaries of Data

It is often desirable to use a few numbers to summarize a distribution. One important aspect of a distribution is where its center is located. Measures of central tendency are discussed first. A second aspect of a distribution is how spread out it is. In other words, how much the data in the distribution vary from one another. The second section describes measures of variability.

9.3   Measures of Central Tendency

Let's begin by trying to find the most "typical" value of a data set.

Note that we just used the word "typical" although in many cases you might think of using the word "average." We need to be careful with the word "average" as it means different things to different people in different contexts. One of the most common uses of the word "average" is what mathematicians and statisticians call the **arithmetic mean**, or just plain old **mean** for short. "Arithmetic mean" sounds rather fancy, but it is likely you have calculated a mean many times without realizing it; the mean is what most people think of when they use the word "average".

> **Mean**
> The **mean** of a set of data is the sum of the data values divided by the number of values.

Example 14

Marci's exam scores for her last math class were: 79, 86, 82, 94. The mean of these values would be

$$\frac{79 + 86 + 82 + 94}{4} = 85.25$$

Typically, we round the mean to one more decimal place than the original data. In this case, we would round 85.25 to 85.3. Thus, we can say Marci's average score on her math exams was 85.25 or about 85.3.

Example 15

The number of touchdown (TD) passes thrown by each of the 31 teams in the National Football League in the 2000 season are shown below.

    37 33 33 32 29 28 28 23 22 22 22 21 21 21 20
    20 19 19 18 18 18 18 16 15 14 14 14 12 12 9 6

Adding these values, we get a sum total of 634 TDs. Dividing by 31, the total number of data values, we get 634/31 = 20.4516.   It would be appropriate to round this to 20.5.

It would be most correct for us to report that *"The mean number of touchdown passes thrown in the NFL in the 2000 season was 20.5 passes,"* but it is not uncommon to see the more casual word "average" used in place of "mean."

Try it Now 4
The price of a jar of peanut butter at 5 stores was: $3.29, $3.59, $3.79, $3.75, and $3.99.
Find the mean price.

Let's look at an example for calculating the mean given a frequency table.

Example 16
___

The one hundred families in a particular neighborhood are asked their annual household income, to the nearest $5 thousand dollars. The results are summarized in a frequency table below.

Income (thousands of dollars)	Frequency
15	6
20	8
25	11
30	17
35	19
40	20
45	12
50	7

Calculating the mean by hand could get tricky if we try to type in all 100 values:

$$\frac{15+15+15+15+15+15+20+20+\ldots+50+50+50+50+50+50+50}{100}$$

That's one long numerator! We could calculate this more efficient by noticing that adding 15 to itself six times is the same as $15 \cdot 6 = 90$. Using this simplification, we get

$$\frac{15 \cdot 6+20 \cdot 8+25 \cdot 11+30 \cdot 17+35 \cdot 19+40 \cdot 20+45 \cdot 12+50 \cdot 7}{100} = \frac{3390}{100} = 33.9$$

The mean household income of our sample is 33.9 thousand dollars ($33,900).

Example 17
___

Extending off the last example, suppose a new family moves into the neighborhood example that has a household income of $5 million ($5000 thousand). Adding this to our sample, our mean is now:

$$\frac{15 \cdot 6+20 \cdot 8+25 \cdot 11+30 \cdot 17+35 \cdot 19+40 \cdot 20+45 \cdot 12+50 \cdot 7+5000 \cdot 1}{101} = \frac{8390}{101} = 83.069$$

While 83.1 thousand dollars ($83,069) is the correct mean household income, it no longer represents a "typical" value.

Imagine the data values on a see-saw or balance scale. The mean is the value that keeps the data in balance, like in the picture below.

If we graph our household data, the $5 million data value is so far out to the right that the mean has to adjust up to keep things in balance

For this reason, when working with data that have **outliers** – values far outside the primary grouping – it is common to use a different measure of center, the **median**.

> **Median**
> The **median** of a set of data is the value in the middle when the data is in order
>
> To find the median, begin by listing the data in order from smallest to largest, or largest to smallest.
>
> If the number of data values, N, is odd, then the median is the middle data value. This value can be found by rounding $N/2$ up to the next whole number.
>
> If the number of data values is even, there is no one middle value, so we find the mean of the two middle values (values $N/2$ and $N/2 + 1$)
>
> *We can interpret the median as "half of the data is less than the median and the other half is more than the median." Of course, we can rewrite this in context of the problem.*

Example 18

Returning to the football touchdown data, we would start by listing the data in order. Luckily, it was already in decreasing order, so we can work with it without needing to reorder it first.

37 33 33 32 29 28 28 23 22 22 22 21 21 21 20
20 19 19 18 18 18 18 16 15 14 14 14 12 12 9 6

Since there are 31 data values, an odd number, the median will be the middle number, the 16th data value ($31/2 = 15.5$, round up to 16, leaving 15 values below and 15 above). The 16th data value is 20, so the median number of touchdown passes in the 2000 season was 20 passes. Notice that for this data, the median is fairly close to the mean we calculated earlier, 20.5. *This means that half of the touchdowns scored were less than 20 and the other half were more than 20.*

Example 19

Find the median of these quiz scores: 5 10 8 6 4 8 2 5 7 7

We start by listing the data in order: 2 4 5 5 6 7 7 8 8 10

Since there are 10 data values, an even number, there is no one middle number. So, we find the mean of the two middle numbers, 6 and 7, and get (6+7)/2 = 6.5.

The median quiz score was 6.5. *We can say, half of the quiz scores were lower than 6.5 and the other half were higher than 6.5.*

Try it Now 5
The price of a jar of peanut butter at 5 stores were: $3.29, $3.59, $3.79, $3.75, and $3.99. Find the median price.

Example 20

Let us return now to our original household income data

Income (thousands of dollars)	Frequency
15	6
20	8
25	11
30	17
35	19
40	20
45	12
50	7

Here we have 100 data values. If we didn't already know that, we could find it by adding the frequencies. Since 100 is an even number, we need to find the mean of the middle two data values - the 50th and 51st data values. To find these, we start counting up from the bottom:

There are 6 data values of $15, so   Values 1 to 6 are $15 thousand
The next 8 data values are $20, so   Values 7 to (6+8)=14 are $20 thousand
The next 11 data values are $25, so   Values 15 to (14+11)=25 are $25 thousand
The next 17 data values are $30, so   Values 26 to (25+17)=42 are $30 thousand
The next 19 data values are $35, so   Values 43 to (42+19)=61 are $35 thousand

From this we can tell that values 50 and 51 will be $35 thousand, and the mean of these two values is $35 thousand. The median income in this neighborhood is $35 thousand. *Thus, half of the households' earned income is less than $35,000 and the other half earned more than $35,000.*

Example 21

If we add in the new neighbor with a $5 million household income, then there will be 101 data values, and the 51st value will be the median. As we discovered in the last example, the 51st value is $35 thousand. Notice that the new neighbor did not affect the median in this case. The median is not swayed as much by outliers as the mean.

Let's think about the previous example. When we added the 101st family's income, the mean was $81,069 from $31,900. That's a big difference in the average household income. We see that the mean is influenced by the values of the data, i.e., the mean could get larger or smaller depending on the values of the data. However, when calculating the median including the 101st family's income, the median wasn't influenced at all. In fact, in general, the median is known as a better statistic for household income since there is a wide spread of incomes among families. Thus, the values of the data influence the mean, but not the median.

In addition to the mean and the median, there is one other common measurement of the "typical" value of a data set: the **mode**.

> **Mode**
> The **mode** is the observed value of the data set that occurs most frequently.

The mode is most commonly used for categorical data, for which the median and mean cannot be computed. Also, the mode is the only central tendency that is used for both categorical and quantitative data. The mean and median are only used with quantitative data.

Example 22

In our vehicle color survey, we collected the data

Color	Frequency
Blue	3
Green	5
Red	4
White	3
Black	2
Grey	3

For this data, Green is the mode, since it is the data value that occurred the most frequently.

It is possible for a data set to have more than one mode if several categories have the same frequency, or no modes if every category occurs only once.

Try it Now 6
Reviewers were asked to rate a product on a scale of 1 to 5. Find
a. The mean rating
b. The median rating
c. The mode rating

Rating	Frequency
1	4
2	8
3	7
4	3
5	1

9.4   Measures of Variation

Consider these three sets of quiz scores:

Section A: 5 5 5 5 5 5 5 5 5 5

Section B: 0 0 0 0 0 10 10 10 10 10

Section C: 4 4 4 5 5 5 5 6 6 6

All three of these sets of data have a mean of 5 and median of 5, yet the sets of scores are clearly quite different. In section A, everyone had the same score; in section B, half the class got no points and the other half got a perfect score, assuming this was a 10-point quiz. Section C was not as consistent as section A, but not as widely varied as section B.

In addition to the mean and median, which are measures of the "typical" or "middle" value, we also need a measure of how "spread out" or varied each data set is.

There are several ways to measure this "spread" of the data. The first is the simplest and is called the **range**.

> **Range**
> The range is the difference between the maximum value and the minimum value of the data set.

Example 23

Using the quiz scores from above,

For section A, the range is 0 since both maximum and minimum are 5 and $5 - 5 = 0$
For section B, the range is 10 since $10 - 0 = 10$
For section C, the range is 2 since $6 - 4 = 2$

In the last example, the range seems to be revealing how spread out the data is. However, suppose we add a fourth section, Section D, with scores 0 5 5 5 5 5 5 5 5 10. This section also has a mean and median of 5. The range is 10, yet, this data set is quite different than Section B. To better illuminate the differences, we'll have to turn to more sophisticated measures of variation.

> **Standard deviation**
> The **standard deviation** is a measure of variation based on measuring the distance each data value deviates, or is different, from the mean. A few important characteristics:
> - Standard deviation is always positive. Standard deviation will be zero if all the data values are equal, and will get larger as the data spreads out.
> - Standard deviation has the same units as the original data.
> - Standard deviation, like the mean, can be highly influenced by outliers.

Using the data from section D, we could compute for each data value the difference between the data value and the mean:

data value	deviation: data value - mean
0	0-5 = -5
5	5-5 = 0
5	5-5 = 0
5	5-5 = 0
5	5-5 = 0
5	5-5 = 0
5	5-5 = 0
5	5-5 = 0
5	5-5 = 0
10	10-5 = 5

We would like to get an idea of the "average" deviation from the mean, but if we find the average of the values in the second column the negative and positive values cancel each other out (this will always happen), so to prevent this we square every value in the second column:

data value	deviation: data value - mean	(deviation)^2
0	0-5 = -5	$(-5)^2 = 25$
5	5-5 = 0	$0^2 = 0$
5	5-5 = 0	$0^2 = 0$
5	5-5 = 0	$0^2 = 0$
5	5-5 = 0	$0^2 = 0$
5	5-5 = 0	$0^2 = 0$
5	5-5 = 0	$0^2 = 0$
5	5-5 = 0	$0^2 = 0$
5	5-5 = 0	$0^2 = 0$
10	10-5 = 5	$(5)^2 = 25$

We then add the squared deviations up to get $25 + 0 + 0 + 0 + 0 + 0 + 0 + 0 + 0 + 25 =$ 50. Ordinarily, we would then divide by the number of scores, n, (in this case, 10) to find the mean of the deviations. But we only do this if the data set represents a population; if the data set represents a sample (as it almost always does), we instead divide by $n - 1$ (in this case, 10 - 1 = 9).[4]

So, in our example, we would have 50/10 = 5 if section D represents a population and 50/9 = about 5.56 if section D represents a sample. These values (5 and 5.56) are called, respectively, the **population variance** and the **sample variance** for section D.

---

[4] The reason we do this is highly technical, but we can see how it might be useful by considering the case of a small sample from a population that contains an outlier, which would increase the average deviation: the outlier very likely won't be included in the sample, so the mean deviation of the sample would underestimate the mean deviation of the population; thus, we divide by a slightly smaller number to get a slightly bigger average deviation.

Variance can be a useful statistical concept, but note that the units of variance in this instance would be points-squared since we squared all of the deviations. What are points-squared? Good question. We would rather deal with the units we started with (points in this case), so to convert back we take the square root and get:

population standard deviation $= \sqrt{\dfrac{50}{10}} = \sqrt{5} \approx 2.2$

or

sample standard deviation $= \sqrt{\dfrac{50}{9}} \approx 2.4$

What does this say about section D? We can say that the average score was 5 give or take 2.4. The "give or take" part is the prefix for standard deviation. In the last chapter, we discuss more about the relationship between the average and standard deviation. For now, we can interpret results as "the average is _____ give or take [standard deviation]."

If we are unsure whether the data set is a sample or a population, we will usually assume it is a sample, and we will round answers to one more decimal place than the original data, as we have done above.

> **To compute standard deviation:**
> 1. Find the deviation of each data from the mean. In other words, subtract the mean from the data value.
> 2. Square each deviation.
> 3. Add the squared deviations.
> 4. Divide by n, the number of data values, if the data represents a whole population; **divide by $n-1$ if the data is from a sample.** (This result is the *sample variance*.)
> 5. Compute the square root of the result. (This result is the *standard deviation*.)

Example 24

Computing the standard deviation for Section B above, we first calculate that the mean is 5. Using a table can help keep track of your computations for the standard deviation:

data value	deviation: data value - mean	(deviation)^2
0	0-5 = -5	$(-5)^2 = 25$
0	0-5 = -5	$(-5)^2 = 25$
0	0-5 = -5	$(-5)^2 = 25$
0	0-5 = -5	$(-5)^2 = 25$
0	0-5 = -5	$(-5)^2 = 25$
10	10-5 = 5	$(5)^2 = 25$
10	10-5 = 5	$(5)^2 = 25$
10	10-5 = 5	$(5)^2 = 25$
10	10-5 = 5	$(5)^2 = 25$
10	10-5 = 5	$(5)^2 = 25$

Assuming this data represents a population, we will add the squared deviations, divide by 10, the number of data values, and compute the square root:

$$\sqrt{\frac{25+25+25+25+25+25+25+25+25+25}{10}} = \sqrt{\frac{250}{10}} = 5$$

Notice that the standard deviation of this data set is much larger than that of section D since the data in this set is more spread out. Thus, the average score was 5 give or take 5.

For comparison, the standard deviations of all four sections are

Section A: 5 5 5 5 5 5 5 5 5 5	Standard deviation: 0
Section B: 0 0 0 0 0 10 10 10 10 10	Standard deviation: 5
Section C: 4 4 4 5 5 5 5 6 6 6	Standard deviation: 0.8
Section D: 0 5 5 5 5 5 5 5 5 10	Standard deviation: 2.2

Try it Now 7
The price of a jar of peanut butter at 5 stores were: $3.29, $3.59, $3.79, $3.75, and $3.99. Find the standard deviation of the prices.

Where standard deviation is a measure of variation based on the mean, **quartiles** are based on the median.

Quartiles
Quartiles are values that divide the data in quarters.

The first quartile (Q_1) is the value so that 25% of the data values are below it; the third quartile (Q_3) is the value so that 75% of the data values are below it. You may have guessed that the second quartile is the same as the median, since the median is the value so that 50% of the data values are below it.

This divides the data into quarters; 25% of the data is between the minimum and Q_1, 25% is between Q_1 and the median, 25% is between the median and Q_3, and 25% is between Q_3 and the maximum value

While quartiles are not a 1-number summary of variation like standard deviation, the quartiles are used with the median, minimum, and maximum values to form a **5-number summary** of the data.

Five-number summary
The five-number summary takes this form

Minimum, Q_1, Median, Q_3, Maximum

To find the first quartile, we need to find the data value so that 25% of the data is below it. If *n* is the number of data values, we compute a locator by finding 25% of *n*. If this locator is a decimal value, we round up, and find the data value in that position. If the locator is a whole number, we find the mean of the data value in that position and the next data value. This is identical to the process we used to find the median, except we use 25% of the data values rather than half the data values as the locator.

> **To find the first quartile, Q_1**
> 1. Begin by ordering the data from smallest to largest.
> 2. Compute the locator: $L = 0.25n$.
> 3. If L is a decimal value:
>   Round up to $L+$
>   Use the data value in the $L+^{\text{th}}$ position.
>
>   If L is a whole number:
>   Find the mean of the data values in the Lth and $L+1^{\text{th}}$ positions.

> **To find the third quartile, Q_3**
> Use the same procedure as for Q_1, but with locator: $L = 0.75n$

Let's look at some examples. We can also calculate the 5-number summary in calculators, or some scientific software like Excel, Minitab, or R. However, in this course, we only get our feet wet with statistics, so we can calculate these values quickly by hand.

Example 25

Suppose we have measured 9 females and their heights (in inches), sorted from smallest to largest are:

59 60 62 64 66 67 69 70 72

To find the first quartile we first compute the locator: 25% of 9 is $L = 0.25(9) = 2.25$. Since this value is not a whole number, we round up to 3. The first quartile will be the third data value: 62 inches. We can say that the 25% of females are shorter than 62 inches and the other 75% is taller than 62 inches.

To find the third quartile, we again compute the locator: 75% of 9 is $0.75(9) = 6.75$. Since this value is not a whole number, we round up to 7. The third quartile will be the seventh data value: 69 inches. We can say that the 75% of females are shorter than 69 inches and the other 25% is taller than 69 inches.

Example 26

Suppose we had measured 8 females and their heights (in inches), sorted from smallest to largest are:

59 60 62 64 66 67 69 70

To find the first quartile we first compute the locator: 25% of 8 is $L = 0.25(8) = 2$. Since this value *is* a whole number, we will find the mean of the 2nd and 3rd data values: $(60+62)/2 = 61$, so the first quartile is 61 inches. We can say that the 25% of females are shorter than 61 inches and the other 75% is taller than 61 inches.

The third quartile is computed similarly, using 75% instead of 25%. $L = 0.75(8) = 6$. This is a whole number, so we will find the mean of the 6th and 7th data values: $(67+69)/2 = 68$, so Q_3 is 68 inches. We can say that the 75% of females are shorter than 68 inches and the other 25% is taller than 68 inches.

Note, the median could be computed the same way, using 50% or a locator of $L = 0.5n$.

The 5-number summary combines the first and third quartile with the minimum, median, and maximum values.

Example 27

In the example with a sample of 9 females, the median is 66, the minimum is 59, and the maximum is 72. Hence, the 5-number summary is: 59, 62, 66, 69, 72.

In the example with a sample of 8 females, the median is 65, the minimum is 59, and the maximum is 70, so the 5-number summary is: 59, 61, 65, 68, 70.

Example 28

Returning to our quiz score data. In each case, the first quartile locator is $0.25(10) = 2.5$, so the first quartile will be the 3rd data value, and the third quartile will be the 8th data value. Creating the five-number summaries:

Section and data	5-number summary
Section A: 5 5 5 5 5 5 5 5 5 5	5, 5, 5, 5, 5
Section B: 0 0 0 0 0 10 10 10 10 10	0, 0, 5, 10, 10
Section C: 4 4 4 5 5 5 5 6 6 6	4, 4, 5, 6, 6
Section D: 0 5 5 5 5 5 5 5 5 10	0, 5, 5, 5, 10

Of course, with a relatively small data set, finding a five-number summary is a bit silly, since the summary contains almost as many values as the original data.

Try it Now 8

The total cost of textbooks for the term was collected from 36 students. Find the 5-number summary of this data.

$140	$160	$160	$165	$180	$220	$235	$240	$250	$260	$280	$285
$285	$285	$290	$300	$300	$305	$310	$310	$315	$315	$320	$320
$330	$340	$345	$350	$355	$360	$360	$380	$395	$420	$460	$460

Example 29

Returning to the household income data from earlier, create the five-number summary.

Income (thousands of dollars)	Frequency
$15	6
$20	8
$25	11
$30	17
$35	19
$40	20
$45	12
$50	7

By adding the frequencies, we can see there are 100 data values represented in the table. In Example 20, we found the median was $35 thousand. We can see in the table that the minimum income is $15 thousand, and the maximum is $50 thousand.

To find Q_1, we calculate the locator: $L = 0.25(100) = 25$. This is a whole number, so Q_1 will be the mean of the 25th and 26th data values.

Counting up in the data as we did before,

There are 6 data values of $15, so   Values 1 to 6 are $15 thousand
The next 8 data values are $20, so   Values 7 to (6+8)=14 are $20 thousand
The next 11 data values are $25, so   Values 15 to (14+11)=25 are $25 thousand
The next 17 data values are $30, so   Values 26 to (25+17)=42 are $30 thousand

The 25th data value is $25 thousand, and the 26th data value is $30 thousand, so Q_1 will be the mean of these: $(25 + 30)/2 = \$27.5$ thousand.

To find Q_3, we calculate the locator: $L = 0.75(100) = 75$. This is a whole number, so Q_3 will be the mean of the 75th and 76th data values. Continuing our counting from earlier,

The next 19 data values are $35, so   Values 43 to (42+19)=61 are $35 thousand
The next 20 data values are $40, so   Values 61 to (61+20)=81 are $40 thousand

Both the 75th and 76th data values lie in this group, so Q_3 will be $40 thousand.

Putting these values together into a five-number summary, we get: 15, 27.5, 35, 40, 50.

Note that the 5-number summary divides the data into four intervals, each of which will contain about 25% of the data. In the previous example, about 25% of households have income between $40 thousand and $50 thousand.

For visualizing data, there is a graphical representation of a 5-number summary called a **box plot**, or box and whisker graph.

> **Box plot**
> A **box plot** is a graphical representation of a five-number summary.
>
> To create a box plot, a number line is first drawn with equidistant tick marks. A box is drawn from the first quartile to the third quartile, and a line is drawn through the box at the median. "Whiskers" are extended out to the minimum and maximum values.

Example 30

The box plot below is based on the 5-number summary from the sample of 9 female heights:

59, 62, 66, 69, 72

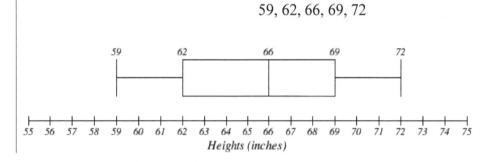

Example 31

The box plot below is based on the 5-number summary from the sample of the household incomes:

15, 27.5, 35, 40, 50

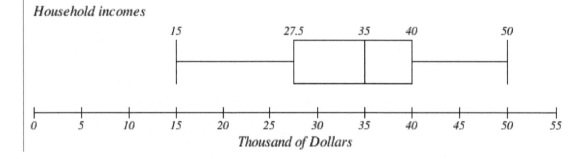

Try it Now 9
Create a boxplot based on the textbook price data from the last *Try it Now*.

Box plots are particularly useful for comparing data from two populations or samples. In fact, when we have two samples to compare, it is always preferred to use box plots.

Example 32

The box plot of service times for two fast-food restaurants is shown below.

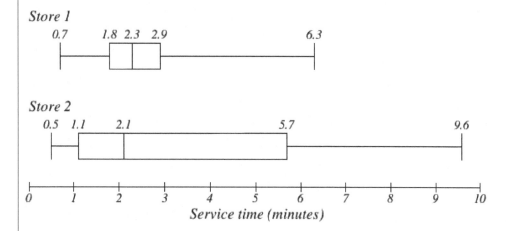

While store 2 had a slightly shorter median service time (2.1 minutes vs. 2.3 minutes), store 2 is less consistent, with a wider spread of the data.

At store 1, 75% of customers were served within 2.9 minutes, while at store 2, 75% of customers were served within 5.7 minutes.

Which store should you go to in a hurry? That depends upon your opinions about luck – 25% of customers at store 2 had to wait between 5.7 and 9.6 minutes.

Example 33

The boxplot below is based on the birth weights of infants with severe idiopathic respiratory distress syndrome (SIRDS)[5]. The boxplot is separated to show the birth weights of infants who survived and those that did not.

Comparing the two groups, the boxplot reveals that the birth weights of the infants that died appear to be, overall, smaller than the weights of infants that survived. In fact, we can see that the median birth weight of infants that survived is the same as the third quartile of the infants that died.

Similarly, we can see that the first quartile of the survivors is larger than the median weight of those that died, meaning that over 75% of the survivors had a birth weight larger than the median birth weight of those that died.

---

[5] van Vliet, P.K. and Gupta, J.M. (1973) Sodium bicarbonate in idiopathic respiratory distress syndrome. *Arch. Disease in Childhood*, **48**, 249–255. As quoted on
http://openlearn.open.ac.uk/mod/oucontent/view.php?id=398296&section=1.1.3

Looking at the maximum value for those that died and the third quartile of the survivors, we can see that over 25% of the survivors had birth weights higher than the heaviest infant that died.

The box plot gives us a quick, albeit informal, way to determine that birth weight is quite likely linked to survival of infants with SIRDS.

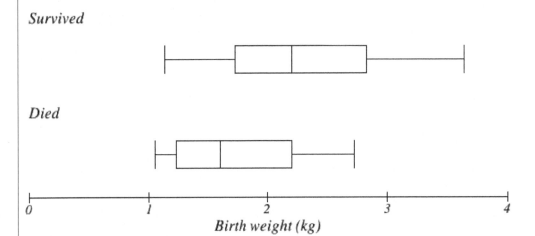

Survived

Died

Birth weight (kg)

Try it Now Answers

1.

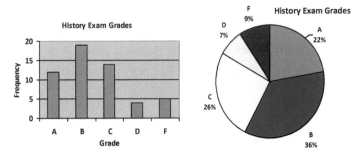

2. While the pie chart accurately depicts the relative size of the people agreeing with each candidate, the chart is confusing, since usually percents on a pie chart represent the percentage of the pie the slice represents.

3. Using a class intervals of size 55, we can group our data into six intervals:

Cost interval	Frequency
$140-194	5
$195-249	3
$250-304	9
$305-359	12
$360-414	4
$415-469	3

We can use the frequency distribution to generate the histogram.

4. Adding the prices and dividing by 5 we get the mean price: $3.682

5. First we put the data in order: $3.29, $3.59, $3.75, $3.79, $3.99. Since there are an odd number of data, the median will be the middle value, $3.75.

6. There are 23 ratings.
   a. The mean is $\dfrac{1 \cdot 4 + 2 \cdot 8 + 3 \cdot 7 + 4 \cdot 3 + 5 \cdot 1}{23} \approx 2.5$
   b. There are 23 data values, so the median will be the 12th data value. Ratings of 1 are the first 4 values, while a rating of 2 are the next 8 values, so the 12th value will be a rating of 2. The median is 2.
   c. The mode is the most frequent rating. The mode rating is 2.

7. Earlier we found the mean of the data was $3.682.

data value	deviation: data value - mean	deviation squared
3.29	3.29 – 3.682 = -0.391	0.153664
3.59	3.59 – 3.682 = -0.092	0.008464
3.79	3.79 – 3.682 = 0.108	0.011664
3.75	3.75 – 3.682 = 0.068	0.004624
3.99	3.99 – 3.682 = 0.308	0.094864

This data is from a sample, so we will add the squared deviations, divide by 4, the number of data values minus 1, and compute the square root:

$$\sqrt{\frac{0.153664 + 0.008464 + 0.011664 + 0.004624 + 0.094864}{4}} \approx \$0.261$$

Thus, the average price of peanut butter is $3.68 give or take $0.26.

8. The data is already in order, so we don't need to sort it first.
   The minimum value is $140 and the maximum is $460.

   There are 36 data values so $n = 36$. $n/2 = 18$, which is a whole number, so the median is the mean of the 18th and 19th data values, $305 and $310. The median is $307.50.

   To find the first quartile, we calculate the locator, $L = 0.25(36) = 9$. Since this is a whole number, we know Q_1 is the mean of the 9th and 10th data values, $250 and $260. $Q_1 = \$255$.

   To find the third quartile, we calculate the locator, $L = 0.75(36) = 27$. Since this is a whole number, we know Q_3 is the mean of the 27th and 28th data values, $345 and $350. $Q_3 = \$347.50$.

   The 5-number summary of this data is: $140, $255, $307.50, $347.50, $460

9. Boxplot of textbook costs:

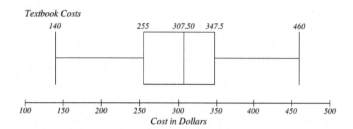

Exercises

Skills

1. The table below shows scores on a Math test.
   a. Complete the frequency table for the Math test scores
   b. Construct a histogram of the data
   c. Construct a pie chart of the data

80	50	50	90	70	70	100	60	70	80	70	50
90	100	80	70	30	80	80	70	100	60	60	50

2. A group of adults where asked how many cars they had in their household
   a. Complete the frequency table for the car number data
   b. Construct a histogram of the data
   c. Construct a pie chart of the data

1	4	2	2	1	2	3	3	1	4	2	2
1	2	1	3	2	2	1	2	1	1	1	2

3. A group of adults were asked how many children they have in their families. The bar graph to the right shows the number of adults who indicated each number of children.
   a. How many adults where questioned?
   b. What percentage of the adults questioned had 0 children?

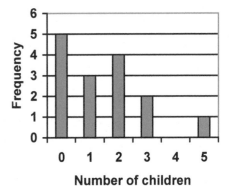

4. Jasmine was interested in how many days it would take an order from Netflix to arrive at her door. The graph below shows the data she collected.
   a. How many movies did she order?
   b. What percentage of the movies arrived in one day?

5. The bar graph below shows the *percentage* of students who received each letter grade on their last English paper. The class contains 20 students. What number of students earned an A on their paper?

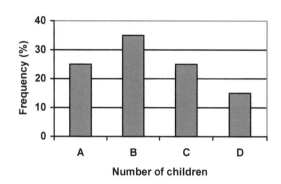

6. Kori categorized her spending for this month into four categories: Rent, Food, Fun, and Other. The percents she spent in each category are pictured here. If she spent a total of $2600 this month, how much did she spend on rent?

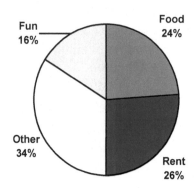

7. A group of diners were asked how much they would pay for a meal. Their responses were: $7.50, $8.25, $9.00, $8.00, $7.25, $7.50, $8.00, $7.00.
   a. Find the mean
   b. Find the median
   c. Write the 5-number summary for this data

8. You recorded the time in seconds it took for 8 participants to solve a puzzle. The times were: 15.2, 18.8, 19.3, 19.7, 20.2, 21.8, 22.1, 29.4.
   a. Find the mean
   b. Find the median
   c. Write the 5-number summary for this data

9. Refer back to the histogram from question #3.
   a. Compute the mean number of children for the group surveyed
   b. Compute the median number of children for the group surveyed
   c. Write the 5-number summary for this data.
   d. Create box plot.

10. Refer back to the histogram from question #4.
    a. Compute the mean number of shipping days
    b. Compute the median number of shipping days
    c. Write the 5-number summary for this data.
    d. Create box plot.

Concepts

11. The box plot below shows salaries for Actuaries and CPAs. Kendra makes the median salary for an Actuary. Kelsey makes the first quartile salary for a CPA. Who makes more money? How much more?

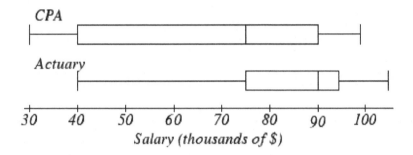

12. Referring to the boxplot above, what percentage of actuaries makes more than the median salary of a CPA?

Exploration

13. Studies are often done by pharmaceutical companies to determine the effectiveness of a treatment program. Suppose that a new AIDS antibody drug is currently under study. It is given to patients once the AIDS symptoms have revealed themselves. Of interest is the average length of time, in months, patients live once starting the treatment. Two researchers each follow a different set of 40 AIDS patients from the start of treatment until their deaths. The following data (in months) are collected.

    Researcher 1: 3; 4; 11; 15; 16; 17; 22; 44; 37; 16; 14; 24; 25; 15; 26; 27; 33; 29; 35; 44; 13; 21; 22; 10; 12; 8; 40; 32; 26; 27; 31; 34; 29; 17; 8; 24; 18; 47; 33; 34

    Researcher 2: 3; 14; 11; 5; 16; 17; 28; 41; 31; 18; 14; 14; 26; 25; 21; 22; 31; 2; 35; 44; 23; 21; 21; 16; 12; 18; 41; 22; 16; 25; 33; 34; 29; 13; 18; 24; 23; 42; 33; 29

    a. Create comparative histograms of the data
    b. Create comparative boxplots of the data

14. A graph appears below showing the number of adults and children who prefer each type of soda. There were 130 adults and kids surveyed. Discuss some ways in which the graph below could be improved.

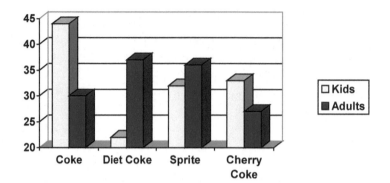

15. Make up three data sets with 5 numbers each that have:
    a. the same mean but different standard deviations.
    b. the same mean but different medians.
    c. the same median but different means.

16. A sample of 30 distance scores measured in yards has a mean of 7, a variance of 16, and a standard deviation of 4.
    a. You want to convert all your distances from yards to feet, so you multiply each score in the sample by 3. What are the new mean, median, variance, and standard deviation?
    b. You then decide that you only want to look at the distance past a certain point. Thus, after multiplying the original scores by 3, you decide to subtract 4 feet from each of the scores. Now what are the new mean, median, variance, and standard deviation?

17. In your class, design a poll on a topic of interest to you and give it to the class.
    a. Summarize the data, computing the mean and five-number summary.
    b. Create a graphical representation of the data.
    c. Write several sentences about the topic, using your computed statistics as evidence in your writing.

Chapter 10:   Probability

Introduction

The probability of a specified event is the chance or likelihood that it will occur. There are several ways of viewing probability. One would be **experimental** in nature, where we repeatedly conduct an experiment. Suppose we flipped a coin over and over and over again and it came up heads about half of the time; we would expect that in the future whenever we flipped the coin it would turn up heads about half of the time. When a weather reporter says "there is a 10% chance of rain tomorrow," she is basing that on prior evidence; that out of all days with similar weather patterns, it has rained on 1 out of 10 of those days.

Another view would be **subjective** in nature, in other words an educated guess. If someone asked you the probability that the Seattle Mariners would win their next baseball game, it would be impossible to conduct an experiment where the same two teams played each other repeatedly, each time with the same starting lineup and starting pitchers, each starting at the same time of day on the same field under the precisely the same conditions. Since there are so many variables to take into account, someone familiar with baseball and with the two teams involved might make an educated guess that there is a 75% chance they will win the game; that is, *if* the same two teams were to play each other repeatedly under identical conditions, the Mariners would win about three out of every four games. But this is just a guess, with no way to verify its accuracy, and depending upon how educated the educated guesser is, a subjective probability may not be worth very much.

We will return to the experimental and subjective probabilities from time to time, but in this course, we will mostly be concerned with **theoretical probability**, which is defined as follows: Suppose there is a situation with n <u>equally likely</u> possible outcomes and that m of those n outcomes correspond to a particular event; then the **probability** of that event is defined as $\dfrac{m}{n}$.

Table of Contents

10.1 Basic Concepts

If you roll a die, pick a card from deck of playing cards, or randomly select a person and observe their hair color, we are executing an experiment or procedure. In probability, we look at the likelihood of different outcomes. We begin with some terminology.

Events and Outcomes
The result of an experiment is called an **outcome**.

An **event** is any particular outcome or group of outcomes.

A **simple event** is an event that cannot be broken down further

The **sample space** is the set of all possible simple events.

Example 1

If we roll a standard 6-sided die, describe the sample space and some simple events.

The sample space is the set of all possible simple events: {1,2,3,4,5,6}

Some examples of simple events:
We roll a 1
We roll a 5

Some compound events:
We roll a number bigger than 4
We roll an even number

Two dice   One die

Basic Probability
Given that all outcomes are equally likely, we can compute the probability of an event E using this formula:

$$P(E) = \frac{\text{Number of outcomes corresponding to the event E}}{\text{Total number of equally - likely outcomes}}$$

Example 2

If we roll a 6-sided die, calculate
a) P(rolling a 1)
b) P(rolling a number bigger than 4)

Recall that the sample space is {1,2,3,4,5,6}

a) There is one outcome corresponding to "rolling a 1", so the probability is $\dfrac{1}{6}$

b) There are two outcomes bigger than a 4, so the probability is $\dfrac{2}{6} = \dfrac{1}{3}$

Probabilities are essentially fractions, and can be reduced to lower terms like fractions.

Example 3

Let's say you have a bag with 20 cherries, 14 sweet and 6 sour. If you pick a cherry at random, what is the probability that it will be sweet?

There are 20 possible cherries that could be picked, so the number of possible outcomes is 20. Of these 20 possible outcomes, 14 are favorable (sweet), so the probability that the cherry will be sweet is $\dfrac{14}{20} = \dfrac{7}{10}$.

There is one potential complication to this example, however. It must be assumed that the probability of picking any of the cherries is the same as the probability of picking any other. This wouldn't be true if (let us imagine) the sweet cherries are smaller than the sour ones. (The sour cherries would come to hand more readily when you sampled from the bag.) Let us keep in mind, therefore, that when we assess probabilities in terms of the ratio of favorable to all potential cases, we rely heavily on the assumption of equal probability for all outcomes.

Try it Now 1
At some random moment, you look at your clock and note the minutes reading.
a. What is probability the minutes reading is 15?
b. What is the probability the minutes reading is 15 or less?

Cards
A standard deck of 52 playing cards consists of four **suits** (hearts, spades, diamonds and clubs). Spades and clubs are black while hearts and diamonds are red. Each suit contains 13 cards, each of a different **rank**: An Ace (which in many games functions as both a low card and a high card), cards numbered 2 through 10, a Jack, a Queen and a King.

The image below gives an example of a complete deck of 52 cards.

	Ace	2	3	4	5	6	7	8	9	10	Jack	Queen	King
Clubs													
Diamonds													
Hearts													
Spades													

Example 4

Compute the probability of randomly drawing one card from a deck and getting an Ace.

There are 52 cards in the deck and 4 Aces so $P(Ace) = \dfrac{4}{52} = \dfrac{1}{13} \approx 0.0769$

We can also think of probabilities as percents: There is a 7.69% chance that a randomly selected card will be an Ace.

Notice that the smallest possible probability is 0 – if there are no outcomes that correspond with the event. The largest possible probability is 1 – if all possible outcomes correspond with the event.

Certain and Impossible events
An *impossible* event has a probability of 0.
A *certain* event has a probability of 1.
The probability of any event must be $0 \leq P(E) \leq 1$, i.e., the probability of any event is between (or equal to) zero and one.

In the course of this chapter, *if you compute a probability and get an answer that is negative or greater than 1, you have made a mistake and should check your work.*

10.2   Working with Events

Complementary Events

Now let us examine the probability that an event does **not** happen. As in the previous section, consider the situation of rolling a six-sided die and first compute the probability of rolling a six: the answer is $P(\text{six}) = 1/6$. Now consider the probability that we do *not* roll a six: there are 5 outcomes that are not a six, so the answer is $P(\text{not a six}) = \dfrac{5}{6}$. Notice that

$$P(\text{six}) + P(\text{not a six}) = \frac{1}{6} + \frac{5}{6} = \frac{6}{6} = 1$$

This is not a coincidence. Consider a generic situation with n possible outcomes and an event E that corresponds to m of these outcomes. Then the remaining n - m outcomes correspond to E not happening, thus

$$P(\text{not } E) = \frac{n-m}{n} = \frac{n}{n} - \frac{m}{n} = 1 - \frac{m}{n} = 1 - P(E)$$

> **Complement of an Event**
> The **complement** of an event is the event "E doesn't happen."
> The notation $\overline{E}$ or E^c is used for the complement of event E.
> We can compute the probability of the complement using $P\left(\overline{E}\right) = 1 - P(E)$.
> Notice also that $P(E) = 1 - P\left(\overline{E}\right)$.

Example 5

> If you pull a random card from a deck of playing cards, what is the probability *it is not a heart*?
>
> There are 13 hearts in the deck, so $P(\text{heart}) = \dfrac{13}{52} = \dfrac{1}{4}$.
>
> The probability of *not* drawing a heart is the complement:
>
> $$P(\text{not heart}) = 1 - P(\text{heart}) = 1 - \frac{1}{4} = \frac{3}{4}$$

Probability of two independent events

Example 6

> Suppose we flipped a coin and rolled a die, and wanted to know the probability of getting a head on the coin and a 6 on the die.
>
> We could list all possible outcomes: {H1,H2,H3,H4,H5,H6,T1,T2,T3,T4,T5,T6}.

Notice there are 2 · 6 = 12 total outcomes. Out of these, only 1 is the desired outcome, so the probability is $\frac{1}{12}$.

The prior example finding the probability for two independent events.

> **Independent Events**
> Events A and B are **independent events** if the probability of Event B occurring is the same whether or not Event A occurs.

Example 7

Are these events independent?

a) A fair coin is tossed two times. The two events are (1) first toss is a head and (2) second toss is a head.

b) The two events (1) "It will rain tomorrow in Houston" and (2) "It will rain tomorrow in Galveston" (a city near Houston).

c) You draw a card from a deck, then draw a second card without replacing the first.

a) The probability that a head comes up on the second toss is 1/2 regardless of whether or not a head came up on the first toss, so these events are *independent*.

b) These events are *not independent* because it is more likely that it will rain in Galveston on days it rains in Houston than on days it does not.

c) The probability of the second card being red depends on whether the first card is red or not, so these events are *not independent*.

When two events are independent, the probability of both occurring is the product of the probabilities of the individual events.

> **$P(A$ and $B)$ for independent events**
> If events A and B are independent, then the probability of both A and B occurring is
>
> $P(A$ and $B) = P(A) \cdot P(B)$
>
> where $P(A$ and $B)$ is the probability of events A and B both occurring, $P(A)$ is the probability of event A occurring, and $P(B)$ is the probability of event B occurring.

If you look back at the coin and die example from earlier, you can see how the number of outcomes of the first event multiplied by the number of outcomes in the second event multiplied to equal the total number of possible outcomes in the combined event.

Example 8

In your drawer, you have 10 pairs of socks, 6 of which are white, and 7 tee shirts, 3 of which are white. If you randomly reach in and pull out a pair of socks and a tee shirt, what is the probability both are white?

The probability of choosing a white pair of socks is $\frac{6}{10}$.

The probability of choosing a white tee shirt is $\frac{3}{7}$.

Since the probability of randomly selecting a white sock is the same whether or not we randomly select a white t-shirt from the drawer, then these are independent events. Hence, we can use the probability formula for independent events.

The probability of both being white is $\frac{6}{10} \cdot \frac{3}{7} = \frac{18}{70} = \frac{9}{35}$.

Try it Now 2

A card is pulled a deck of cards and noted. The card is then replaced, the deck is shuffled, and a second card is removed and noted. What is the probability that both cards are Aces?

The previous examples looked at the probability of *both* events occurring. Now we will look at the probability of *either* event occurring.

Probability of either two events occurring

Example 9

Suppose we flipped a coin and rolled a die, and wanted to know the probability of getting a head on the coin *or* a 6 on the die.

Here, there are still 12 possible outcomes: {H1,H2,H3,H4,H5,H6,T1,T2,T3,T4,T5,T6}

By simply counting, we can see that 7 of the outcomes have a head on the coin *or* a 6 on the die *or* both – we use *or* inclusively here (these 7 outcomes are H1, H2, H3, H4, H5, H6, T6), so the probability is $\frac{7}{12}$. How could we have found this from the individual probabilities?

As we would expect, $\frac{1}{2}$ of these outcomes have a head, and $\frac{1}{6}$ of these outcomes have a 6 on the die. If we add these, $\frac{1}{2} + \frac{1}{6} = \frac{6}{12} + \frac{2}{12} = \frac{8}{12}$, which is not the correct probability. Looking at the outcomes we can see why: the outcome H6 would have been counted twice, since it contains both a head and a 6; the probability of both a head *and* rolling a 6 is $\frac{1}{12}$.

If we subtract out this double count, we have the correct probability: $\frac{8}{12} - \frac{1}{12} = \frac{7}{12}$.

P(A or B)
The probability of either A or B occurring (or both) is

$$P(A \text{ or } B) = P(A) + P(B) - P(A \text{ and } B)$$

Example 10

Suppose we draw one card from a standard deck. What is the probability that we get a Queen or a King?

There are 4 Queens and 4 Kings in the deck. Hence, 8 outcomes corresponding to a Queen or King out of 52 possible outcomes. Thus, the probability of drawing a Queen or a King is

$$P(\text{King or Queen}) = \frac{8}{52}$$

Note that in this case, there are no cards that are both a Queen and a King, so $P(\text{King and Queen}) = 0$. Using the probability rule, we get

$$P(\text{King or Queen}) = P(\text{King}) + P(\text{Queen}) - P(\text{King and Queen}) = \frac{4}{52} + \frac{4}{52} - 0 = \frac{8}{52}.$$

In the last example, the events were **mutually exclusive**, so $P(A \text{ or } B) = P(A) + P(B)$.

Mutually Exclusive
Two events are mutually exclusive if the events cannot occur at the same time. If this is the case, then we can use the formula

$$P(A \text{ or } B) = P(A) + P(B)$$

to find the probability.

Some examples:
- If we toss a coin, the coin lands on heads or tails, but not both.
- If we draw a card, the card cannot be a queen and king at the same time (or any two faces). *Careful, though, because if you draw a card from a deck, then the card could be a heart and an 8 at the same time, making them not mutually exclusive.*

Example 11

Suppose we draw one card from a standard deck. What is the probability that we get a red card or a King?

Half the cards are red, so $P(\text{red}) = \frac{26}{52}$

There are four kings, so $P(\text{King}) = \dfrac{4}{52}$

There are two red kings, so $P(\text{Red and King}) = \dfrac{2}{52}$

We can then calculate

$$P(\text{Red or King}) = P(\text{Red}) + P(\text{King}) - P(\text{Red and King}) = \frac{26}{52} + \frac{4}{52} - \frac{2}{52} = \frac{28}{52}$$

Probability of at least one

> *P***(at least one)**
> The probability of at least one event occurring is
>
> $P(\text{at least one}) = 1 - P(\text{none})$

Example

In a large population, 70% of the people have been vaccinated. If 5 people are randomly selected, what is the probability that at least one of them has been vaccinated?

Since we need to find the probability that at least one of the 5 selected people is vaccinated, we get

$P(\text{at least one}) = 1 - P(\text{none})$
$P(\text{at least one is vaccinated}) = 1 - P(\text{none of the 5 are vaccinated})$

If 70% of the people are vaccinated, then this means 30% are not vaccinated. Hence,

$P(\text{at least one is vaccinated}) \quad = 1 - P(\text{none of the 5 are vaccinated})$
$= 1 - P(1^{\text{st}} \text{ isn't vaccinated } \textbf{and } 2^{\text{nd}} \text{ isn't vaccinated}\ldots)$
$= 1 - (0.30)(0.30)(0.30)(0.30)(0.30)$
$= 1 - (0.30)^5$
$= 0.99757$

Thus, there is a 0.99757 or 99.757% chance that at least one of the selected 5 people is vaccinated.

Try it Now 3

In your drawer, you have 10 pairs of socks, 6 of which are white, and 7 tee shirts, 3 of which are white. If you reach in and randomly grab a pair of socks and a tee shirt, what the probability at least one is white?

Example 12

The table below shows the number of survey subjects who have received and not received a speeding ticket in the last year, and the color of their car. Find the probability that a randomly chosen person:

a) Has a red car *and* got a speeding ticket
b) Has a red car *or* got a speeding ticket.

	Speeding ticket	No speeding ticket	Total
Red car	15	135	150
Not red car	45	470	515
Total	60	605	665

We can see that 15 people of the 665 surveyed had both a red car and got a speeding ticket, so the probability is $\frac{15}{665} \approx 0.0226$.

Notice that having a red car and getting a speeding ticket are not independent events, so the probability of both of them occurring is not simply the product of probabilities of each one occurring.

We could answer this question by simply adding up the numbers: 15 people with red cars and speeding tickets + 135 with red cars but no ticket + 45 with a ticket but no red car = 195 people. So, the probability is $\frac{195}{665} \approx 0.2932$.

We also could have found this probability by:

P(had a red car) + P(got a speeding ticket) – P(had a red car and got a speeding ticket)
$= \frac{150}{665} + \frac{60}{665} - \frac{15}{665} = \frac{195}{665}$.

Conditional Probability

Often it is required to compute the probability of an event given that another event has occurred.

Example 13

What is the probability that two cards drawn at random from a deck of playing cards will both be aces?

It might seem that you could use the formula for the probability of two independent events and simply multiply $\frac{4}{52} \cdot \frac{4}{52} = \frac{1}{169}$. This would be incorrect, however, because the two events are not independent. If the first card drawn is an ace, then the probability that the second card is also an ace would be lower because there would only be three aces left in the deck.

Once the first card chosen is an ace, the probability that the second card chosen is also an ace is called the **conditional probability** of drawing an ace. In this case, the "condition" is that the first card is an ace. Symbolically, we write this as:

P(ace on second draw | an ace on the first draw).

The vertical bar "|" is read as "given," so the above expression is short for "The probability that an ace is drawn on the second draw given that an ace was drawn on the first draw." What is this probability? After an ace is drawn on the first draw, there are 3 aces out of 51 total cards left. This means that the conditional probability of drawing an ace after one ace has already been drawn is $\dfrac{3}{51} = \dfrac{1}{17}$.

Thus, the probability of both cards being aces is $\dfrac{4}{52} \cdot \dfrac{3}{51} = \dfrac{12}{2652} = \dfrac{1}{221}$.

Conditional Probability
The probability the event *B* occurs, given that event *A* has happened, is represented as $P(B \mid A)$.

This is read as "the probability of *B* given *A*"

Example 14

Find the probability that a die rolled shows a 6, given that a flipped coin shows a head.

These are two independent events, so the probability of the die rolling a 6 is $\dfrac{1}{6}$, regardless of the result of the coin flip.

Example 15

The table below shows the number of survey subjects who have received and not received a speeding ticket in the last year, and the color of their car. Find the probability that a randomly chosen person:

a) Has a speeding ticket *given* they have a red car
b) Has a red car *given* they have a speeding ticket

	Speeding ticket	No speeding ticket	Total
Red car	15	135	150
Not red car	45	470	515
Total	60	605	665

a) Since we know the person has a red car, we are only considering the 150 people in the first row of the table. Of those, 15 have a speeding ticket, so

$$P(\text{ticket} \mid \text{red car}) = \frac{15}{150} = \frac{1}{10} = 0.1$$

b) Since we know the person has a speeding ticket, we are only considering the 60 people in the first column of the table. Of those, 15 have a red car, so

$$P(\text{red car} \mid \text{ticket}) = \frac{15}{60} = \frac{1}{4} = 0.25 \,.$$

Notice from the last example that *P(B | A)* is **not** equal to *P(A | B)*.

These kinds of conditional probabilities are what insurance companies use to determine your insurance rates. They look at the conditional probability of you having accident, given your age, your car, your car color, your driving history, etc., and price your policy based on that likelihood.

Conditional Probability Formula
If events *A* and *B* are not independent, then

$$P(A \text{ and } B) = P(A) \cdot P(B \mid A)$$

Example 16

If you pull 2 cards out of a deck, what is the probability that both are spades?

The probability that the first card is a spade is $\frac{13}{52}$.

The probability that the second card is a spade, given the first was a spade, is $\frac{12}{51}$, since there is one less spade in the deck, and one less total cards.

The probability that both cards are spades is $\frac{13}{52} \cdot \frac{12}{51} = \frac{156}{2652} \approx 0.0588$

Example 17

If you draw two cards from a deck, what is the probability that you will get the Ace of Diamonds and a black card?

You can satisfy this condition by having Case A or Case B, as follows:

Case A) you can get the Ace of Diamonds first and then a black card or
Case B) you can get a black card first and then the Ace of Diamonds.

Let's calculate the probability of Case A. The probability that the first card is the Ace of Diamonds is $\frac{1}{52}$. The probability that the second card is black given that the first card is the

Ace of Diamonds is $\frac{26}{51}$ because 26 of the remaining 51 cards are black. The probability is therefore $\frac{1}{52} \cdot \frac{26}{51} = \frac{1}{102}$.

Now for Case B: the probability that the first card is black is $\frac{26}{52} = \frac{1}{2}$. The probability that the second card is the Ace of Diamonds given that the first card is black is $\frac{1}{51}$. The probability of Case B is therefore $\frac{1}{2} \cdot \frac{1}{51} = \frac{1}{102}$, the same as the probability of Case 1.

Recall that the probability of A or B is $P(A) + P(B) - P(A \text{ and } B)$. In this problem, $P(A \text{ and } B) = 0$ since the first card cannot be the Ace of Diamonds and be a black card. Therefore, the probability of Case A or Case B is $\frac{1}{101} + \frac{1}{101} = \frac{2}{101}$. The probability that you will get the Ace of Diamonds and a black card when drawing two cards from a deck is $\frac{2}{101}$.

Try it Now 4
In your drawer, you have 10 pairs of socks, 6 of which are white. If you reach in and randomly grab two pairs of socks, what is the probability that both are white?

Example 18

A home pregnancy test was given to women, then pregnancy was verified through blood tests. The following table shows the home pregnancy test results. Find
a) $P(\text{not pregnant} \mid \text{positive test result})$
b) $P(\text{positive test result} \mid \text{not pregnant})$

	Positive test	Negative test	Total
Pregnant	70	4	74
Not Pregnant	5	14	19
Total	75	18	93

a) Since we know the test result was positive, we're limited to the 75 women in the first column, of which 5 were not pregnant. $P(\text{not pregnant} \mid \text{positive test result}) = \frac{5}{75} \approx 0.067$.

b) Since we know the woman is not pregnant, we are limited to the 19 women in the second row, of which 5 had a positive test. $P(\text{positive test result} \mid \text{not pregnant}) = \frac{5}{19} \approx 0.263$

The second result is what is usually called a false positive: A positive result when the woman is not actually pregnant.

10.3   Bayes Theorem

In this section, we concentrate on the more complex conditional probability problems we began looking at in the last section.

Example 19

Suppose a certain disease has an incidence rate of 0.1% (that is, it afflicts 0.1% of the population). A test has been devised to detect this disease. The test does not produce false negatives (that is, anyone who has the disease will test positive for it), but the false positive rate is 5% (that is, about 5% of people who take the test will test positive, even though they do not have the disease). Suppose a randomly selected person takes the test and tests positive. What is the probability that this person actually has the disease?

There are two ways to approach the solution to this problem. One involves an important result in probability theory called Bayes' theorem. We will discuss this theorem a bit later, but for now we will use an alternative and, we hope, much more intuitive approach.

Let's break down the information in the problem piece by piece.

Suppose a certain disease has an incidence rate of 0.1% (that is, it afflicts 0.1% of the population). The percentage 0.1% can be converted to a decimal number by moving the decimal place two places to the left, to get 0.001. In turn, 0.001 can be rewritten as a fraction: 1/1000. This tells us that about 1 in every 1000 people has the disease. (If we wanted we could write P(disease)=0.001.)

A test has been devised to detect this disease. The test does not produce false negatives (that is, anyone who has the disease will test positive for it). This part is fairly straightforward: everyone who has the disease will test positive, or alternatively everyone who tests negative does not have the disease. (We could also say P(positive | disease)=1.)

The false positive rate is 5% (that is, about 5% of people who take the test will test positive, even though they do not have the disease). This is even more straightforward. Another way of looking at it is that of every 100 people who are tested and do not have the disease, 5 will test positive even though they do not have the disease. (We could also say that P(positive | no disease)=0.05.)

Suppose a randomly selected person takes the test and tests positive. What is the probability that this person actually has the disease? Here we want to compute P(disease|positive). We already know that P(positive|disease)=1, but remember that conditional probabilities are not equal if the conditions are switched.

Rather than thinking in terms of all these probabilities we have developed, let's create a hypothetical situation and apply the facts as set out above. First, suppose we randomly select 1000 people and administer the test. How many do we expect to have the disease? Since about 1/1000 of all people are afflicted with the disease, 1/1000 of 1000 people is 1. (Now you know why we chose 1000.) Only 1 of 1000 test subjects actually has the disease; the other 999 do not.

We also know that 5% of all people who do not have the disease will test positive. There are 999 disease-free people, so we would expect (0.05)(999)=49.95 (so, about 50) people to test positive who do not have the disease.

Now back to the original question, computing *P*(disease|positive). There are 51 people who test positive in our example (the one unfortunate person who actually has the disease, plus the 50 people who tested positive but don't). Only one of these people has the disease, so

$$P(\text{disease} \mid \text{positive}) \approx \frac{1}{51} \approx 0.0196$$

or less than 2%. Does this surprise you? This means that of all people who test positive, over 98% *do not have the disease.*

The answer we got was slightly approximate, since we rounded 49.95 to 50. We could redo the problem with 100,000 test subjects, 100 of whom would have the disease and (0.05)(99,900)=4995 test positive but do not have the disease, so the exact probability of having the disease if you test positive is

$$P(\text{disease} \mid \text{positive}) \approx \frac{100}{5095} \approx 0.0196$$

which is pretty much the same answer.

But back to the surprising result. *Of all people who test positive, over 98% do not have the disease.* If your guess for the probability a person who tests positive has the disease was wildly different from the right answer (2%), don't feel bad. The exact same problem was posed to doctors and medical students at the Harvard Medical School 25 years ago and the results revealed in a 1978 *New England Journal of Medicine* article. Only about 18% of the participants got the right answer. Most of the rest thought the answer was closer to 95% (perhaps they were misled by the false positive rate of 5%).

So at least you should feel a little better that a bunch of doctors didn't get the right answer either (assuming you thought the answer was much higher). But the significance of this finding and similar results from other studies in the intervening years lies not in making math students feel better but in the possibly catastrophic consequences it might have for patient care. If a doctor thinks the chances that a positive test result nearly guarantees that a patient has a disease, they might begin an unnecessary and possibly harmful treatment regimen on a healthy patient. Or worse, as in the early days of the AIDS crisis when being HIV-positive was often equated with a death sentence, the patient might take a drastic action and commit suicide.

As we have seen in this hypothetical example, the most responsible course of action for treating a patient who tests positive would be to counsel the patient that they most likely do *not* have the disease and to order further, more reliable, tests to verify the diagnosis.

One of the reasons that the doctors and medical students in the study did so poorly is that such problems, when presented in the types of statistics courses that medical students often take, are solved by use of Bayes' theorem, which is stated as follows:

> **Bayes' Theorem**
> $$P(A \mid B) = \frac{P(A)P(B \mid A)}{P(A)P(B \mid A) + P(\overline{A})P(B \mid \overline{A})}$$

In our earlier example, this translates to

$$P(\text{disease} \mid \text{positive}) = \frac{P(\text{disease})P(\text{positive} \mid \text{disease})}{P(\text{disease})P(\text{positive} \mid \text{disease}) + P(\text{no disease})P(\text{positive} \mid \text{no disease})}$$

Plugging in the numbers gives

$$P(\text{disease} \mid \text{positive}) = \frac{(0.001)(1)}{(0.001)(1) + (0.999)(0.05)} \approx 0.0196$$

which is exactly the same answer as our original solution.

The problem is that you (or the typical medical student, or even the typical math professor) are much more likely to be able to remember the original solution than to remember Bayes' theorem. Psychologists, such as Gerd Gigerenzer, author of *Calculated Risks: How to Know When Numbers Deceive You*, have advocated that the method involved in the original solution (which Gigerenzer calls the method of "natural frequencies") be employed in place of Bayes' Theorem. Gigerenzer performed a study and found that those educated in the natural frequency method were able to recall it far longer than those who were taught Bayes' theorem. When one considers the possible life-and-death consequences associated with such calculations it seems wise to heed his advice.

Example 20

A certain disease has an incidence rate of 2%. If the false negative rate is 10% and the false positive rate is 1%, compute the probability that a person who tests positive actually has the disease.

Imagine 10,000 people who are tested. Of these 10,000, 200 will have the disease; 10% of them, or 20, will test negative and the remaining 180 will test positive. Of the 9800 who do not have the disease, 98 will test positive. So, of the 278 total people who test positive, 180 will have the disease. Thus,

$$P(\text{disease} \mid \text{positive}) = \frac{180}{278} \approx 0.647$$

so about 65% of the people who test positive will have the disease.

Using Bayes theorem directly would give the same result:

$$P(\text{disease} \mid \text{positive}) = \frac{(0.02)(0.90)}{(0.02)(0.90) + (0.98)(0.01)} = \frac{0.018}{0.0278} \approx 0.647$$

Try it Now 5

A certain disease has an incidence rate of 0.5%. If there are no false negatives and if the false positive rate is 3%, compute the probability that a person who tests positive actually has the disease.

10.4 Counting

Counting? You already know how to count or you wouldn't be taking a college-level math class, right? Well yes, but what we'll really be investigating here are ways of counting *efficiently*. When we get to the probability situations a bit later in this chapter, we will need to count some *very* large numbers, like the number of possible winning lottery tickets. One way to do this would be to write down every possible set of numbers that might show up on a lottery ticket, but believe me: you don't want to do this.

Basic Counting

We will start, however, with some more reasonable sorts of counting problems in order to develop the ideas that we will soon need.

Example 21

Suppose at a particular restaurant you have three choices for an appetizer (soup, salad or breadsticks) and five choices for a main course (hamburger, sandwich, quiche, fajita or pizza). If you are allowed to choose exactly one item from each category for your meal, how many different meal options do you have?

Solution 1: One way to solve this problem would be to systematically list each possible meal:

soup + hamburger	soup + sandwich	soup + quiche
soup + fajita	soup + pizza	salad + hamburger
salad + sandwich	salad + quiche	salad + fajita
salad + pizza	breadsticks + hamburger	breadsticks + sandwich
breadsticks + quiche	breadsticks + fajita	breadsticks + pizza

Assuming that we did this systematically and that we neither missed any possibilities nor listed any possibility more than once, the answer would be 15. Thus, you could go to the restaurant 15 nights in a row and have a different meal each night.

Solution 2: Another way to solve this problem would be to list all the possibilities in a table:

	hamburger	sandwich	quiche	fajita	pizza
soup	soup+burger				
salad	salad+burger				
bread	*etc.*				

In each of the cells in the table we could list the corresponding meal: soup + hamburger in the upper left corner, salad + hamburger below it, etc. But if we didn't really care *what* the possible meals are, only *how many* possible meals there are, we could just count the number of cells and arrive at an answer of 15, which matches our answer from the first solution. (It's always good when you solve a problem two different ways and get the same answer!)

Solution 3: We already have two perfectly good solutions. Why do we need a third? The first method was not very systematic, and we might easily have made an omission. The second method was better, but suppose that in addition to the appetizer and the main course we further complicated the problem by adding desserts to the menu: we've used the rows of the table for the appetizers and the columns for the main courses—where will the desserts go? We would need a third dimension, and since drawing 3-D tables on a 2-D page or computer screen isn't terribly easy, we need a better way in case we have three categories to choose form instead of just two.

So, back to the problem in the example. What else can we do? Let's draw a **tree diagram**:

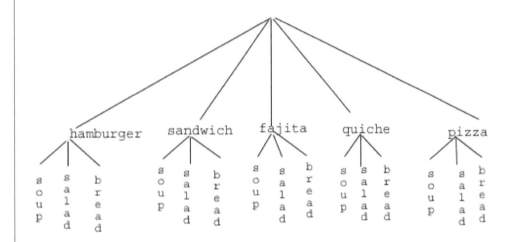

This is called a "tree" diagram because at each stage we branch out, like the branches on a tree. In this case, we first drew five branches (one for each main course) and then for each of those branches we drew three more branches (one for each appetizer). We count the number of branches at the final level and get (surprise, surprise!) 15.

If we wanted, we could instead draw three branches at the first stage for the three appetizers and then five branches (one for each main course) branching out of each of those three branches.

Ok, so now we know how to count possibilities using tables and tree diagrams. These methods will continue to be useful in certain cases, but imagine a game where you have two decks of cards (with 52 cards in each deck) and you select one card from each deck. Would you really want to draw a table or tree diagram to determine the number of outcomes of this game?

Let's go back to the previous example that involved selecting a meal from three appetizers and five main courses, and look at the second solution that used a table. Notice that one way to count the number of possible meals is simply to number each of the appropriate cells in the table, as we have done above. But another way to count the number of cells in the table would be multiply the number of rows (3) by the number of columns (5) to get 15. Notice that we could have arrived at the same result without making a table at all by simply multiplying the number of choices for the appetizer (3) by the number of choices for the main course (5). We generalize this technique as the *basic counting rule*:

> **Basic Counting Rule**
>
> If we are asked to choose one item from each of two separate categories where there are m items in the first category and n items in the second category, then the total number of available choices is $m \cdot n$.
>
> This is sometimes called the multiplication rule for probabilities.

Example 22

There are 21 novels and 18 volumes of poetry on a reading list for a college English course. How many different ways can a student select one novel and one volume of poetry to read during the quarter?

There are 21 choices from the first category and 18 for the second, so there are $21 \cdot 18 = 378$ possibilities.

The Basic Counting Rule can be extended when there are more than two categories by applying it repeatedly, as we see in the next example.

Example 23

Suppose at a particular restaurant you have three choices for an appetizer (soup, salad or breadsticks), five choices for a main course (hamburger, sandwich, quiche, fajita or pasta) and two choices for dessert (pie or ice cream). If you are allowed to choose exactly one item from each category for your meal, how many different meal options do you have?

There are 3 choices for an appetizer, 5 for the main course and 2 for dessert, so there are $3 \cdot 5 \cdot 2 = 30$ possibilities.

Example 24

A quiz consists of 3 true-or-false questions. In how many ways can a student answer the quiz?

There are 3 questions. Each question has 2 possible answers (true or false), so the quiz may be answered in $2 \cdot 2 \cdot 2 = 8$ different ways. Recall that another way to write $2 \cdot 2 \cdot 2$ is 2^3, which is much more compact.

Try it Now 6

Suppose at a particular restaurant you have eight choices for an appetizer, eleven choices for a main course and five choices for dessert. If you are allowed to choose exactly one item from each category for your meal, how many different meal options do you have?

Permutations

In this section, we will develop an even faster way to solve some of the problems we have already learned to solve by other means. Let's start with a couple examples.

Example 25

> How many different ways can the letters of the word MATH be rearranged to form a four-letter code word?
>
> This problem is a bit different. Instead of choosing one item from each of several different categories, we are repeatedly choosing items from the *same* category (the category is: the letters of the word MATH) and each time we choose an item we *do not replace* it, so there is one fewer choice at the next stage: we have 4 choices for the first letter (say we choose A), then 3 choices for the second (M, T and H; say we choose H), then 2 choices for the next letter (M and T; say we choose M) and only one choice at the last stage (T). Thus, there are $4 \cdot 3 \cdot 2 \cdot 1 = 24$ ways to spell a code worth with the letters MATH.

In this example, we needed to calculate $n \cdot (n-1) \cdot (n-2) \cdots 3 \cdot 2 \cdot 1$. This calculation shows up often in mathematics, and is called the **factorial**, and is notated $n!$

> **Factorial**
> A factorial is when we take a positive integer and find the product of all descending positive integers, including itself, all the way to 1:
>
> $$n! = n \cdot (n-1) \cdot (n-2) \cdots 3 \cdot 2 \cdot 1$$
>
> We say $n!$ as "n factorial." E.g., given $5!$, then we say "5 factorial."

Example 26

> How many ways can five different door prizes be distributed among five people?
>
> There are 5 choices of prize for the first person, 4 choices for the second, and so on. The number of ways the prizes can be distributed will be $5! = 5 \cdot 4 \cdot 3 \cdot 2 \cdot 1 = 120$ ways.

Now we will consider some slightly different examples.

Example 27

> A charity benefit is attended by 25 people and three gift certificates are given away as door prizes: one gift certificate is in the amount of $100, the second is worth $25 and the third is worth $10. Assuming that no person receives more than one prize, how many different ways can the three gift certificates be awarded?
>
> Using the Basic Counting Rule, there are 25 choices for the person who receives the $100 certificate, 24 remaining choices for the $25 certificate and 23 choices for the $10 certificate, so there are $25 \cdot 24 \cdot 23 = 13,800$ ways in which the prizes can be awarded.

Example 28

> Eight sprinters have made it to the Olympic finals in the 100-meter race. In how many different ways can the gold, silver and bronze medals be awarded?

Using the Basic Counting Rule, there are 8 choices for the gold medal winner, 7 remaining choices for the silver, and 6 for the bronze, so there are $8 \cdot 7 \cdot 6 = 336$ ways the three medals can be awarded to the 8 runners.

Note that in these preceding examples, the gift certificates and the Olympic medals were awarded *without replacement*; that is, once we have chosen a winner of the first door prize or the gold medal, they are not eligible for the other prizes. Thus, at each succeeding stage of the solution there is one fewer choice (25, then 24, then 23 in the first example; 8, then 7, then 6 in the second). Contrast this with the situation of a multiple-choice test, where there might be five possible answers — A, B, C, D or E — for each question on the test.

Note also that *the order of selection was important* in each example: for the three door prizes, being chosen first means that you receive substantially more money; in the Olympics example, coming in first means that you get the gold medal instead of the silver or bronze. In each case, if we had chosen the same three people in a different order there might have been a different person who received the $100 prize, or a different gold medalist. (Contrast this with the situation where we might draw three names out of a hat to each receive a $10 gift certificate; in this case, the order of selection is *not* important since each of the three people receive the same prize. Situations where the order is *not* important will be discussed in the next section.)

We can generalize the situation in the two examples above to any problem *without replacement* where the *order of selection is important*. If we are arranging in order r items out of n possibilities (instead of 3 out of 25 or 3 out of 8 as in the previous examples), the number of possible arrangements will be given by

$$n \cdot (n-1) \cdot (n-2) \cdots (n-r+1)$$

If you don't see why $(n-r+1)$ is the right number to use for the last factor, just think back to the first example in this section, where we calculated $25 \cdot 24 \cdot 23$ to get 13,800. In this case $n = 25$ and $r = 3$, so $n - r + 1 = 25 - 3 + 1 = 23$, which is exactly the right number for the final factor.

Now, why would we want to use this complicated formula when it's actually easier to use the Basic Counting Rule, as we did in the first two examples? Well, we won't actually use this formula all that often, we only developed it so that we could attach a special notation and a special definition to this situation where we are choosing r items out of n possibilities *without replacement* and where the *order of selection is important*.

Permutations

$$_nP_r = n \cdot (n-1) \cdot (n-2) \cdots (n-r+1)$$

We say that there are $_nP_r$ **permutations** of size r that may be selected from among n choices *without replacement* when *order matters*.

It turns out that we can express this result more simply using factorials.

$$_nP_r = \frac{n!}{(n-r)!}$$

In practicality, we usually use technology rather than factorials or repeated multiplication to compute permutations.

Example 29

I have nine paintings and have room to display only four of them at a time on my wall. How many different ways could I do this?

Since we are choosing 4 paintings out of 9 *without replacement* where the *order of selection is important* there are $_9P_4 = 9 \cdot 8 \cdot 7 \cdot 6 = 3{,}024$ permutations.

Example 30

How many ways can a four-person executive committee (president, vice-president, secretary, treasurer) be selected from a 16-member board of directors of a non-profit organization?

We want to choose 4 people out of 16 without replacement and where the order of selection is important. So, the answer is $_{16}P_4 = 16 \cdot 15 \cdot 14 \cdot 13 = 43{,}680$.

Try it Now 7
How many 5 character passwords can be made using the letters A through Z
a. if repeats are allowed
b. if no repeats are allowed

Combinations

In the previous section, we considered the situation where we chose r items out of n possibilities *without replacement* and where the *order of selection was important*. We now consider a similar situation in which the order of selection is *not* important.

Example 31

A charity benefit is attended by 25 people at which three $50 gift certificates are given away as door prizes. Assuming no person receives more than one prize, how many different ways can the gift certificates be awarded?

Using the Basic Counting Rule, there are 25 choices for the first person, 24 remaining choices for the second person and 23 for the third, so there are $25 \cdot 24 \cdot 23 = 13{,}800$ ways to choose three people. Suppose for a moment that Abe is chosen first, Bea second and Cindy third; this is one of the 13,800 possible outcomes. Another way to award the prizes would be to choose Abe first, Cindy second and Bea third; this is another of the 13,800 possible outcomes. But either way Abe, Bea and Cindy each get $50, so it doesn't really matter the order in which we select them. In how many different orders can Abe, Bea and Cindy be selected? It turns out there are 6:

ABC   ACB   BAC   BCA   CAB   CBA

How can we be sure that we have counted them all? We are really just choosing 3 people out of 3, so there are $3 \cdot 2 \cdot 1 = 6$ ways to do this; we didn't really need to list them all, we can just use permutations!

So, out of the 13,800 ways to select 3 people out of 25, six of them involve Abe, Bea and Cindy. The same argument works for any other group of three people (say Abe, Bea and David or Frank, Gloria and Hildy) so each three-person group is counted *six times*. Thus the 13,800 figure is six times too big. The number of distinct three-person groups will be $13,800/6 = 2300$.

We can generalize the situation in this example above to any problem of choosing a collection of items *without replacement* where the *order of selection is **not** important*. If we are choosing *r* items out of *n* possibilities (instead of 3 out of 25 as in the previous examples), the number of possible choices will be given by $\dfrac{{}_nP_r}{{}_rP_r}$, and we could use this formula for computation. However, this situation arises so frequently that we attach a special notation and a special definition to this situation where we are choosing *r* items out of *n* possibilities *without replacement* where the *order of selection is **not** important*.

> **Combinations**
>
> $${}_nC_r = \dfrac{{}_nP_r}{{}_rP_r}$$
>
> We say that there are ${}_nC_r$ **combinations** of size *r* that may be selected from among *n* choices *without replacement* where *order doesn't matter*.
>
> We can also write the combinations formula in terms of factorials:
>
> $${}_nC_r = \dfrac{n!}{(n-r)!r!}$$

Example 32

A group of four students is to be chosen from a 35-member class to represent the class on the student council. How many ways can this be done?

Since we are choosing 4 people out of 35 *without replacement* where the *order of selection is **not** important* there are ${}_{35}C_4 = \dfrac{35 \cdot 34 \cdot 33 \cdot 32}{4 \cdot 3 \cdot 2 \cdot 1} = 52{,}360$ combinations.

Try it Now 8
The United States Senate Appropriations Committee consists of 29 members; the Defense Subcommittee of the Appropriations Committee consists of 19 members. Disregarding party affiliation or any special seats on the Subcommittee, how many different 19-member subcommittees may be chosen from among the 29 Senators on the Appropriations Committee?

In the preceding Try it Now problem we assumed that the 19 members of the Defense Subcommittee were chosen without regard to party affiliation. In reality this would never happen: if Republicans are in the majority they would never let a majority of Democrats sit on (and thus control) any subcommittee. (The same of course would be true if the Democrats were in control.) So, let's consider the example again, in a slightly more complicated form.

Example 33

The United States Senate Appropriations Committee consists of 29 members, 15 Republicans and 14 Democrats. The Defense Subcommittee consists of 19 members, 10 Republicans and 9 Democrats. How many different ways can the members of the Defense Subcommittee be chosen from among the 29 Senators on the Appropriations Committee?

In this case, we need to choose 10 of the 15 Republicans and 9 of the 14 Democrats. There are $_{15}C_{10} = 3003$ ways to choose the 10 Republicans and $_{14}C_9 = 2002$ ways to choose the 9 Democrats. But now what? How do we finish the problem?

Suppose we listed all of the possible 10-member Republican groups on 3003 slips of red paper and all of the possible 9-member Democratic groups on 2002 slips of blue paper. How many ways can we choose one red slip and one blue slip? This is a job for the Basic Counting Rule! We are simply making one choice from the first category and one choice from the second category, just like in the restaurant menu problems from earlier.

There must be $3003 \cdot 2002 = 6,012,006$ possible ways of selecting the members of the Defense Subcommittee.

Probability using Permutations and Combinations

We can use permutations and combinations to help us answer more complex probability questions

Example 34

A 4-digit PIN number is selected. What is the probability that there are no repeated digits?

There are 10 possible values for each digit of the PIN (namely: 0, 1, 2, 3, 4, 5, 6, 7, 8, 9), so there are $10 \cdot 10 \cdot 10 \cdot 10 = 10^4 = 10,000$ total possible PIN numbers.

To have no repeated digits, all four digits would have to be different, which is selecting without replacement. We could either compute $10 \cdot 9 \cdot 8 \cdot 7$, or notice that this is the same as the permutation $_{10}P_4 = 5040$.

The probability of no repeated digits is the number of 4-digit PIN numbers with no repeated digits divided by the total number of 4-digit PIN numbers. This probability is

$$\frac{_{10}P_4}{10^4} = \frac{5040}{10000} = 0.504$$

Hence, there is a 50.4% chance that the 4-digit PIN has no repeated digits.

Example 35

In a certain state's lottery, 48 balls numbered 1 through 48 are placed in a machine and six of them are drawn at random. If the six numbers drawn match the numbers that a player had chosen, the player wins $1,000,000. In this lottery, the order the numbers are drawn in doesn't matter. Compute the probability that you win the million-dollar prize if you purchase a single lottery ticket.

In order to compute the probability, we need to count the total number of ways six numbers can be drawn, and the number of ways the six numbers on the player's ticket could match the six numbers drawn from the machine. Since there is no stipulation that the numbers be in any particular order, the number of possible outcomes of the lottery drawing is $_{48}C_6 =$ 12,271,512. Of these possible outcomes, only one would match all six numbers on the player's ticket, so the probability of winning the grand prize is:

$$\frac{_6C_6}{_{48}C_6} = \frac{1}{12271512} \approx 0.0000000815$$

Thus, there is less than a 1% chance of winning the grand prize.

Example 36

In the state lottery from the previous example, if five of the six numbers drawn match the numbers that a player has chosen, the player wins a second prize of $1,000. Compute the probability that you win the second prize if you purchase a single lottery ticket.

As above, the number of possible outcomes of the lottery drawing is $_{48}C_6 =$ 12,271,512. In order to win the second prize, five of the six numbers on the ticket must match five of the six winning numbers; in other words, we must have chosen five of the six winning numbers and one of the 42 losing numbers. The number of ways to choose 5 out of the 6 winning numbers is given by $_6C_5 = 6$ and the number of ways to choose 1 out of the 42 losing numbers is given by $_{42}C_1 = 42$. Thus, the number of favorable outcomes is then given by the Basic Counting Rule: $_6C_5 \cdot {}_{42}C_1 = 6 \cdot 42 = 252$. So, the probability of winning the second prize is.

$$\frac{\left(_6C_5\right)\left(_{42}C_1\right)}{_{48}C_6} = \frac{252}{12271512} \approx 0.0000205$$

This means there is less than a 1% chance of winning second prize. Wow! We now can see why some people call it "luck" when winning the lottery because the chances of winning are so low.

Try it Now 9

A multiple-choice question on an economics quiz contains 10 questions with five possible answers each. Compute the probability of randomly guessing the answers and getting 9 questions correct.

Example 37

Compute the probability of randomly drawing five cards from a deck and getting exactly one Ace.

In many card games (such as poker) the order in which the cards are drawn is not important (since the player may rearrange the cards in his hand any way he chooses); in the problems that follow, we will assume that this is the case unless otherwise stated. Thus, we use combinations to compute the possible number of 5-card hands, $_{52}C_5$. This number will go in the denominator of our probability formula, since it is the number of possible outcomes.

For the numerator, we need the number of ways to draw one Ace and four other cards (none of them Aces) from the deck. Since there are four Aces and we want exactly one of them, there will be $_4C_1$ ways to select one Ace; since there are 48 non-Aces and we want 4 of them, there will be $_{48}C_4$ ways to select the four non-Aces. Now we use the Basic Counting Rule to calculate that there will be $_4C_1 \cdot {_{48}C_4}$ ways to choose one ace and four non-Aces.

Putting this all together, we have

$$P(\text{one Ace}) = \frac{\left(_4C_1\right)\left(_{48}C_4\right)}{_{52}C_5} = \frac{778320}{2598960} \approx 0.299$$

Thus, there is about a 30% chance of drawing exactly one Ace in a hand of 5 cards.

Example 38

Compute the probability of randomly drawing five cards from a deck and getting exactly two Aces.

The solution is similar to the previous example, except now we are choosing 2 Aces out of 4 and 3 non-Aces out of 48; the denominator remains the same:

$$P(\text{two Aces}) = \frac{\left(_4C_2\right)\left(_{48}C_3\right)}{_{52}C_5} = \frac{103776}{2598960} \approx 0.0399$$

There is about a 4% chance of drawing 2 Aces in a hand of 5 cards. Notice from example 37, the probability of drawing exactly one Ace is much higher that drawing two.

It is useful to note that these card problems are remarkably similar to the lottery problems discussed earlier.

Try it Now 10

Compute the probability of randomly drawing five cards from a deck of cards and getting three Aces and two Kings.

Birthday Problem

Let's take a pause to consider a famous problem in probability theory:

> Suppose you have a room full of 30 people. What is the probability that there is at least one shared birthday?

Take a guess at the answer to the above problem. Was your guess fairly low, like around 10%? That seems to be the intuitive answer (30/365, perhaps?). Let's see if we should listen to our intuition. Let's start with a simpler problem, however.

Example 39

Suppose three people are in a room. What is the probability that there is at least one shared birthday among these three people?

There are a lot of ways there could be at least one shared birthday. Fortunately, there is an easier way. We ask ourselves "What is the alternative to having at least one shared birthday?" In this case, the alternative is that there are **no** shared birthdays. In other words, the alternative to "at least one" is having **none**. In other words, since this is a complementary event,

P(at least one) = 1 – P(none)

We will start, then, by computing the probability that there is no shared birthday. Let's imagine that you are one of these three people. Your birthday can be anything without conflict, so there are 365 choices out of 365 for your birthday. What is the probability that the second person does not share your birthday? There are 365 days in the year (let's ignore leap years) and removing your birthday from contention, there are 364 choices that will guarantee that you do not share a birthday with this person, so the probability that the second person does not share your birthday is 364/365. Now we move to the third person. What is the probability that this third person does not have the same birthday as either you or the second person? There are 363 days that will not duplicate your birthday or the second person's, so the probability that the third person does not share a birthday with the first two is 363/365.

We want the second person not to share a birthday with you *and* the third person not to share a birthday with the first two people, so we use the multiplication rule:

$$P(\text{no shared birthday}) = \frac{365}{365} \cdot \frac{364}{365} \cdot \frac{363}{365} \approx 0.9918$$

and then subtract from 1 to get

P(shared birthday) = 1 – P(no shared birthday) = 1 – 0.9918 = 0.0082. This means there is less than a 1% chance that a person in a room of 3 has at least one shared birthday.

This is a pretty small number, so maybe it makes sense that the answer to our original problem will be small. Let's make our group a bit bigger.

Example 40

Suppose five people are in a room. What is the probability that there is at least one shared birthday among these five people?

Continuing the pattern of the previous example, the answer should be

$$P(\text{shared birthday}) = 1 - \frac{365}{365} \cdot \frac{364}{365} \cdot \frac{363}{365} \cdot \frac{362}{365} \cdot \frac{361}{365} \approx 0.0271$$

Note that we could rewrite this more compactly as

$$P(\text{shared birthday}) = 1 - \frac{_{365}P_5}{365^5} \approx 0.0271$$

which makes it a bit easier to type into a calculator or computer, and which suggests a nice formula as we continue to expand the population of our group.

Example 41

Suppose 30 people are in a room. What is the probability that there is at least one shared birthday among these 30 people?

Here we can calculate

$$P(\text{shared birthday}) = 1 - \frac{_{365}P_{30}}{365^{30}} \approx 0.706$$

which gives us the surprising result that when you are in a room with 30 people there is a 70% chance that there will be at least one shared birthday.

If you like to bet, and if you can convince 30 people to reveal their birthdays, you might be able to win some money by betting a friend that there will be at least two people with the same birthday in the room anytime you are in a room of 30 or more people. (Of course, you would need to make sure your friend hasn't studied probability!) You wouldn't be guaranteed to win, but you should win more than half the time.

This is one of many results in probability theory that is counterintuitive; that is, it goes against our gut instincts. If you still don't believe the math, you can carry out a simulation. Just so you won't have to go around rounding up groups of 30 people, someone has kindly developed a Java applet so that you can conduct a computer simulation. Go to this web page: http://www-stat.stanford.edu/~susan/surprise/Birthday.html, and once the applet has loaded, select 30 birthdays and then keep clicking Start and Reset. If you keep track of the number of times that there is a repeated birthday, you should get a repeated birthday about 7 out of every 10 times you run the simulation.

Try it Now 11

Suppose 10 people are in a room. What is the probability that there is at least one shared birthday among these 10 people?

10.5 Expected Value

Expected value is perhaps the most useful probability concept we will discuss. It has many applications, from insurance policies to making financial decisions, and it's one thing that the casinos and government agencies that run gambling operations and lotteries hope most people never learn about.

Example 42

[1]In the casino game roulette, a wheel with 38 spaces (18 red, 18 black, and 2 green) is spun. In one possible bet, the player bets $1 on a single number. If that number is spun on the wheel, then they receive $36 (their original $1 + $35). Otherwise, they lose their $1. On average, how much money should a player expect to win or lose if they play this game repeatedly?

Suppose you bet $1 on each of the 38 spaces on the wheel, for a total of $38 bet. When the winning number is spun, you are paid $36 on that number. While you won on that one number, overall, you've lost $2. On a per-space basis, you have "won" -$2/$38 ≈ -$0.053. In other words, on average you lose 5.3 cents per space you bet on.

We call this average gain or loss the expected value of playing roulette. Notice that no one ever loses exactly 5.3 cents: most people (in fact, about 37 out of every 38) lose $1 and a very few people (about 1 person out of every 38) gain $35 (the $36 they win minus the $1 they spent to play the game).

There is another way to compute expected value without imagining what would happen if we play every possible space. There are 38 possible outcomes when the wheel spins, so the probability of winning is $\frac{1}{38}$. The complement, the probability of losing, is $\frac{37}{38}$.

Summarizing these along with the values, we get this table:

Outcome	Probability of outcome
$35	$\frac{1}{38}$
-$1	$\frac{37}{38}$

Notice that if we multiply each outcome by its corresponding probability we get

$\$35 \cdot \frac{1}{38} = 0.9211$ and $-\$1 \cdot \frac{37}{38} = -0.9737$, and if we add these numbers, we get

$0.9211 + (-0.9737) \approx -0.053$, which is the expected value we computed above.

Expected Value

Expected Value is the average gain or loss of an event if the procedure is repeated many times.

We can compute the expected value by multiplying each outcome by the probability of that outcome, then adding up the products.

Try it Now 12

You purchase a raffle ticket to help out a charity. The raffle ticket costs $5. The charity is selling 2000 tickets. One of them will be drawn and the person holding the ticket will be given a prize worth $4000. Compute the expected value for this raffle.

Example 43
--

In a certain state's lottery, 48 balls numbered 1 through 48 are placed in a machine and six of them are drawn at random. If the six numbers drawn match the numbers that a player had chosen, the player wins $1,000,000. If they match 5 numbers, then win $1,000. It costs $1 to buy a ticket. Find the expected value.

Earlier, we calculated the probability of matching all 6 numbers and the probability of matching 5 numbers:

$$\frac{_6C_6}{_{48}C_6} = \frac{1}{12271512} \approx 0.0000000815 \text{ for all 6 numbers,}$$

$$\frac{\left(_6C_5\right)\left(_{42}C_1\right)}{_{48}C_6} = \frac{252}{12271512} \approx 0.0000205 \text{ for 5 numbers.}$$

Our probabilities and outcome values are

Outcome	Probability of outcome
$999,999	$\dfrac{1}{12271512}$
$999	$\dfrac{252}{12271512}$
-$1	$1 - \dfrac{253}{12271512} = \dfrac{12271259}{12271512}$

The expected value is

$$(\$999,999) \cdot \frac{1}{12271512} + (\$999) \cdot \frac{252}{12271512} + (-\$1) \cdot \frac{12271259}{12271512} \approx -\$0.898$$

On average, one can expect to lose about 90 cents on a lottery ticket. Of course, most players will lose $1.

In general, if the expected value of a game is negative, it is not a good idea to play the game, since on average you will lose money. It would be better to play a game with a positive expected value (good luck trying to find one!), although keep in mind that even if the *average* winnings are positive it could be the case that most people lose money and one very fortunate individual wins a great deal of money. If the expected value of a game is 0, we call it a **fair game**, since neither side has an advantage.

Not surprisingly, the expected value for casino games is negative for the player, which is positive for the casino. It must be positive or they would go out of business. Players just need to keep in mind that when they play a game repeatedly, their expected value is negative. That is fine so long as you enjoy playing the game and think it is worth the cost. But it would be wrong to expect to come out ahead.

Try it Now 13
A friend offers to play a game, in which you roll 3 standard 6-sided dice. If all the dice roll different values, you give him $1. If any two dice match values, you get $2. What is the expected value of this game? Would you play?

Expected value also has applications outside of gambling. Expected value is very common in making insurance decisions.

Example 44

A 40-year-old man in the U.S. has a 0.242% risk of dying during the next year[2]. An insurance company charges $275 for a life-insurance policy that pays a $100,000 death benefit. What is the expected value for the person buying the insurance?

The probabilities and outcomes are

Outcome	Probability of outcome
$100,000 - $275 = $99,725	0.00242

The expected value is ($99,725)(0.00242) + (-$275)(0.99758) = -$33.

Not surprisingly, the expected value is negative; the insurance company can only afford to offer policies if they, on average, make money on each policy. They can afford to pay out the occasional benefit because they offer enough policies that those benefit payouts are balanced by the rest of the insured people.

For people buying the insurance, there is a negative expected value, but there is a security that comes from insurance that is worth that cost.

[2] According to the estimator at http://www.numericalexample.com/index.php?view=article&id=91

Try it Now Answers

1. There are 60 possible readings, from 00 to 59. a. $\dfrac{1}{60}$ b. $\dfrac{16}{60}$ (counting 00 through 15)

2. Since the second draw is made after replacing the first card, these events are independent.
   The probability of an ace on each draw is $\dfrac{4}{52} = \dfrac{1}{13}$, so the probability of an Ace on both
   draws is $\dfrac{1}{13} \cdot \dfrac{1}{13} = \dfrac{1}{169}$

3. P(white sock and white tee) = $\dfrac{6}{10} \cdot \dfrac{3}{7} = \dfrac{9}{35}$

   P(white sock or white tee) = $\dfrac{6}{10} + \dfrac{3}{7} - \dfrac{9}{35} = \dfrac{27}{35}$

4. a. $\dfrac{6}{10} \cdot \dfrac{5}{9} = \dfrac{30}{90} = \dfrac{1}{3}$

5. Out of 100,000 people, 500 would have the disease. Of those, all 500 would test positive.
   Of the 99,500 without the disease, 2,985 would falsely test positive and the other 96,515
   would test negative.
   P(disease | positive) = $\dfrac{500}{500 + 2985} = \dfrac{500}{3485} \approx 14.3\%$

6. $8 \cdot 11 \cdot 5 = 440$ menu combinations

7. There are 26 characters. a. $26^5 = 11{,}881{,}376$. b. $_{26}P_5 = 26 \cdot 25 \cdot 24 \cdot 23 \cdot 22 = 7{,}893{,}600$

8. Order does not matter. $_{29}C_{19} = 20{,}030{,}010$ possible subcommittees

9. There are $5^{10} = 9{,}765{,}625$ different ways the exam can be answered. There are 9 possible
   locations for the one missed question, and in each of those locations there are 4 wrong
   answers, so there are 36 ways the test could be answered with one wrong answer.
   P(9 answers correct) = $\dfrac{36}{5^{10}} \approx 0.0000037$ chance

10. P(three Aces and two Kings) = $\dfrac{\left(_4C_3\right)\left(_4C_2\right)}{_{52}C_5} = \dfrac{24}{2598960} \approx 0.0000092$

11. P(shared birthday) = $1 - \dfrac{_{365}P_{10}}{365^{10}} \approx 0.117$

12. $(\$3,995) \cdot \dfrac{1}{2000} + (-\$5) \cdot \dfrac{1999}{2000} \approx -\$3.00$

13. Suppose you roll the first die. The probability the second will be different is $\dfrac{5}{6}$. The probability that the third roll is different than the previous two is $\dfrac{4}{6}$, so the probability that the three dice are different is $\dfrac{5}{6} \cdot \dfrac{4}{6} = \dfrac{20}{36}$. The probability that two dice will match is the complement, $1 - \dfrac{20}{36} = \dfrac{16}{36}$.

The expected value is: $(\$2) \cdot \dfrac{16}{36} + (-\$1) \cdot \dfrac{20}{36} = \dfrac{12}{36} \approx \$0.33$. Yes, it is in your advantage to play. On average, you'd win $0.33 per play.

Exercises

1. A ball is drawn randomly from a jar that contains 6 red balls, 2 white balls, and 5 yellow balls. Find the probability of the given event.
    a. A red ball is drawn
    b. A white ball is drawn

2. Suppose you write each letter of the alphabet on a different slip of paper and put the slips into a hat. What is the probability of drawing one slip of paper from the hat at random and getting:
    a. A consonant
    b. A vowel

3. A group of people were asked if they had run a red light in the last year. 150 responded "yes", and 185 responded "no." Find the probability that if a person is chosen at random, they have run a red light in the last year.

4. In a survey, 205 people indicated they prefer cats, 160 indicated they prefer dots, and 40 indicated they don't enjoy either pet. Find the probability that if a person is chosen at random, they prefer cats.

5. Compute the probability of tossing a six-sided die (with sides numbered 1 through 6) and getting a 5.

6. Compute the probability of tossing a six-sided die and getting a 7.

7. Giving a test to a group of students, the grades and gender are summarized below. If one student was chosen at random, find the probability that the student was female.

	A	B	C	Total
Male	8	18	13	39

Female	10	4	12	26
Total	18	22	25	65

8. The table below shows the number of credit cards owned by a group of individuals. If one person was chosen at random, find the probability that the person had no credit cards.

	Zero	One	Two or more	Total
Male	9	5	19	33
Female	18	10	20	48
Total	27	15	39	81

9. Compute the probability of tossing a six-sided die and getting an even number.

10. Compute the probability of tossing a six-sided die and getting a number less than 3.

11. If you pick one card at random from a standard deck of cards, what is the probability it will be a King?

12. If you pick one card at random from a standard deck of cards, what is the probability it will be a Diamond?

13. Compute the probability of rolling a 12-sided die and getting a number other than 8.

14. If you pick one card at random from a standard deck of cards, what is the probability it is not the Ace of Spades?

15. Referring to the grade table from question #7, what is the probability that a student chosen at random did NOT earn a C?

16. Referring to the credit card table from question #8, what is the probability that a person chosen at random has at least one credit card?

17. A six-sided die is rolled twice. What is the probability of showing a 6 on both rolls?

18. A fair coin is flipped twice. What is the probability of showing heads on both flips?

19. A die is rolled twice. What is the probability of showing a 5 on the first roll and an even number on the second roll?

20. Suppose that 21% of people own dogs. If you pick two people at random, what is the probability that they both own a dog?

21. Suppose a jar contains 17 red marbles and 32 blue marbles. If you reach in the jar and pull out 2 marbles at random, find the probability that both are red.

22. Suppose you write each letter of the alphabet on a different slip of paper and put the slips into a hat. If you pull out two slips at random, find the probability that both are vowels.

23. Bert and Ernie each have a well-shuffled standard deck of 52 cards. They each draw one card from their own deck. Compute the probability that:
    a. Bert and Ernie both draw an Ace.
    b. Bert draws an Ace but Ernie does not.
    c. neither Bert nor Ernie draws an Ace.
    d. Bert and Ernie both draw a heart.
    e. Bert gets a card that is not a Jack and Ernie draws a card that is not a heart.

24. Bert has a well-shuffled standard deck of 52 cards, from which he draws one card; Ernie has a 12-sided die, which he rolls at the same time Bert draws a card. Compute the probability that:
    a. Bert gets a Jack and Ernie rolls a five.
    b. Bert gets a heart and Ernie rolls a number less than six.
    c. Bert gets a face card (Jack, Queen or King) and Ernie rolls an even number.
    d. Bert gets a red card and Ernie rolls a fifteen.
    e. Bert gets a card that is not a Jack and Ernie rolls a number that is not twelve.

25. Compute the probability of drawing a King from a deck of cards and then drawing a Queen.

26. Compute the probability of drawing two spades from a deck of cards.

27. A math class consists of 25 students, 14 female and 11 male. Two students are selected at random to participate in a probability experiment. Compute the probability that
    a. a male is selected, then a female.
    b. a female is selected, then a male.
    c. two males are selected.
    d. two females are selected.
    e. no males are selected.

28. A math class consists of 25 students, 14 female and 11 male. Three students are selected at random to participate in a probability experiment. Compute the probability that
    a. a male is selected, then two females.
    b. a female is selected, then two males.
    c. two females are selected, then one male.
    d. three males are selected.
    e. three females are selected.

29. Giving a test to a group of students, the grades and gender are summarized below. If one student was chosen at random, find the probability that the student was female and earned an A.

	A	B	C	Total
Male	8	18	13	39
Female	10	4	12	26
Total	18	22	25	65

30. The table below shows the number of credit cards owned by a group of individuals. If one person was chosen at random, find the probability that the person was male and had two or more credit cards.

	Zero	One	Two or more	Total
Male	9	5	19	33
Female	18	10	20	48
Total	27	15	39	81

31. A jar contains 6 red marbles numbered 1 to 6 and 8 blue marbles numbered 1 to 8. A marble is drawn at random from the jar. Find the probability the marble is red or odd-numbered.

32. A jar contains 4 red marbles numbered 1 to 4 and 10 blue marbles numbered 1 to 10. A marble is drawn at random from the jar. Find the probability the marble is blue or even-numbered.

33. Referring to the table from #29, find the probability that a student chosen at random is female or earned a B.

34. Referring to the table from #30, find the probability that a person chosen at random is male or has no credit cards.

35. Compute the probability of drawing the King of hearts or a Queen from a deck of cards.

36. Compute the probability of drawing a King or a heart from a deck of cards.

37. A jar contains 5 red marbles numbered 1 to 5 and 8 blue marbles numbered 1 to 8. A marble is drawn at random from the jar. Find the probability the marble is
    a. Even-numbered given that the marble is red.
    b. Red given that the marble is even-numbered.

38. A jar contains 4 red marbles numbered 1 to 4 and 8 blue marbles numbered 1 to 8. A marble is drawn at random from the jar. Find the probability the marble is
    a. Odd-numbered given that the marble is blue.
    b. Blue given that the marble is odd-numbered.

39. Compute the probability of flipping a coin and getting heads, given that the previous flip was tails.

40. Find the probability of rolling a "1" on a fair die, given that the last 3 rolls were all ones.

41. Suppose a math class contains 25 students, 14 females (three of whom speak French) and 11 males (two of whom speak French). Compute the probability that a randomly selected student speaks French, given that the student is female.

42. Suppose a math class contains 25 students, 14 females (three of whom speak French) and 11 males (two of whom speak French). Compute the probability that a randomly selected student is male, given that the student speaks French.

43. A certain virus infects one in every 400 people. A test used to detect the virus in a person is positive 90% of the time if the person has the virus and 10% of the time if the person does not have the virus. Let A be the event "the person is infected" and B be the event "the person tests positive."
    a. Find the probability that a person has the virus given that they have tested positive, i.e. find P(A | B).
    b. Find the probability that a person does not have the virus given that they test negative, i.e. find P(not A | not B).

44. A certain virus infects one in every 2000 people. A test used to detect the virus in a person is positive 96% of the time if the person has the virus and 4% of the time if the person does not have the virus. Let A be the event "the person is infected" and B be the event "the person tests positive."
    a. Find the probability that a person has the virus given that they have tested positive, i.e. find P(A | B).
    b. Find the probability that a person does not have the virus given that they test negative, i.e. find P(not A | not B).

45. A certain disease has an incidence rate of 0.3%. If the false negative rate is 6% and the false positive rate is 4%, compute the probability that a person who tests positive actually has the disease.

46. A certain disease has an incidence rate of 0.1%. If the false negative rate is 8% and the false positive rate is 3%, compute the probability that a person who tests positive actually has the disease.

47. A certain group of symptom-free women between the ages of 40 and 50 are randomly selected to participate in mammography screening. The incidence rate of breast cancer among such women is 0.8%. The false negative rate for the mammogram is 10%. The false positive rate is 7%. If a the mammogram results for a particular woman are positive (indicating that she has breast cancer), what is the probability that she actually has breast cancer?

48. About 0.01% of men with no known risk behavior are infected with HIV. The false negative rate for the standard HIV test 0.01% and the false positive rate is also 0.01%. If a randomly selected man with no known risk behavior tests positive for HIV, what is the probability that he is actually infected with HIV?

49. A boy owns 2 pairs of pants, 3 shirts, 8 ties, and 2 jackets. How many different outfits can he wear to school if he must wear one of each item?

50. At a restaurant you can choose from 3 appetizers, 8 entrees, and 2 desserts. How many different three-course meals can you have?

51. How many three-letter words can be made from 4 letters FGHI if
    a. repetition of letters is allowed
    b. repetition of letters is not allowed

52. How many four-letter words can be made from 6 letters AEBWDP if
    a. repetition of letters is allowed
    b. repetition of letters is not allowed

53. All of the license plates in a particular state feature three letters followed by three digits (e.g. ABC 123). How many different license plate numbers are available to the state's Department of Motor Vehicles?

54. A computer password must be eight characters long. How many passwords are possible if only the 26 letters of the alphabet are allowed?

55. A pianist plans to play 4 pieces at a recital. In how many ways can she arrange these pieces in the program?

56. In how many ways can first, second, and third prizes be awarded in a contest with 210 contestants?

57. Seven Olympic sprinters are eligible to compete in the 4 x 100 m relay race for the USA Olympic team. How many four-person relay teams can be selected from among the seven athletes?

58. A computer user has downloaded 25 songs using an online file-sharing program and wants to create a CD-R with ten songs to use in his portable CD player. If the order that the songs are placed on the CD-R is important to him, how many different CD-Rs could he make from the 25 songs available to him?

59. In western music, an octave is divided into 12 pitches. For the film *Close Encounters of the Third Kind*, director Steven Spielberg asked composer John Williams to write a five-note theme, which aliens would use to communicate with people on Earth. Disregarding rhythm and octave changes, how many five-note themes are possible if no note is repeated?

60. In the early twentieth century, proponents of the Second Viennese School of musical composition (including Arnold Schönberg, Anton Webern and Alban Berg) devised the twelve-tone technique, which utilized a tone row consisting of all 12 pitches from the chromatic scale in any order, but with not pitches repeated in the row. Disregarding rhythm and octave changes, how many tone rows are possible?

61. In how many ways can 4 pizza toppings be chosen from 12 available toppings?

62. At a baby shower 17 guests are in attendance and 5 of them are randomly selected to receive a door prize. If all 5 prizes are identical, in how many ways can the prizes be awarded?

63. In the 6/50 lottery game, a player picks six numbers from 1 to 50. How many different choices does the player have if order doesn't matter?

64. In a lottery daily game, a player picks three numbers from 0 to 9. How many different choices does the player have if order doesn't matter?

65. A jury pool consists of 27 people. How many different ways can 11 people be chosen to serve on a jury and one additional person be chosen to serve as the jury foreman?

66. The United States Senate Committee on Commerce, Science, and Transportation consists of 23 members, 12 Republicans and 11 Democrats. The Surface Transportation and Merchant Marine Subcommittee consists of 8 Republicans and 7 Democrats. How many ways can members of the Subcommittee be chosen from the Committee?

67. You own 16 CDs. You want to randomly arrange 5 of them in a CD rack. What is the probability that the rack ends up in alphabetical order?

68. A jury pool consists of 27 people, 14 men and 13 women. Compute the probability that a randomly selected jury of 12 people is all male.

69. In a lottery game, a player picks six numbers from 1 to 48. If 5 of the 6 numbers match those drawn, they player wins second prize. What is the probability of winning this prize?

70. In a lottery game, a player picks six numbers from 1 to 48. If 4 of the 6 numbers match those drawn, they player wins third prize. What is the probability of winning this prize?

71. Compute the probability that a 5-card poker hand is dealt to you that contains all hearts.

72. Compute the probability that a 5-card poker hand is dealt to you that contains four Aces.

73. A bag contains 3 gold marbles, 6 silver marbles, and 28 black marbles. Someone offers to play this game: You randomly select on marble from the bag. If it is gold, you win $3. If it is silver, you win $2. If it is black, you lose $1. What is your expected value if you play this game?

74. A friend devises a game that is played by rolling a single six-sided die once. If you roll a 6, he pays you $3; if you roll a 5, he pays you nothing; if you roll a number less than 5, you pay him $1. Compute the expected value for this game. Should you play this game?

75. In a lottery game, a player picks six numbers from 1 to 23. If the player matches all six numbers, they win 30,000 dollars. Otherwise, they lose $1. Find the expected value of this game.

76. A game is played by picking two cards from a deck. If they are the same value, then you win $5, otherwise you lose $1. What is the expected value of this game?

77. A company estimates that 0.7% of their products will fail after the original warranty period but within 2 years of the purchase, with a replacement cost of $350. If they offer a 2-year extended warranty for $48, what is the company's expected value of each warranty sold?

78. An insurance company estimates the probability of an earthquake in the next year to be 0.0013. The average damage done by an earthquake it estimates to be $60,000. If the company offers earthquake insurance for $100, what is their expected value of the policy?

Exploration

Some of these questions were adapted from puzzles at mindyourdecisions.com.

79. A small college has been accused of gender bias in its admissions to graduate programs.
    a. Out of 500 men who applied, 255 were accepted. Out of 700 women who applied, 240 were accepted. Find the acceptance rate for each gender. Does this suggest bias?
    b. The college then looked at each of the two departments with graduate programs, and found the data below. Compute the acceptance rate within each department by gender. Does this suggest bias?

Department	Men		Women	
	Applied	Admitted	Applied	Admitted
Dept A	400	240	100	90
Dept B	100	15	600	150

    c. Looking at our results from Parts *a* and *b*, what can you conclude? Is there gender bias in this college's admissions? If so, in which direction?

80. A bet on "black" in Roulette has a probability of 18/38 of winning. If you win, you double your money. You can bet anywhere from $1 to $100 on each spin.
    a. Suppose you have $10, and are going to play until you go broke or have $20. What is your best strategy for playing?
    b. Suppose you have $10, and are going to play until you go broke or have $30. What is your best strategy for playing?

81. Your friend proposes a game: You flip a coin. If it's heads, you win $1. If it's tails, you lose $1. However, you are worried the coin might not be fair coin. How could you change the game to make the game fair, without replacing the coin?

82. Fifty people are in a line. The first person in the line to have a birthday matching someone in front of them will win a prize. Of course, this means the first person in the line has no chance of winning. Which person has the highest likelihood of winning?

83. Three people put their names in a hat, then each draw a name, as part of a randomized gift exchange. What is the probability that no one draws their own name? What about with four people?

84. How many different "words" can be formed by using all the letters of each of the following words exactly once?
    a. "ALICE"
    b. "APPLE"

85. How many different "words" can be formed by using all the letters of each of the following words exactly once?
    a. "TRUMPS"
    b. "TEETER"

86. The *Monty Hall problem* is named for the host of the game show *Let's make a Deal*. In this game, there would be three doors, behind one of which there was a prize. The contestant was asked to choose one of the doors. Monty Hall would then open one of the other doors to show there was no prize there. The contestant was then asked if they wanted to stay with their original door, or switch to the other unopened door. Is it better to stay or switch, or does it matter?

87. Suppose you have two coins, where one is a fair coin, and the other coin comes up heads 70% of the time. What is the probability you have the fair coin given each of the following outcomes from a series of flips?
    a. 5 Heads and 0 Tails
    b. 8 Heads and 3 Tails
    c. 10 Heads and 10 Tails
    d. 3 Heads and 8 Tails

88. Suppose you have six coins, where five are fair coins, and one coin comes up heads 80% of the time. What is the probability you have a fair coin given each of the following outcomes from a series of flips?
    a. 5 Heads and 0 Tails
    b. 8 Heads and 3 Tails
    c. 10 Heads and 10 Tails
    d. 3 Heads and 8 Tails

89. In this problem, we will explore probabilities from a series of events.
    a. If you flip 20 coins, how many would you *expect* to come up "heads", on average? Would you expect *every* flip of 20 coins to come up with exactly that many heads?
    b. If you were to flip 20 coins, what would you consider a "usual" result? An "unusual" result?
    c. Flip 20 coins (or one coin 20 times) and record how many come up "heads". Repeat this experiment 9 more times. Collect the data from the entire class.
    d. When flipping 20 coins, what is the theoretic probability of flipping 20 heads?
    e. Based on the class's experimental data, what appears to be the probability of flipping 10 heads out of 20 coins?
    f. The formula $_nC_x p^x (1-p)^{n-x}$ will compute the probability of an event with probability p occurring x times out of n, such as flipping x heads out of n coins where the probability of heads is $p = \frac{1}{2}$. Use this to compute the theoretic probability of flipping 10 heads out of 20 coins.
    g. If you were to flip 20 coins, based on the class's experimental data, what range of values would you consider a "usual" result? What is the combined probability of these results? What would you consider an "unusual" result? What is the combined probability of these results?

h. We'll now consider a simplification of a case from the 1960s. In the area, about 26% of the jury eligible population was black. In the court case, there were 100 men on the juror panel, of which 8 were black. Does this provide evidence of racial bias in jury selection?

Chapter 11: Normal Distribution

Introduction

Most high schools have a set amount of time in-between classes during which students must get to their next class. If you were to stand at the door of your statistics class and watch the students coming in, think about how the students would enter. Usually, one or two students enter early, then more students come in, then a large group of students enter, and finally, the number of students entering decreases again, with one or two students barely making it on time, or perhaps even coming in late!

Now consider this. Have you ever popped popcorn in a microwave? Think about what happens in terms of the rate at which the kernels pop. For the first few minutes, nothing happens, and then, after a while, a few kernels start popping. This rate increases to the point at which you hear most of the kernels popping, and then it gradually decreases again until just a kernel or two pops.

Here's something else to think about. Try measuring the height, shoe size, or the width of the hands of the students in your class. In most situations, you will probably find that there are a couple of students with very low measurements and a couple with very high measurements, with the majority of students centered on a particular value.

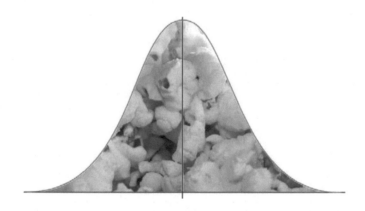

All of these examples show a typical pattern that seems to be a part of many real-life phenomena. In statistics, because this pattern is so pervasive, it seems to fit to call it normal, or more formally, the normal distribution. The normal distribution is an extremely important concept, because it occurs so often in the data we collect from the natural world, as well as in many of the more theoretical ideas that are the foundation of statistics. This chapter explores the details of the normal distribution.

Table of Contents

11.1   The Standard Normal Probability Distribution

Learning Objectives

1.   Identify the characteristics of a normal distribution.
2.   Identify and use the Empirical Rule (68-95-99.7 Rule) for normal distributions.
3.   Calculate a z-score and relate it to probability.
4.   Determine if a data set corresponds to a normal distribution.

The Characteristics of a Normal Distribution

Shape

When graphing the data from each of the examples in the introduction, the distributions from each of these situations would be mound-shaped and mostly symmetric. A *normal distribution* is a perfectly symmetric, mound-shaped distribution. It is commonly referred to the as a normal curve, or bell curve.

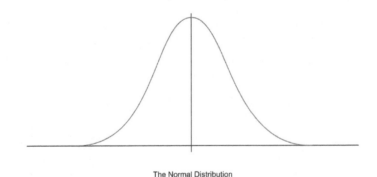

The Normal Distribution

Because so many real data sets closely approximate a normal distribution, we can use the idealized normal curve to learn a great deal about such data. With a practical data collection, the distribution will never be exactly symmetric, so just like situations involving probability, a true normal distribution only results from an infinite collection of data. Also, it is important to note that the normal distribution describes a continuous random variable.

Center

Due to the exact symmetry of a normal curve, the center of a normal distribution, or a data set that approximates a normal distribution, is located at the highest point of the distribution, and all the statistical measures of center we have already studied (the mean, median, and mode) are equal.

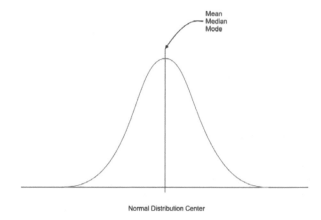

Normal Distribution Center

It is also important to realize that this center peak divides the data into two equal parts.

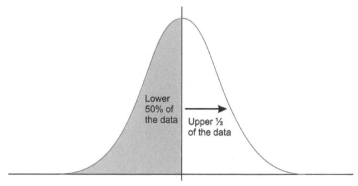

Normal Distribution divided into equal halves by the center

Spread

Let's go back to our popcorn example. The bag advertises a certain time, beyond which you risk burning the popcorn. From experience, the manufacturers know when most of the popcorn will stop popping, but there is still a chance that there are those rare kernels that will require more (or less) time to pop than the time advertised by the manufacturer. The directions usually tell you to stop when the time between popping is a few seconds, but aren't you tempted to keep going so you don't end up with a bag full of un-popped kernels? Because this is a real, and not theoretical, situation, there will be a time when the popcorn will stop popping and start burning, but there is always a chance, no matter how small, that one more kernel will pop if you keep the microwave going. In an idealized normal distribution of a continuous random variable, the distribution continues infinitely in both directions.

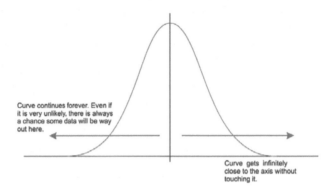

Because of this infinite spread, the range would not be a useful statistical measure of spread. The most common way to measure the spread of a normal distribution is with the standard deviation, or the typical distance away from the mean. Because of the symmetry of a normal distribution, the standard deviation indicates how far away from the maximum peak the data will be. Here are two normal distributions with the same center (mean):

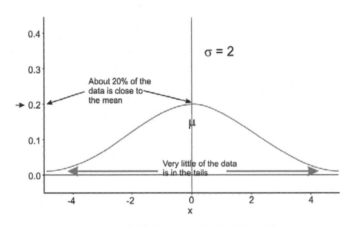

normal distribution, standard deviation = 2

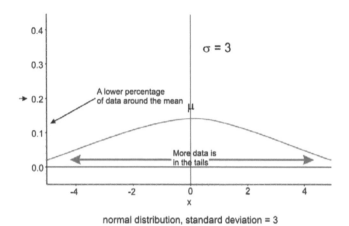

normal distribution, standard deviation = 3

The first distribution pictured above has a smaller standard deviation, and so more of the data are heavily concentrated around the mean than in the second distribution. Also, in the first distribution, there are fewer data values at the extremes than in the second distribution. Because the second distribution has a larger standard deviation, the data are spread farther from the mean value, with more of the data appearing in the tails.

Technology Note: Investigating the Normal Distribution on a TI-83/84 Graphing Calculator

We can graph a normal curve for a probability distribution on the TI-83/84 calculator. To do so, first press [Y=]. To create a normal distribution, we will draw an idealized curve using something called a density function. The command is called 'normalpdf(', and it is found by pressing [2nd][DISTR][1]. Enter an X to represent the random variable, followed by the mean and the standard deviation, all separated by commas. For this example, choose a mean of 5 and a standard deviation of 1.

Adjust your window to match the following settings and press [GRAPH].

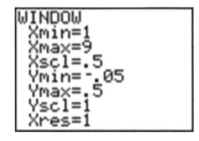

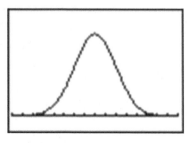

Press [2ND][QUIT] to go to the home screen. We can draw a vertical line at the mean to show it is in the center of the distribution by pressing [2ND][DRAW] and choosing 'Vertical'. Enter the mean, which is 5, and press [ENTER].

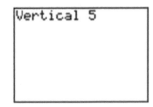 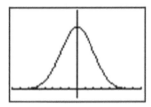

Remember that even though the graph appears to touch the *x*-axis, it is actually just very close to it.

In your Y= Menu, enter the following to graph 3 different normal distributions, each with a different standard deviation:

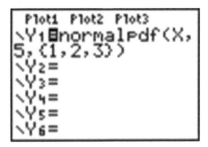

This makes it easy to see the change in spread when the standard deviation changes.

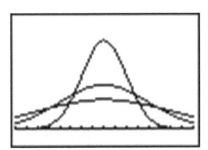

The Empirical Rule for normal distributions

Because of the similar shape of all normal distributions, we can measure the percentage of data that is a certain distance from the mean no matter what the standard deviation of the data set is. The following graph shows a normal distribution with $\mu = 0$ and $\sigma = 1$. This curve is called a *standard normal curve*. In this case, the values of *x* represent the number of standard deviations away from the mean.

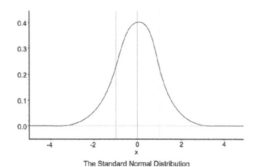

The Standard Normal Distribution

Notice that vertical lines are drawn at points that are exactly one standard deviation to the left and right of the mean. We have consistently described standard deviation as a measure of the typical distance away from the mean. How much of the data is actually within one standard deviation of the mean? To answer this question, think about the space, or area, under the curve. The entire data set, or 100% of it, is contained under the whole curve. What percentage would you estimate is between the two lines? To help estimate the answer, we can use a graphing calculator. Graph a standard normal distribution over an appropriate window.

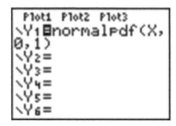
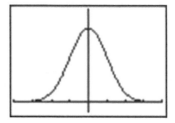

Now press [2ND][DISTR], go to the DRAW menu, and choose 'ShadeNorm('. Insert '−1, 1' after the 'Shade-Norm(' command and press [ENTER]. It will shade the area within one standard deviation of the mean.

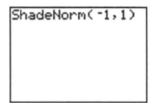
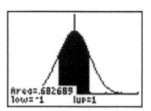

The calculator also gives a very accurate estimate of the area. We can see from the rightmost screenshot above that approximately 68% of the area is within one standard deviation of the mean. If we venture to 2 standard deviations away from the mean, how much of the data should we expect to capture? Make the following changes to the 'ShadeNorm(' command to find out:

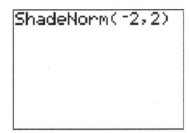

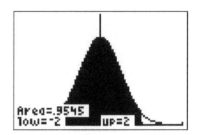

Notice from the shading that almost all of the distribution is shaded, and the percentage of data is close to 95%. If you were to venture to 3 standard deviations from the mean, 99.7%, or virtually all of the data, is captured, which tells us that very little of the data in a normal distribution is more than 3 standard deviations from the mean.

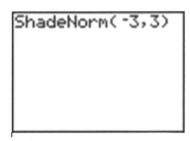

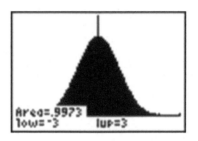

Notice that the calculator actually makes it look like the entire distribution is shaded because of the limitations of the screen resolution, but as we have already discovered, there is still some area under the curve further out than that. These three approximate percentages, 68%, 95%, and 99.7%, are extremely important and are part of what is called the *Empirical Rule*.

The Empirical Rule states that the percentages of data in a normal distribution within 1, 2, and 3 standard deviations of the mean are approximately 68%, 95%, and 99.7%, respectively.

On the Web

http://tinyurl.com/2ue78u Explore the Empirical Rule.

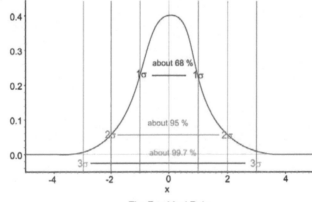

The Empirical Rule

Calculating and interpreting z-Scores

A *z-score* is a measure of the number of standard deviations a particular data point is away from the mean. For example, let's say the mean score on a test for your statistics class was an 82, with a standard deviation of 7 points. If your score was an 89, it is exactly one standard deviation to the right of the mean; therefore, your z-score would be 1. If, on the other hand, you scored a 75, your score would be exactly one standard deviation below the mean, and your z-score would be −1. All values that are below the mean have negative z-scores, while all values that are above the mean have positive z-scores. A z-score of −2 would represent a value that is exactly 2 standard deviations below the mean, so in this case, the value would be 82 − 14 = 68.

To calculate a z-score for which the numbers are not so obvious, you take the deviation and divide it by the standard deviation.

$$z = \frac{\text{deviation}}{\text{standard deviation}}$$

You may recall that deviation is the mean value of the variable subtracted from the observed value, so in symbolic terms, the z-score would be:

$$z = \frac{x - \mu}{\sigma}$$

As previously stated, since σ is always positive, z will be positive when x is greater than μ and negative when x is less than μ. A z-score of zero means that the term has the same value as the mean. The value of z represents the number of standard deviations the given value of x is above or below the mean.

Example 1

What is the z-score for an *A* on the test described above, which has a mean score of 82? (Assume that an *A* is a 93.)

The z-score can be calculated as follows:

$$z = \frac{x - \mu}{\sigma}$$

$$z = \frac{93 - 82}{7}$$

$$z \approx 1.57$$

If we know that the test scores from the last example are distributed normally, then a z-score can tell us something about how our test score relates to the rest of the class. From the Empirical Rule, we know that about 68% of the students would have scored between a z-score of −1 and 1, or between a 75 and an 89, on the test. If 68% of the data is between these two values, then that leaves the remaining 32% in the tail areas. Because of symmetry, half of this, or 16%, would be in each individual tail.

Example 2

On a nationwide math test, the mean was 65 and the standard deviation was 10. If Robert scored 81, what was his z-score?

$$z = \frac{x - \mu}{\sigma}$$

$$z = \frac{81 - 65}{10}$$

$$z \approx 1.60$$

Example 3

On a college entrance exam, the mean was 70, and the standard deviation was 8. If Helen's z-score was -1.5, what was her exam score?

Since $z = \frac{x-\mu}{\sigma}$, then we can rewrite this formula solving for x:

$$x = \mu + z\sigma$$

Now, we can obtain Helen's exam score with the given parameters:

$$x = \mu + z\sigma$$
$$x = 70 + (-1.5)(8)$$
$$x = 58$$

Thus, Helen's exam score was 58; notice a score of 58 is below the mean and this makes sense since her z-score was negative.

Assessing Normality

The best way to determine if a data set approximates a normal distribution is to look at a visual representation. Histograms and box plots can be useful indicators of normality, but they are not always definitive. It is often easier to tell if a data set is *not* normal from these plots.

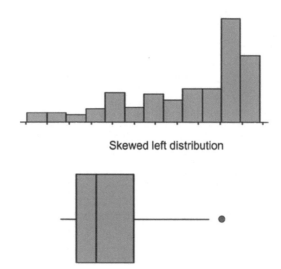

Skewed left distribution

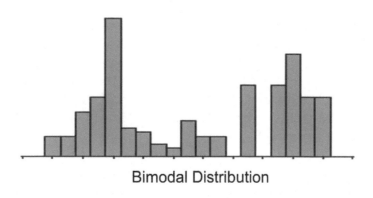

Skewed right distribution with outliers

Bimodal Distribution

If a data set is skewed right, it means that the right tail is significantly longer than the left. Similarly, skewed left means the left tail has more weight than the right. A bimodal distribution, on the other hand, has two modes, or peaks. For instance, with a histogram of the heights of American 30-year-old adults, you will see a bimodal distribution—one mode for males and one mode for females.

There is a plot we can use to determine if a distribution is normal called a *normal probability plot* or *normal quantile plot*. To make this plot by hand, first order your data from smallest to largest. Then, determine the quantile of each data point. Finally, using a table of standard normal probabilities, determine the closest z-score for each quantile. Plot these z-scores against the actual data values. To make a normal probability plot using your calculator, enter your data into a list, then use the last type of graph in the STAT PLOT menu, as shown below:

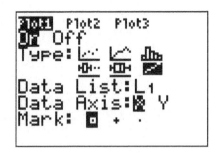

If the data set is normal, then this plot will be perfectly linear. The closer to being linear the normal probability plot is, the more closely the data set approximates a normal distribution.

Look below at the histogram and the normal probability plot for the same data.

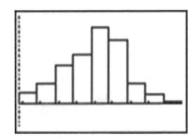   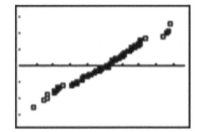

The histogram is fairly symmetric and mound-shaped and appears to display the characteristics of a normal distribution. When the z-scores of the quantiles of the data are plotted against the actual data values, the normal probability plot appears strongly linear, indicating that the data set closely approximates a normal distribution. The following example will allow you to see how a normal probability plot is made in more detail.

Example 4

The following data set tracked high school seniors' involvement in traffic accidents. The participants were asked the following question: "During the last 12 months, how many accidents have you had while you were driving (whether or not you were responsible)?"

Year	Percentage of high school seniors who said they were
	involved in no traffic accidents
1991	75.7
1992	76.9
1993	76.1
1994	75.7
1995	75.3
1996	74.1
1997	74.4
1998	74.4

1999	75.1
2000	75.1
2001	75.5
2002	75.5
2003	75.8

Figure: Percentage of high school seniors who said they were involved in no traffic accidents. *Source*: Sourcebook of Criminal Justice Statistics: http://www.albany.edu/sourcebook/pdf/t352.pdf

Here is a histogram and a box plot of this data:

 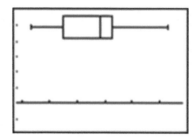

The histogram appears to show a roughly mound-shaped and symmetric distribution. The box plot does not appear to be significantly skewed, but the various sections of the plot also do not appear to be overly symmetric, either. In the following chart, the data has been reordered from smallest to largest, the quantiles have been determined, and the closest corresponding z-scores have been found using a table of standard normal probabilities.

Year	Percentage	Quantile	z-score
1996	74.1	$\frac{1}{13} = 0.078$	-1.42
1997	74.4	$\frac{2}{13} = 0.154$	-1.02
1998	74.4	$\frac{3}{13} = 0.231$	-0.74
1999	75.1	$\frac{4}{13} = 0.286$	-0.56
2000	75.1	$\frac{5}{13} = 0.385$	-0.29
1995	75.3	$\frac{6}{13} = 0.462$	-0.09
2001	75.5	$\frac{7}{13} = 0.538$	0.1
2002	75.5	$\frac{8}{13} = 0.615$	0.29
1991	75.7	$\frac{9}{13} = 0.692$	0.50
1994	75.7	$\frac{10}{13} = 0.769$	0.74
2003	75.8	$\frac{11}{13} = 0.846$	1.02
1993	76.1	$\frac{12}{13} = 0.923$	1.43
1992	76.9	$\frac{13}{13} = 1$	3.49

Figure: Table of quantiles and corresponding *z*-scores for senior no-accident data.

Here is a plot of the percentages versus the *z*-scores of their quantiles, or the normal probability plot:

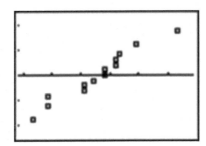

Remember that you can simplify this process by simply entering the percentages into a $L1$ in your calculator and selecting the normal probability plot option (the last type of plot) in STAT PLOT.

While not perfectly linear, this plot does have a strong linear pattern, and we would, therefore, conclude that the distribution is reasonably normal.

Exercises for 11.1

1. Which of the following data sets is most likely to be normally distributed? For the other choices, explain why you believe they would not follow a normal distribution.
   a) The hand span (measured from the tip of the thumb to the tip of the extended 5th finger) of a random sample of high school seniors
   b) The annual salaries of all employees of a large shipping company
   c) The annual salaries of a random sample of 50 CEOs of major companies, 25 women and 25 men
   d) The dates of 100 pennies taken from a cash drawer in a convenience store

2. The grades on a statistics mid-term for a high school are normally distributed, with $\mu = 81$ and $\sigma = 6.3$. Calculate the z-scores for each of the following exam grades. Draw and label a sketch for each example. 65, 83, 93, 100

3. Assume that the mean weight of 1-year-old girls in the USA is normally distributed, with a mean of about 9.5 kilograms and a standard deviation of approximately 1.1 kilograms. Without using a calculator, estimate the percentage of 1-year-old girls who meet the following conditions. Draw a sketch and shade the proper region for each problem.
   a) Less than 8.4 kg
   b) Between 7.3 kg and 11.7 kg
   c) More than 12.8 kg

4. For a standard normal distribution, place the following in order from smallest to largest.
   a) The percentage of data below 1
   b) The percentage of data below -1
   c) The mean
   d) The standard deviation
   e) The percentage of data above 2

5. The 2007 AP Statistics examination scores were not normally distributed, with $\mu = 2.8$ and $\sigma = 1.34$. What is the approximate z-score that corresponds to an exam score of 5? (The scores range from 1 to 5.)
   a) 0.786
   b) 1.46
   c) 1.64
   d) 2.20
   e) A z-score cannot be calculated because the distribution is not normal.

[1] Data available on the College Board Website: http://professionals.collegeboard.com/data-reports-research/ap/archi ved/2007

6. The heights of 5th grade boys in the USA is approximately normally distributed, with a mean height of 143.5 cm and a standard deviation of about 7.1 cm. What is the probability that a randomly chosen 5th grade boy would be taller than 157.7 cm?

7. A statistics class bought some sprinkle (or jimmies) doughnuts for a treat and noticed that the number of sprinkles seemed to vary from doughnut to doughnut, so they counted the sprinkles on each doughnut. Here are the results: 241, 282, 258, 223, 133, 335, 322, 323, 354, 194, 332, 274, 233, 147, 213, 262, 227, and 366. Create a histogram, dot plot, or box plot for this data. Comment on the shape, center and spread of the distribution.

Answers for Exercises 11.1:
1. a) Normal
   b) Management and corporate salaries would most likely skew the data right
   c) Individually, the gender data might be normal
   d) Most pennies usually are from the current or previous year making the data skew left
2. -2.54; 0.32, 1.90; 3.02
3. a) 16%, b) 95%, c) 0.15%
4. c,e,b,a,d
5. c) 1.64
6. 2.5%
7. 5-number summary: 133,224,260,322.75,366; the data is fairly normal since the mean is approximately the median

11.2   The Density Curve of a Normal Distribution

Learning Objectives

1. Identify the properties of a normal density curve and the relationship between concavity and standard deviation.
2. Convert between z-scores and areas under a normal probability curve.
3. Calculate probabilities that correspond to left, right, and middle areas from a z-score table.
4. Calculate probabilities that correspond to left, right, and middle areas using a graphing calculator.
5. Calculate for unknown values other than the z-score and area.

Introduction

In this section, we will continue our investigation of normal distributions to include density curves and learn various methods for calculating probabilities from the normal density curve.

Density Curves

A *density curve* is an idealized representation of a distribution in which the area under the curve is defined to be 1. Density curves need not be normal, but the normal density curve will be the most useful to us.

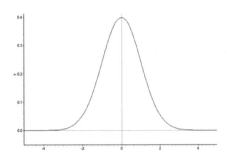

Inflection Points on a Normal Density Curve

We already know from the Empirical Rule that approximately $^2/_3$ of the data in a normal distribution lies within 1 standard deviation of the mean. With a normal density curve, this means that about 68% of the total area under the curve is within z-scores of ± 1. Look at the following three density curves:

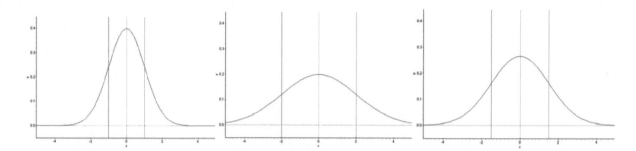

Notice that the curves are spread increasingly wider. Lines have been drawn to show the points that are one standard deviation on either side of the mean. Look at where this happens on each density curve.

Here is a normal distribution with an even larger standard deviation.

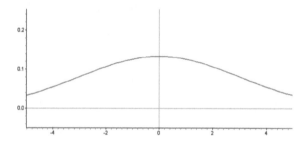

Is it possible to predict the standard deviation of this distribution by estimating the *x*-coordinate of a point on the density curve? Read on to find out!

You may have noticed that the density curve changes shape at two points in each of our examples. These are the points where the curve changes concavity. Starting from the mean and heading outward to the left and right, the curve is *concave down*. (It looks like a mountain, or '*n*' shape.) After passing these points, the curve is *concave up*. (It looks like a valley, or '*u*' shape.) The points at which the curve changes from being concave up to being concave down are called the *inflection points*. On a normal density curve, these inflection points are always exactly one standard deviation away from the mean.

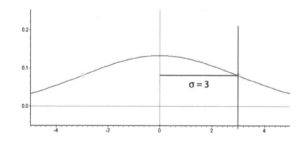

In this example, the standard deviation is 3 units. We can use this concept to estimate the standard deviation of a normally distributed data set.

Example 5

Estimate the standard deviation of the distribution represented by the following histogram.

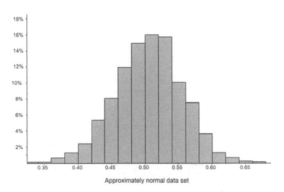

Approximately normal data set

This distribution is fairly normal, so we could draw a density curve to approximate it as follows:

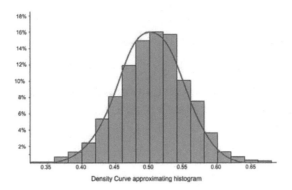

Density Curve approximating histogram

Now estimate the inflection points as shown below:

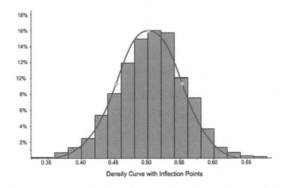

Density Curve with Inflection Points

It appears that the mean is about 0.5 and that the *x*-coordinates of the inflection points are about 0.45 and 0.55, respectively. This would lead to an estimate of about 0.05 for the standard deviation.

The actual statistics for this distribution are as follows:

$s \approx 0.04988$
$\bar{x} \approx 0.4997$

We can verify these figures by using the expectations from the Empirical Rule. In the following graph, we have highlighted the bins that are contained within one standard deviation of the mean.

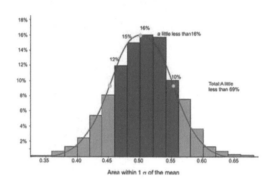

If you estimate the relative frequencies from each bin, their total is remarkably close to 68%. Make sure to divide the relative frequencies from the bins on the ends by 2 when performing your calculation.

Convert between z-scores and areas

While it is convenient to estimate areas under a normal curve using the Empirical Rule, we often need more precise methods to calculate these areas. Luckily, we can use formulas or technology to help us with the calculations.

z-scores

All normal distributions have the same basic shape, and therefore, rescaling and re-centering can be implemented to change any normal distributions to one with a mean of 0 and a standard deviation of 1. This configuration is referred to as a *standard normal distribution*. In a standard normal distribution, the variable along the horizontal axis is the *z*-score. This score is another measure of the performance of an individual score in a population. To review, the *z*-score measures how many standard deviations a score is away from the mean. The *z*-score of the term x in a population distribution whose mean is μ and whose standard deviation is σ is given by: $= \frac{x-\mu}{\sigma}$.
Since σ is always positive, z will be positive when x is greater than μ and negative when x is less than μ. A *z*-score of 0 means that the term has the same value as the mean. The value of z is the number of standard deviations the given value of x is above or below the mean.

Example 6

On a nationwide math test, the mean was 65 and the standard deviation was 10. If Robert scored 81, what was his *z*-score?

$$z = \frac{x - \mu}{\sigma}$$

$$z = \frac{81 - 65}{1.6}$$

$$z = 1.6$$

Example 7

On a college entrance exam, the mean was 70 and the standard deviation was 8. If Helen's z-score was -1.5, what was her exam score?

Recall, the equation to obtain x is

$$x = \mu + z\sigma$$

Using this equation, we can find Helen's score:

$$x = \mu + z\sigma$$
$$x = 70 + (-1.5)(8)$$
$$x = 58$$

Now you will see how z-scores are used to determine the probability of an event.

Example 8

Suppose you were to toss 8 coins 256 times. The following figure shows the histogram and the approximating normal curve for the experiment. The random variable represents the number of tails obtained.

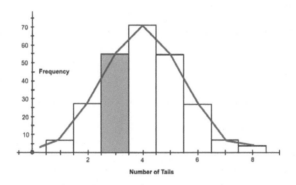

The blue section of the graph represents the probability that exactly 3 of the coins turned up tails. Geometrically, this probability represents the area of the blue shaded bar divided by the total area of the bars. The area of the blue shaded bar is approximately equal to the area under the normal curve from 2.5 to 3.5.

Since areas under normal curves correspond to the probability of an event occurring, a special normal distribution table is used to calculate the probabilities. ***This table can be found at the end of this section where the area is given from the mean.*** The following is an example of a table of z-scores and a brief explanation of how it works: http://tinyurl.com/2ce9o gv.

The values inside the given table represent the areas under the standard normal curve for values between 0 and the relative *z*-score. For example, to determine the area under the curve between *z*-scores of 0 and 2.36, look in the intersecting cell for the row labeled 2.3 and the column labeled 0.06. The area under the curve is 0.4909. To determine the area between 0 and a negative value, look in the intersecting cell of the row and column which sums to the absolute value of the number in question. For example, the area under the curve between −1.3 and 0 is equal to the area under the curve between 1.3 and 0, so look at the cell that is the intersection of the 1.3 row and the 0.00 column. (The area is 0.4032.)

Calculate probabilities that correspond to z-scores and areas

It is extremely important, especially when you first start with these calculations, that you get in the habit of relating it to the normal distribution by drawing a sketch of the situation. In this case, simply draw a sketch of a standard normal curve with the appropriate region shaded and labeled.

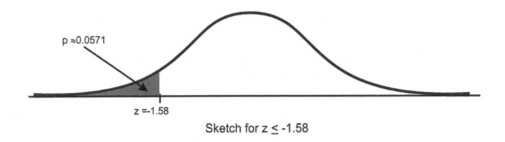

Sketch for z ≤ -1.58

Example 9

Find the probability of choosing a value that is greater than $z = -0.528$, or $P(z > -0.528)$.

Before even using the table, first draw a figure with the shaded region. This *z*-score is just below the mean, so the answer should be more than 0.5.

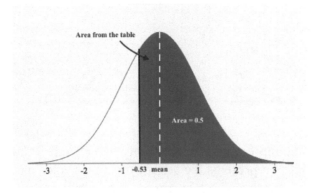

Next, read the table to find the correct probability for the data below this *z*-score. We must first round this *z*-score to −0.53, so this will slightly under-estimate the probability, but it is the best we can do using the table. Looking up a *z*-score of −0.53, we see

z	0.00	0.01	0.02	0.03
0.00	0.00000	0.00399	0.00798	0.01197
0.10	0.03983	0.04380	0.04776	0.05172
0.20	0.07926	0.08317	0.08706	0.09095
0.30	0.11791	0.12172	0.12552	0.12930
0.40	0.15542	0.15910	0.16276	0.16640
0.50	0.19146	0.19497	0.19847	0.20194

The table returns an area of 0.20194. Since the area from the mean to $z = -0.53$ is 0.20194 and the area on the right of the mean is 0.5, then the area of the shaded region is

$$0.5 + 0.20194 = 0.70194$$

Thus, the probability of choosing a value that is greater than $z = -0.528$ is 0.7019.

What about values between two z-scores? While it is an interesting and worthwhile exercise to do this using a table, we can also use statistical software or a graphing calculator.

Example 10

Find $P(-2.60 < z < 1.30)$.

First, we draw a figure with the shaded region:

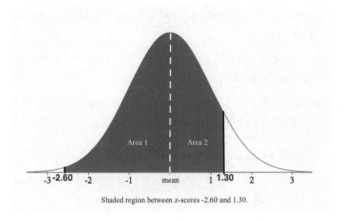

Shaded region between z-scores -2.60 and 1.30.

Since the table gives us the area from the mean to a z-score, we can see that we will add the areas, Area 1 + Area 2, to obtain the area of the shaded region, resulting in the probability. Let's look up the z-scores on the table to find the area from the mean to each z-score:

z	0.00
1.30	0.40320
2.60	0.49534

Area 1 is 0.49534 and Area 2 is 0.40320. Adding these two together, we get

$P(-2.60 < z < 1.30)$= Area 1 + Area 2 = 0.49534 + 0.40320 = 0.89854

Thus, the probability $P(-2.60 < z < 1.30) = 0.89854$.

The probability can also be found using the TI-83/84 calculator. Use the 'normalcdf(−2.60, 1.30, 0, 1)' command, and the calculator will return the result 0.898538. The syntax for this command is 'normalcdf(min, max, μ, σ)'. When using this command, you do not need to first standardize. You can use the mean and standard deviation of the given distribution.

Technology Note: The 'normalcdf(' Command on the TI-83/84 Calculator

Your graphing calculator has already been programmed to calculate probabilities for a normal density curve using what is called a *cumulative density function*. The command you will use is found in the DISTR menu, which you can bring up by pressing [2ND][DISTR].

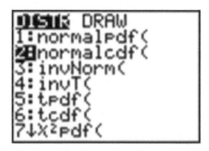

Press [2] to select the 'normalcdf(' command, which has a syntax of 'normalcdf(lower bound, upper bound, mean, standard deviation)'.

The command has been programmed so that if you do not specify a mean and standard deviation, it will default to the standard normal curve, with $\mu = 0$ and $\sigma = 1$.

For example, entering 'normalcdf(−1, 1)' will specify the area within one standard deviation of the mean, which we already know to be approximately 0.68.

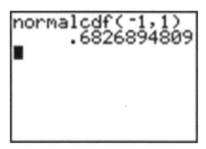

Try verifying the other values from the Empirical Rule.

Summary:

'Normalcdf (a,b,μ,σ)' gives values of the cumulative normal density function. In other words, it gives the probability of an event occurring between $x = a$ and $x = b$, or the area under the probability density curve between the vertical lines $x = a$ and $x = b$, where the normal distribution has a mean of μ and a standard deviation of σ. If μ and σ are not specified, it is assumed that $\mu = 0$ and $\sigma = 1$.

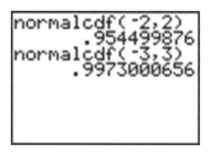

Example 11

Find the probability $P(z < -1.58)$.

First, we draw a figure with the shaded region:

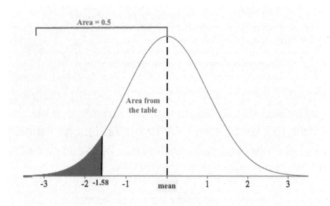

Since the table gives us the area from the mean to a z-score and the total area to the left of the mean is 0.5, we can see that we will subtract the area given in the table from 0.5 to obtain the area of the shaded region, resulting in the probability. Let's look up the z-score on the table to find the area from the mean to the z-score:

z	0.08
1.50	0.44295

The area from the mean to $z = -1.58$ is 0.44295. Subtracting this from 0.5, we get

$P(z < -1.58) = 0.5 - 0.44295 = 0.05705$.

Doing this on the calculator, we must have both an upper and lower bound. Technically, though, the density curve does not have a lower bound, as it continues infinitely in both directions. We do know, however, that a very small percentage of the data is below 3 standard deviations to the left of the mean. Use -3 as the lower bound and see what answer you get.

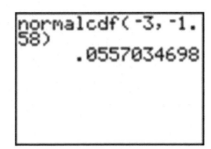

The answer is fairly accurate, but you must remember that there is really still some area under the probability density curve, even though it is just a little, that we are leaving out if we stop at −3. If you look at the z-table, you can see that we are, in fact, leaving out about 0.5 − 0.4987 = 0.0013. Next, try going out to −4 and −5.

```
normalcdf( -4, -1.
58)
          .057021751
normalcdf( -5, -1.
58)
          .0570531499
```

Once we get to −5, the answer is quite accurate. Since we cannot really capture all the data, entering a sufficiently small value should be enough for any reasonable degree of accuracy. A quick and easy way to handle this is to enter −99999 (or "a bunch of nines"). It really doesn't matter exactly how many nines you enter. The difference between five and six nines will be beyond the accuracy that even your calculator can display.

```
normalcdf( -99999
, -1.58)
          .057053437
normalcdf( -99999
9, -1.58)
          .057053437
```

Example 12
_____

Find the probability $P(0 < z < 1.78)$.

First, we draw a figure with the shaded region:

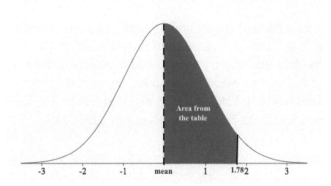

Since the table gives us the area from the mean to a *z*-score, we can see that whatever area is given from the table results in the probability. Let's look up the *z*-score on the table to find the area from the mean to the *z*-score.

z	0.08
1.70	0.46246

The area from the mean to $z = 1.78$ is 0.46246. Thus,

$P(z < 1.78) = 0.46246$.

We are at an advantage using the calculator, because we do not have to round off the *z*-score in this example. Let's try this example with the calculator. Enter the 'normalcdf(' command, using −0.528 to "a bunch of nines." The nines represent a ridiculously large upper bound that will insure that the unaccounted-for probability will be so small that it will be virtually undetectable.

```
normalcdf(-.528,
9999999)
          .7012503533
```

Remember that because of rounding, our answer from the table was slightly too small, so when we subtracted it from 1, our final answer was slightly too large. The calculator answer of about 0.70125 is a more accurate approximation than the answer arrived at by using the table.

Standardizing

In most practical problems involving normal distributions, the curve will not be as we have seen so far, with $\mu = 0$ and $\sigma = 1$. When using a *z*-table, you will first have to *standardize* the distribution by calculating the *z*-score(s).

Example 13

A candy company sells small bags of candy and attempts to keep the number of pieces in each bag the same, though small differences due to random variation in the packaging process lead to different amounts in individual packages. A quality control expert from the company has determined that the mean number of pieces in each bag is normally distributed, with a mean of 57.3 and a standard deviation of 1.2. Endy opened a bag of candy and felt he was cheated. His bag contained only 55 candies. Does Endy have reason to complain?

To determine if Endy was cheated, we need to find the probability of selecting a bag of candy with 55 or fewer candies, i.e., we let $x = 55$. Let's calculate the z-score for 55:

$$z = \frac{x - \mu}{\sigma}$$

$$z = \frac{55 - 57.3}{1.2}$$

$$z \approx -1.91$$

Next, we can draw a figure to see the shaded region:

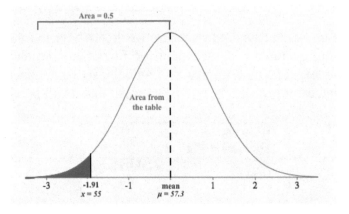

Using a table, we obtain a value 0.47193. This is the area from the mean to $z = -1.91$. We can subtract this value from 0.5, since the area on the left of the mean is 0.5:

$$0.5 - 0.47193 = 0.02807$$

Hence, there is about a 3% chance that he would get a bag of candy with 55 or fewer pieces, so Endy should feel cheated because the chances of getting a bag with 55 or fewer candies is so low.

Using a graphing calculator, the results would look as follows (the 'Ans' function has been used to avoid rounding off the z-score):

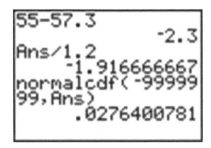

However, one of the advantages of using a calculator is that it is unnecessary to standardize. We can simply enter the mean and standard deviation from the original population distribution of candy, avoiding the z-score calculation completely.

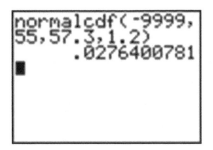

Unknown Value Problems

If you understand the relationship between the area under a density curve and mean, standard deviation, and z-scores, you should be able to solve problems in which you are provided all but one of these values and are asked to calculate the remaining value. In the last lesson, we found the probability that a variable is within a particular range, or the area under a density curve within that range. What if you are asked to find a value that gives a particular probability? We rewrite the z-score formula $z = \frac{x-\mu}{\sigma}$ as

$$x = \mu + z \cdot \sigma \quad \text{or} \quad \mu = x - z \cdot \sigma$$

Unknown Original Value, x

Example 14

Given the normally-distributed random variable x, with $\mu = 35$ and $\sigma = 7.4$, what is the value of x where the probability of experiencing a value less than it is 80%?

As suggested before, it is important and helpful to sketch the distribution.

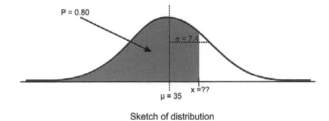

Sketch of distribution

We need to find a *z*-score from the table that corresponds to the area from the mean. Since the area on the left of the mean is 0.5, we see that the area from the mean to *x* is 0.30, i.e.,

$$P(z < x) = 0.8$$

and this implies that

$$P(0 < z < x) = 0.8 - 0.5 = 0.3$$

We need to find, somewhere in the areas given in the table, an area of 0.3 (or the closest to it) and its corresponding *z*-score. Let's take a look:

z	0.04	0.05
0.80	0.29955	0.30234

We see the closest area to 0.3, given in the table, is 0.29955, which has a corresponding *z*-score of 0.84. Hence, we use $z = 0.84$ for the *z*-score in the formula to obtain *x*. Given $\mu = 35$ and $\sigma = 7.4$, we get

$$x = \mu + z \cdot \sigma$$
$$x = 35 + 0.84(7.4)$$
$$x = 41.216$$

Thus, the value of *x* where the probability of experiencing a value less than it is 80% is 41.216. In general, when we want to obtain an *x* value from a given probability, we find the *z*-score first, then plug-n-chug this into the rewritten *z*-score formula.

When we were given a value of the variable and were asked to find the percentage or probability, the 'normalcdf(' command on a graphing calculator. But how do we find a value given the percentage? Graphing calculators and computer software are much more convenient and accurate. The command on the TI-83/84 calculator is 'invNorm('. You may have seen it already in the DISTR menu.

The syntax for this command is as follows:

'InvNorm(percentage or probability to the left, mean, standard deviation)'

Make sure to enter the values in the correct order, such as in the example below:

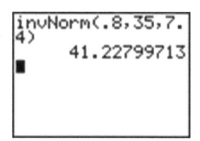

Unknown Mean or Standard Deviation

Example 15

For a normally distributed random variable, $\sigma = 4.5$, $x = 20$, and $P = 0.05$, find μ.

To solve this problem, first draw a sketch:

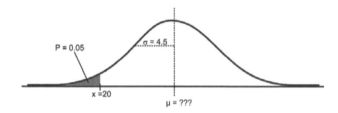

We need to find a z-score from the table that corresponds to the area from the mean. Since the area on the left of $x = 20$ is 0.05, we see that the area from the mean to x is 0.45, i.e.,

$$P(z < x) = 0.05$$

and this implies that

$$P(x < z < 0) = 0.5 - 0.05 = 0.45$$

We need to find, somewhere in the areas given in the table, an area of 0.45 (or the closest to it) and its corresponding negative z-score, since the x value lies below the mean. Let's take a look:

z	0.04	0.05
1.60	0.44950	0.45053

We see the closest area to 0.45, given in the table, is 0.44950, which has a corresponding z-score of -1.64. Recall, the z-score is negative because the x value lies below the mean. Hence, we use $z = -1.64$ for the z-score in the formula to obtain x. Given $\sigma = 4.5$ and $x = 20$, we get

$$\mu = x - z \cdot \sigma$$
$$\mu = 20 - (-1.64)(4.5)$$
$$\mu = 27.38$$

Thus, the mean is 27.38.

We could also use the 'invNorm(' command on the calculator. The result, −1.645, confirms the prediction that the value is less than 2 standard deviations from the mean.

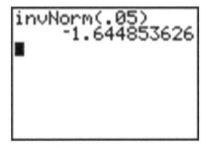

Now, plug in the known quantities into the *z*-score formula and solve for μ as follows:

$$z = \frac{x - \mu}{\sigma}$$
$$\mu = x - z\sigma$$
$$\mu \approx 20 - (-1.645)(4.5)$$
$$\mu \approx 27.402$$

We can see there was little discrepancy from using the table and using the calculator. However, since we were eye-balling from the table, the calculator gives more accurate results.

Example 16

For a normally-distributed random variable, $\mu = 83$, $x = 94$, and $P = 0.90$. Find σ. Again, let's first look at a sketch of the distribution.

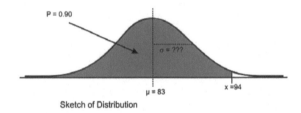

Sketch of Distribution

Since about 97.5% of the data is below 2 standard deviations, it seems reasonable to estimate that the *x* value is less than two standard deviations away from the mean and that σ might be around 7 or 8.

Again, the first step to see if our prediction is right is to use 'invNorm(' to calculate the *z*-score. Remember that since we are not entering a mean or standard deviation, the result is based on the assumption that $\mu = 0$ and $\sigma = 1$.

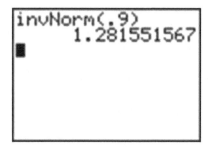

Now, use the *z*-score formula and solve for σ as follows:

$$z = \frac{x - \mu}{\sigma}$$

$$\sigma = \frac{x - \mu}{z}$$

$$\sigma \approx \frac{94 - 83}{1.282}$$

$$\sigma \approx 8.583$$

Technology Note: Drawing a Distribution on the TI-83/84 Calculator

The TI-83/84 calculator will draw a distribution for you, but before doing so, we need to set an appropriate window (see screen below) and delete or turn off any functions or plots. Let's use the last example and draw the shaded region below 94 under a normal curve with $\mu = 83$ and $\sigma = 8.583$. Remember from the Empirical Rule that we probably want to show about 3 standard deviations away from 83 in either direction. If we use 9 as an estimate for σ, then we should open our window 27 units above and below 83. The *y* settings can be a bit tricky, but with a little practice, you will get used to determining the maximum percentage of area near the mean.

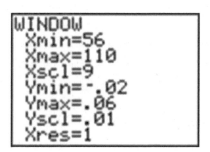

The reason that we went below the *x*-axis is to leave room for the text, as you will see.

Now, press [2ND][DISTR] and arrow over to the DRAW menu.

Choose the 'ShadeNorm(' command. With this command, you enter the values just as if you were doing a 'normal cdf(' calculation. The syntax for the 'ShadeNorm(' command is as follows: 'ShadeNorm(lower bound, upper bound, mean, standard deviation)'

Enter the values shown in the following screenshot:

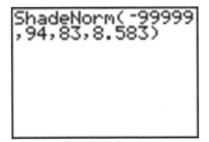

Next, press [ENTER] to see the result. It should appear as follows:

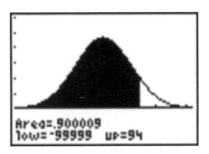

Technology Note: The 'normalpdf(' Command on the TI-83/84 Calculator

You may have noticed that the first option in the DISTR menu is 'normalpdf(', which stands for a normal probability density function. It is the option you used in lesson 5.1 to draw the graph of a normal distribution. Many students wonder what this function is for and occasionally even use it by mistake to calculate what they think are cumulative probabilities, but this function is actually the mathematical formula for drawing a normal distribution. You can find this formula in the resources at the end of the lesson if you are interested. The numbers this function returns are not really useful to us statistically. The primary purpose for this function is to draw the normal curve.

To do this, first be sure to turn off any plots and clear out any functions. Then press [Y=], insert 'normalpdf(', enter 'X', and close the parentheses as shown. Because we did not specify a mean and standard deviation, the standard normal curve will be drawn. Finally, enter the following window settings, which are necessary to fit most of the curve on the screen (think about the Empirical Rule when deciding on settings), and press [GRAPH]. The normal curve below should appear on your screen.

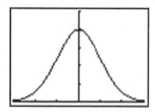

Important links: Tables and calculators

- This link leads you to a *z*-score table and an explanation of how to use it: http://tinyurl.com/2ce9ogv

- Here's a normal distribution calculator. Use this calculator to verify your answers and drawing a figure: http://tinyurl.com/n6uwo5m

Exercises 11.2

1. Estimate the standard deviation of the following distribution.

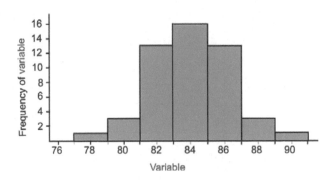

2. Calculate the following probabilities using only the z-table. Show all your work.
   a) $P(z \geq -0.79)$
   b) $P(-1 \leq z \leq 1)$ Show all work.
   c) $P(-1.56 < z < 0.32)$

3. Brielle's statistics class took a quiz, and the results were normally distributed, with a mean of 85 and a standard deviation of 7. She wanted to calculate the percentage of the class that got a *B* (between 80 and 90). She used her calculator and was puzzled by the result. Here is a screen shot of her calculator:

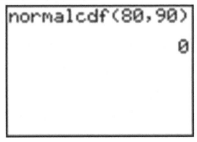

   Explain her mistake and the resulting answer on the calculator, and then calculate the correct answer.

4. Which grade is better: A 78 on a test whose mean is 72 and standard deviation is 6.5, or an 83 on a test whose mean is 77 and standard deviation is 8.4. Justify your answer and draw sketches of each distribution.

5. Teachers A and B have final exam scores that are approximately normally distributed, with the mean for Teacher A equal to 72 and the mean for Teacher B equal to 82. The standard deviation of Teacher A's scores is 10, and the standard deviation of Teacher B's scores is 5.

   a) With which teacher is a score of 90 more impressive? Support your answer with appropriate probability calculations and with a sketch.

b) With which teacher is a score of 60 more discouraging? Again, support your answer with appropriate probability calculations and with a sketch.

6. For each of the following problems, X is a continuous random variable with a normal distribution and the given mean and standard deviation. P is the probability of a value of the distribution being less than x. Find the missing value and sketch and shade the distribution.

Mean	Standard Deviation	x	P, probability
85	4.5		0.68
	1	16	0.05
73		85	0.91
93	5		0.90

7. What is the z-score for the lower quartile in a standard normal distribution?

Answers for Exercises 11.2:

1. 2

2. a) 0.78524

   b) 0.68268

   c) 0.56614

3. She didn't input the mean and standard deviation after 90; 0.525 or about 53%

4. $z = 0.92$, $z = 0.71$; the grade with 78 did relatively better

5. a) Teacher A: $z = 1.8$; Teacher B: $z = 0.8$; it is more impressive for a student to earn a 90 in Teacher A's class

   b) Teacher A: $z = 1.2$; Teacher B: $z = -2.2$; it is more discouraging to earn a 60 in Teacher B's class

6. $x = 87.115$; mean = 17.64; standard deviation = 8.96; $x = 99.4$

7. $z = -0.67$

11.3   Applications of the Normal Distributions

Learning Objective

• Apply the characteristics of a normal distribution to solving applications.

Introduction

The normal distribution is the foundation for statistical inference and will be an essential part of many of those topics in later chapters. In the meantime, this section will cover some of the types of questions that can be answered using the properties of a normal distribution. The first examples deal with more theoretical questions that will help you master basic understandings and computational skills, while the later problems will provide examples with real data, or at least a real context.

Normal Distributions with Real Data

The foundation of performing experiments by collecting surveys and samples is most often based on the normal distribution, as you will learn in greater detail in later chapters. Here are two examples to get you started.

Example 17

The Information Centre of the National Health Service in Britain collects and publishes a great deal of information and statistics on health issues affecting the population. One such comprehensive data set tracks information about the health of children[1]. According to its statistics, in 2006, the mean height of 12-year-old boys was 152.9 cm, with a standard deviation estimate of approximately 8.5 cm. (These are not the exact figures for the population, and in later chapters, we will learn how they are calculated and how accurate they may be, but for now, we will assume that they are a reasonable estimate of the true parameters.) If 12-year-old Cecil is 158 cm, approximately what percentage of all 12-year-old boys in Britain is he taller than?

We first must assume that the height of 12-year-old boys in Britain is normally distributed, and this seems like a reasonable assumption to make. As always, draw a sketch and estimate a reasonable answer prior to calculating the percentage. In this case, let's use the calculator to sketch the distribution and the shading. First decide on an appropriate window that includes about 3 standard deviations on either side of the mean. In this case, 3 standard deviations is about 25.5 cm, so add and subtract this value to/from the mean to find the horizontal extremes. Then enter the appropriate 'ShadeNorm(' command as shown:

WINDOW
Xmin=127.4
Xmax=178.4
Xscl=5
Ymin=-.025
Ymax=.05
Yscl=.05
Xres=1

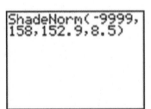

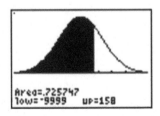

From this data, we would estimate that Cecil is taller than about 73% of 12-year-old boys. We could also phrase our assumption this way: the probability of a randomly selected British 12-year-old boy being shorter than Cecil is about 0.73. Often with data like this, we use percentiles. We would say that Cecil is in the 73rd percentile for height among 12-year-old boys in Britain.

How tall would Cecil need to be in order to be in the top 1% of all 12-year-old boys in Britain?

Here is a sketch:

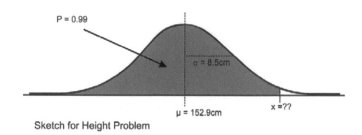

Sketch for Height Problem

In this case, we are given the percentage, so we need to use the 'invNorm(' command as shown.

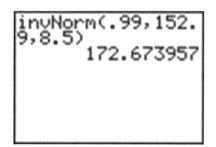

Our results indicate that Cecil would need to be about 173 cm tall to be in the top 1% of 12-year-old boys in Britain.

Example 18

Suppose that the distribution of the masses of female marine iguanas in Puerto Villamil in the Galapagos Islands is approximately normal, with a mean mass of 950 g and a standard deviation of 325 g. There are very few young marine iguanas in the populated areas of the islands, because feral cats tend to kill them. How rare is it that we would find a female marine iguana with a mass less than 400 g in this area?

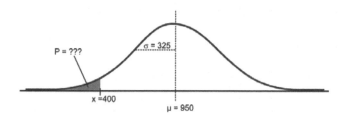

Using a graphing calculator, we can approximate the probability of a female marine iguana being less than 400 grams as follows:

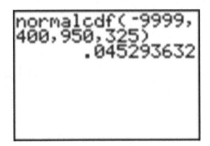

With a probability of approximately 0.045, or only about 5%, we could say it is rather unlikely that we would find an iguana this small.

Example 19

The physical plant at the main campus of a large state university receives daily requests to replace florescent lightbulbs. The distribution of the number of daily requests is bell-shaped and has a mean of 59 and a standard deviation of 9. Using the Empirical rule, what is the approximate percentage of lightbulb replacement requests numbering between 59 and 77?

Since we want to use the Empirical Rule, we should draw a figure reflecting the Empirical Rule given the mean is 59 and the standard deviation is 9. Recall, 1 standard deviation from the mean is 59 ± 9, two standard deviations from the mean is $59 \pm 2 \cdot 9$, and 3 standard deviations from the mean is $59 \pm 3 \cdot 9$.

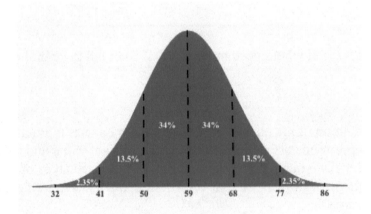

Once we make this figure, we can easily the percentage of lightbulb replacement requests numbering between 59 and 77:

$$34\% + 13.5\% = 47.5\%$$

Thus, 47.5% of lightbulb replacement requests numbering between 59 and 77.

Example 20

A company has a policy of retiring company cars; this policy looks at number of miles driven, purpose of trips, style of car and other features. The distribution of the number of months in service for the fleet of cars is bell-shaped and has a mean of 41 months and a standard deviation of 5 months. Using the Empirical Rule, what is the approximate percentage of cars that remain in service between 46 and 56 months?

Since we want to use the Empirical Rule, we should draw a figure reflecting the Empirical Rule given the mean is 41 and the standard deviation is 5:

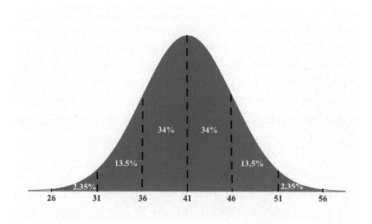

Once we make this figure, we can easily the approximate percentage of cars that remain in service between 46 and 56 months:

13.5% + 2.35% = 15.85%

Thus, 15.85% of cars remain in service between 46 and 56 months.

Exercises 11.3

1. Which of the following intervals contains the middle 95% of the data in a standard normal distribution?

    a) $z < 2$
    b) $z \leq 1.645$
    c) $z \leq 1.96$
    d) $-1.645 \leq z \leq 1.645$
    e) $-1.96 \leq z \leq 1.96$

2. The manufacturing process at a metal-parts factory produces some slight variation in the diameter of metal ball bearings. The quality control experts claim that the bearings produced have a mean diameter of 1.4 cm. If the diameter is more than 0.0035 cm too wide or too narrow, they will not work properly. In order to maintain its reliable reputation, the company wishes to ensure that no more than one-tenth of 1% of the bearings that are made are ineffective. What would the standard deviation of the manufactured bearings need to be in order to meet this goal?

3. Suppose that the wrapper of a certain candy bar lists its weight as 2.13 ounces. Naturally, the weights of individual bars vary somewhat. Suppose that the weights of these candy bars vary according to a normal distribution, with $\mu = 2.2$ ounces and $\sigma = 0.04$ ounces.
    a) What proportion of the candy bars weigh less than the advertised weight?
    b) What proportion of the candy bars weight between 2.2 and 2.3 ounces?
    c) A candy bar of what weight would be heavier than all but 1% of the candy bars out there?
    d) If the manufacturer wants to adjust the production process so that no more than 1 candy bar in 1000 weighs less than the advertised weight, what would the mean of the actual weights need to be? (Assume the standard deviation remains the same.)
    e) If the manufacturer wants to adjust the production process so that the mean remains at 2.2 ounces and no more than 1 candy bar in 1000 weighs less than the advertised weight, how small does the standard deviation of the weights need to be?

4. The Acme Company manufactures widgets. The distribution of widget weights is bell-shaped. The widget weights have a mean of 51 ounces and a standard deviation of 4 ounces. Use the Empirical Rule to answer the following questions.
    a) 99.7% of the widget weights lie between what two weights?
    b) What percentage of the widget weights lie between 43 and 63 ounces?
    c) What percentage of the widget weights lie above 47?

Answers to Exercises 11.3:
1. e
2. 0.00106
3. a) 0.04006; b) 0.49379; c) 2.1068 ounces; d) 2.2536 ounces e) 0.02265 ounces
4. a) 39 and 63; b) 97.35%; c) 84%

Standard Normal Table
Area under the Normal Curve from 0 to z

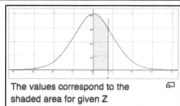

The values correspond to the shaded area for given Z

z	0.00	0.01	0.02	0.03	0.04	0.05	0.06	0.07	0.08	0.09
0.00	0.00000	0.00399	0.00798	0.01197	0.01595	0.01994	0.02392	0.02790	0.03188	0.03586
0.10	0.03983	0.04380	0.04776	0.05172	0.05567	0.05962	0.06356	0.06749	0.07142	0.07535
0.20	0.07926	0.08317	0.08706	0.09095	0.09483	0.09871	0.10257	0.10642	0.11026	0.11409
0.30	0.11791	0.12172	0.12552	0.12930	0.13307	0.13683	0.14058	0.14431	0.14803	0.15173
0.40	0.15542	0.15910	0.16276	0.16640	0.17003	0.17364	0.17724	0.18082	0.18439	0.18793
0.50	0.19146	0.19497	0.19847	0.20194	0.20540	0.20884	0.21226	0.21566	0.21904	0.22240
0.60	0.22575	0.22907	0.23237	0.23565	0.23891	0.24215	0.24537	0.24857	0.25175	0.25490
0.70	0.25804	0.26115	0.26424	0.26730	0.27035	0.27337	0.27637	0.27935	0.28230	0.28524
0.80	0.28814	0.29103	0.29389	0.29673	0.29955	0.30234	0.30511	0.30785	0.31057	0.31327
0.90	0.31594	0.31859	0.32121	0.32381	0.32639	0.32894	0.33147	0.33398	0.33646	0.33891
1.00	0.34134	0.34375	0.34614	0.34849	0.35083	0.35314	0.35543	0.35769	0.35993	0.36214
1.10	0.36433	0.36650	0.36864	0.37076	0.37286	0.37493	0.37698	0.37900	0.38100	0.38298
1.20	0.38493	0.38686	0.38877	0.39065	0.39251	0.39435	0.39617	0.39796	0.39973	0.40147
1.30	0.40320	0.40490	0.40658	0.40824	0.40988	0.41149	0.41308	0.41466	0.41621	0.41774
1.40	0.41924	0.42073	0.42220	0.42364	0.42507	0.42647	0.42785	0.42922	0.43056	0.43189
1.50	0.43319	0.43448	0.43574	0.43699	0.43822	0.43943	0.44062	0.44179	0.44295	0.44408
1.60	0.44520	0.44630	0.44738	0.44845	0.44950	0.45053	0.45154	0.45254	0.45352	0.45449
1.70	0.45543	0.45637	0.45728	0.45818	0.45907	0.45994	0.46080	0.46164	0.46246	0.46327
1.80	0.46407	0.46485	0.46562	0.46638	0.46712	0.46784	0.46856	0.46926	0.46995	0.47062
1.90	0.47128	0.47193	0.47257	0.47320	0.47381	0.47441	0.47500	0.47558	0.47615	0.47670
2.00	0.47725	0.47778	0.47831	0.47882	0.47932	0.47982	0.48030	0.48077	0.48124	0.48169
2.10	0.48214	0.48257	0.48300	0.48341	0.48382	0.48422	0.48461	0.48500	0.48537	0.48574
2.20	0.48610	0.48645	0.48679	0.48713	0.48745	0.48778	0.48809	0.48840	0.48870	0.48899
2.30	0.48928	0.48956	0.48983	0.49010	0.49036	0.49061	0.49086	0.49111	0.49134	0.49158
2.40	0.49180	0.49202	0.49224	0.49245	0.49266	0.49286	0.49305	0.49324	0.49343	0.49361
2.50	0.49379	0.49396	0.49413	0.49430	0.49446	0.49461	0.49477	0.49492	0.49506	0.49520
2.60	0.49534	0.49547	0.49560	0.49573	0.49585	0.49598	0.49609	0.49621	0.49632	0.49643
2.70	0.49653	0.49664	0.49674	0.49683	0.49693	0.49702	0.49711	0.49720	0.49728	0.49736
2.80	0.49744	0.49752	0.49760	0.49767	0.49774	0.49781	0.49788	0.49795	0.49801	0.49807
2.90	0.49813	0.49819	0.49825	0.49831	0.49836	0.49841	0.49846	0.49851	0.49856	0.49861
3.00	0.49865	0.49869	0.49874	0.49878	0.49882	0.49886	0.49889	0.49893	0.49896	0.49900
3.10	0.49903	0.49906	0.49910	0.49913	0.49916	0.49918	0.49921	0.49924	0.49926	0.49929
3.20	0.49931	0.49934	0.49936	0.49938	0.49940	0.49942	0.49944	0.49946	0.49948	0.49950
3.30	0.49952	0.49953	0.49955	0.49957	0.49958	0.49960	0.49961	0.49962	0.49964	0.49965
3.40	0.49966	0.49968	0.49969	0.49970	0.49971	0.49972	0.49973	0.49974	0.49975	0.49976
3.50	0.49977	0.49978	0.49978	0.49979	0.49980	0.49981	0.49981	0.49982	0.49983	0.49983
3.60	0.49984	0.49985	0.49985	0.49986	0.49986	0.49987	0.49987	0.49988	0.49988	0.49989
3.70	0.49989	0.49990	0.49990	0.49990	0.49991	0.49991	0.49992	0.49992	0.49992	0.49992
3.80	0.49993	0.49993	0.49993	0.49994	0.49994	0.49994	0.49994	0.49995	0.49995	0.49995
3.90	0.49995	0.49995	0.49996	0.49996	0.49996	0.49996	0.49996	0.49996	0.49997	0.49997
4.00	0.49997	0.49997	0.49997	0.49997	0.49997	0.49997	0.49998	0.49998	0.49998	0.49998

Exercises

1. The physical plant at the main campus of a large state university receives daily requests to replace florescent lightbulbs. The distribution of the number of daily requests is bell-shaped and has a mean of 56 and a standard deviation of 4. Using the Empirical Rule, what is the approximate percentage of lightbulb replacement requests numbering between 56 and 68?

2. A company has a policy of retiring company cars; this policy looks at number of miles driven, purpose of trips, style of car and other features. The distribution of the number of months in service for the fleet of cars is bell-shaped and has a mean of 65 months and a standard deviation of 4 months. Using the Empirical Rule, what is the approximate percentage of cars that remain in service between 57 and 61 months?

3. The Acme Company manufactures widgets. The distribution of widget weights is bell-shaped. The widget weights have a mean of 48 ounces and a standard deviation of 11 ounces. Suggestion: sketch the distribution in order to answer these questions.

    a. 99.7% of the widget weights lie between _____ and _____.
    b. b) What percentage of the widget weights lie between 26 and 81 ounces?
    c. What percentage of the widget weights lie above 37?

4. For a standard normal distribution, find the following probabilities:

    a. $P(z < 1.42)$
    b. $P(z > -2.52)$
    c. $P(-2.06 < z < 2.81)$

5. For a standard normal distribution, given $P(z < c) = 0.7055$, find c.

6. For a standard normal distribution, given $P(z > c) = 0.7109$, find c.

7. On a nationwide math test, the mean was 72 and the standard deviation was 10. If Roberto scored 70, what was his z-score?

8. On a nationwide math test, the mean was 66 and the standard deviation was 4. If Roberto scored 75, what was his z-score?

9. On a nationwide math test, the mean was 57 and the standard deviation was 4. If Roberto scored 85, what was his z-score?

10. A quick survey of peanut butter prices had standard deviation and mean of $0.26 and $3.68, respectively. Compute the area for a peanut butter jar costing less than $3.50.

11. A quick survey of peanut butter prices had standard deviation and mean of $0.26 and $3.68, respectively. Compute the area for a peanut butter jar costing more than $4.25.

12. A quick survey of peanut butter prices had standard deviation and mean of $0.26 and $3.68, respectively. Compute the area for a peanut butter jar costing between $3.50 and $4.25.

13. A quick survey of peanut butter prices had standard deviation and mean of $0.81 and $3.22, respectively. Compute the price for a peanut butter jar costing given the area from the mean is 0.48422.

14. A quick survey of peanut butter prices had standard deviation and mean of $1.53 and $2.22, respectively. Compute the price for a peanut butter jar costing given the area from the mean is 0.13683.

Solutions to Selected Exercises

Table of Contents

Chapter 1: Problem Solving

1. $18/230 = 0.07826 =$ about 7.8%

3. €250(0.23) = € 57.50 in VAT

5. $15000(5.57) = \$83,550$

7. absolute increase: 1050. Relative: $1050/3250 = 0.323 = 32.3\%$ increase

9. a. $2200 - 2200(0.15) = 2200(0.85) = \$1870$
  b. Yes, their goal was to decrease by at least 15%. They exceeded their goal.

11. Dropping by 6% is the same as keeping 94%. $a(0.94) = 300$. $a = 319.15$. Attendance was about 319 before the drop.

13. a) Kaplan's enrollment was 64.3% larger than Walden's. 30510
  b) Walden's enrollment was 39.1% smaller than Kaplan's.
  c) Walden's enrollment was 60.9% of Kaplan's.

15. If the original price was $100, the basic clearance price would be $100 – \$100(0.60) =$ $40. The additional markdown would bring it to $40 - \$40(0.30) = \$28$. This is 28% of the original price.

17. These are not comparable; "a" is using a base of all Americans and is talking about health insurance from any source, while "b" is using a base of adults and is talking specifically about health insurance provided by employers.

21. These statements are equivalent, if we assume the claim in "a" is a percentage point increase, not a relative change. Certainly, these messages are phrased to convey different opinions of the levy. We are told the new rate will be $9.33 per $1000, which is 0.933% tax rate. If the original rate was 0.833% (0.1 percentage point lower), then this would indeed be a 12% relative increase.

23. 20% of 30% is 30%(0.20) = 6%, a 6-percentage point decrease.

25. Probably not, unless the final is worth 50% of the overall class grade. If the final was worth 25% of the overall grade, then a 100% would only raise her average to 77.5%

27. $4/10 pounds = $0.40 per pound   (or 10 pounds/$4 = 2.5 pounds per dollar)

29. $x = 15$ 31. 2.5 cups   33. 74 turbines

35. 96 inches   37. $6000   39. 55.6 meters

43. The population density of the US is 84 people per square mile. The density of India is about 933 people per square mile. The density of India is about 11 times greater than that of the U.S.

49. The oil in the spill could produce 93.1 million gallons of gasoline. Each car uses about 600 gallons a year. That would fuel 155,167 cars for a year.

53. An answer around 100-300 gallons would be reasonable

57. 156 million miles

59. The time it takes the light to reach you is so tiny for any reasonable distance that we can safely ignore it. 750 miles/hr is about 0.21 miles/sec. If the sound takes 4 seconds to reach you, the lightning is about 0.84 miles away. In general, the lightning will be $0.21n$ miles away, which is often approximated by dividing the number of seconds by 5.

61. About 8.2 minutes

63. Four cubic yards (or 3.7 if they sell partial cubic yards)

Chapter 2: Historical Counting

1. Partial answer: Jars: 3 singles, 3 @ x3, 2 @ x6, 1 @ x12. 3+9+12+12 = 36

3. 113

5. 3022

7. 53

9. 1100100

11. 332

13. 111100010

15. 7,1,10 base 12 = 1030 base 10

17. 6,4,2 base 12 = 914 base 10

19. 175 base 10 = 1,2,7 base 12 = ♉ ♊ ♏

21. 10000 base 10 = 5,9,5,4 base 12 = ♍ ♑ ♍ ♌

23. 135 = 6,15 base 20 =

25. 360 = 18,0 base 20 =

27. 10500 = 1,6,5,0 base 20

29. 1,2,12 base 20 = 452 base 10

31. 3,0,3 base 20 = 1203 base 10

33.

35.

58. 21 five

59. 232 five

60. 33 five

61. 1000 two

62. 100011_{two} 64. 43_{five}

63. 112_{five} 65. 10_{five}

Chapter 3: Sets

1. {m, i, s, p}   3. One possibility is: Multiples of 3 between 1 and 10

5. Yes   7. True

9. True   11. False

13. $A \cup B = \{1, 2, 3, 4, 5\}$ 15. $A \cap C = \{4\}$

17. $A^c = \{6, 7, 8, 9, 10\}$ 19. $D^c \cap E = \{t, s\}$

21. $(D \cap E) \cup F = \{k, b, a, t, h\}$ 23. $(F \cap E)^c \cap D = \{b, c, k\}$

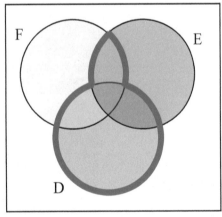

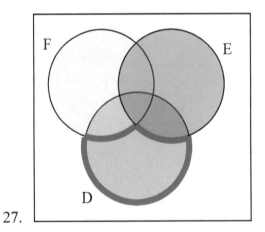

25.   27.

29. One possible answer: $(A \cap B) \cup (B \cap C)$

31. $(A \cap B^c) \cup C$ 33. 5

35. 6   37. $n(A \cap C) = 5$

39. $n(A \cap B \cap C^c) = 3$ 41. $n(G \cup H) = 45$

43. 136 use Redbox

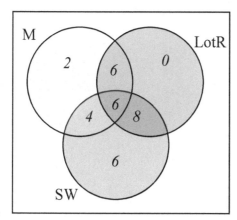

45.   a) 8 had seen exactly one b) 6 had only seen SW

Chapter 4: Logic

1.   1,2,3,4,5,6,7,9
2. $= IF(A1 < 40000, 0.13 * A1, 0.20 * A1)$
3. $F \sim F \sim F \sim T$
4. $T \sim T \sim T \sim F$
5. $a\ b\ c\ d$
6.

A	B	C	$(A \vee B) \wedge C$
T	T	T	T
T	T	F	F
T	F	T	T
T	F	F	F
F	T	T	T
F	T	F	F
F	F	T	F
F	F	F	F

7.

A	B	C	$\sim(A \vee B) \wedge C$
T	T	T	F
T	T	F	F
T	F	T	F
T	F	F	F
F	T	T	F
F	T	F	F
F	F	T	T
F	F	F	F

8.

A	B	C	$\sim(A \wedge B) \rightarrow C$
T	T	T	T

T	*T*	*F*	*T*
T	*F*	*T*	*T*
T	*F*	*F*	*F*
F	*T*	*T*	*T*
F	*T*	*F*	*F*
F	*F*	*T*	*T*
F	*F*	*F*	*F*

9. Invalid argument
10. Invalid argument
11. Invalid argument
12. Valid argument
13. Valid argument
14. Valid argument
15. Ad hominem
16. Appeal to consequence
17. Ad hominem

Chapter 5: Measurement

1. 15
2. 216
3. 3
4. 1356960
5. 1335840
6. 128
7. 0.012
8. 0.013
9. 7
10. 136
11. 224
12. 0.28125
13. 800
14. 4.58
15. 3.52
16. 0.002
17. 1.71
18. 5.18
19. 2.61
20. 2000
21. 5.83
22. 342900
23. 29.505
24. 57.2
25. 5
26. 53.6
27. 37.222222222222

Chapter 6: Geometry

1. Obtuse
2. Right
3. Right
4. Angle COE ~ Angle AOB ~ 40 ~ 130 ~ 140
5. 44
6. 23
7. 72
8. 6
9. 40 ~ 112
10. 98
11. 19.1
12. $\sqrt{168}$
13. 42
14. 74
15. 212
16. 31.42 or 31.4
17. 62.83   or   62.8
18. 432
19. 50 ~ square cm
20. 2200
21. 44 ~ square meters
22. 113.1 or 113.04
23. 40.8 or 40.80
24. a c b
25. 615.8 ~ cubic feet
26. 254.5 ~ cubic feet
27. 9202.77 ~ cubic cm

Chapter 7: Finance

1. A = 200 + .05(200) = \$210

3. I=200. t = 13/52 (13 weeks out of 52 in a year). P_0 = 9800
   200 = 9800(r)(13/52) r = 0.0816 = 8.16% annual rate

5. $P_{10} = 300(1+.05/1)^{10(1)}$ = \$488.67

7. a. $P_{20} = 2000(1+.03/12)^{20(12)}$ = \$3641.51 in 20 years
   b. 3641.51 − 2000 = \$1641.51 in interest

9. $P_8 = P_0(1+.06/12)^{8(12)} = 6000$. $P_0 = \$3717.14$ would be needed

11. a. $P_{30} = \dfrac{200\left((1+0.03/12)^{30(12)} - 1\right)}{0.03/12} = \$116,547.38$
    b. $200(12)(30) = \$72,000$
    c. $\$116,547.40 - \$72,000 = \$44,547.38$ of interest

13. a. $P_{30} = 800,000 = \dfrac{d\left((1+0.06/12)^{30(12)} - 1\right)}{0.06/12}$ $d = \$796.40$ each month
    b. $\$796.40(12)(30) = \$286,704$
    c. $\$800,000 - \$286,704 = \$513,296$ in interest

15. a. $P_0 = \dfrac{30000\left(1 - (1+0.08/1)^{-25(1)}\right)}{0.08/1} = \$320,253.29$
    b. $30000(25) = \$750,000$
    c. $\$750,000 - \$320,253.29 = \$429,756.71$

17. $P_0 = 500,000 = \dfrac{d\left(1 - (1+0.06/12)^{-20(12)}\right)}{0.06/12}$ $d = \$3582.16$ each month

19. a. $P_0 = \dfrac{700\left(1 - (1+0.05/12)^{-30(12)}\right)}{0.05/12} = $ a $\$130,397.13$ loan
    b. $700(12)(30) = \$252,000$
    c. $\$252,200 - \$130,397.13 = \$121,602.87$ in interest

21. $P_0 = 25,000 = \dfrac{d\left(1 - (1+0.02/12)^{-48}\right)}{0.02/12} = \$542.38$ a month

23. a. Down payment of 10% is $\$20,000$, leaving $\$180,000$ as the loan amount
    b. $P_0 = 180,000 = \dfrac{d\left(1 - (1+0.05/12)^{-30(12)}\right)}{0.05/12}$ $d = \$966.28$ a month
    c. $P_0 = 180,000 = \dfrac{d\left(1 - (1+0.06/12)^{-30(12)}\right)}{0.06/12}$ $d = \$1079.19$ a month

25. First we find the monthly payments:
$P_0 = 24,000 = \dfrac{d\left(1 - (1+0.03/12)^{-5(12)}\right)}{0.03/12}$. $d = \$431.25$

Remaining balance: $P_0 = \dfrac{431.25\left(1 - (1+0.03/12)^{-2(12)}\right)}{0.03/12} = \$10,033.45$

27. $6000(1 + 0.04/12)^{12N} = 10000$

$$(1.00333)^{12N} = 1.667$$

$$\log\!\left((1.00333)^{12N}\right) = \log(1.667)$$

$$12N\log(1.00333) = \log(1.667)$$

$$N = \frac{\log(1.667)}{12\log(1.00333)} = \text{about 12.8 years}$$

29. $3000 = \dfrac{60\!\left(1-(1+0.14/12)^{-12N}\right)}{0.14/12}$

$$3000(0.14/12) = 60\!\left(1-(1.0117)^{-12N}\right)$$

$$\frac{3000(0.14/12)}{60} = 0.5833 = 1-(1.0117)^{-12N}$$

$$0.5833 - 1 = -(1.0117)^{-12N}$$

$$-(0.5833-1) = (1.0117)^{-12N}$$

$$\log(0.4167) = \log\!\left((1.0117)^{-12N}\right)$$

$$\log(0.4167) = -12N\log(1.0117)$$

$$N = \frac{\log(0.4167)}{-12\log(1.0117)} = \text{about 6.3 years}$$

31. First 5 years: $P_5 = \dfrac{50\!\left((1+0.08/12)^{5(12)}-1\right)}{0.08/12} = \$3673.84$

   Next 25 years: $3673.84(1+.08/12)^{25(12)} = \$26,966.65$

33. Working backwards, $P_0 = \dfrac{10000\!\left(1-(1+0.08/4)^{-10(4)}\right)}{0.08/4} = \$273,554.79$ needed at

   retirement. To end up with that amount of money, $273,554.70 = \dfrac{d\!\left((1+0.08/4)^{15(4)}-1\right)}{0.08/4}$.

   He'll need to contribute d = \$2398.52 a quarter.

Chapter 8: Statistics, Collecting Data

1. a. Population is the current representatives in the state's congress
   b. 106
   c. the 28 representatives surveyed
   d. 14 out of 28 = ½ = 50%
   e. We might expect 50% of the 106 representatives = 53 representatives

3. This suffers from leading question bias

5. This question would likely suffer from a perceived lack of anonymity

7. This suffers from leading question bias

9. Quantitative

11. Observational study

13. Stratified sample

15. a. Group 1, receiving the vaccine
    b. Group 2 is acting as a control group. They are not receiving the treatment (new vaccine).
    c. The study is at least blind. We are not provided enough information to determine if it is double-blind.
    d. This is a controlled experiment

17. a. Census
    b. Observational study

Chapter 9: Statistics, Describing Data
1. a. Different tables are possible

Score	Frequency
30	1
40	0
50	4
60	3
70	6
80	5
90	2
100	3

b. This is technically a bar graph, not a histogram:

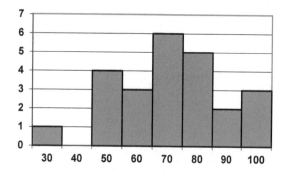

c.

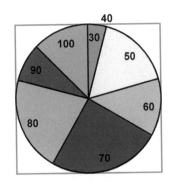

3. a. 5+3+4+2+1 = 15
   b. 5/15 = 0.3333 = 33.33%

5. Bar is at 25%. 25% of 20 = 5 students earned an A

7. a. (7.25+8.25+9.00+8.00+7.25+7.50+8.00+7.00)/8 = $7.781
   b. In order, 7.50 and 8.00 are in middle positions. Median = $7.75
   c. 0.25*8 = 2. Q1 is average of 2nd and 3rd data values: $7.375
      0.75*8 = 6. Q3 is average of 6th and 7th data values: $8.125
      5-number summary: $7.00, $7.375, $7.75, $8.125, $9.00

9. a. (5*0 + 3*1 + 4*2 + 2*3 + 1*5)/15 = 1.4667
   b. Median is 8th data value: 1 child
   c. 0.25*15 = 3.75. Q1 is 4th data value: 0 children
      0.75*15 = 11.25. Q3 is 12th data value: 2 children
      5-number summary: 0, 0, 1, 2, 5

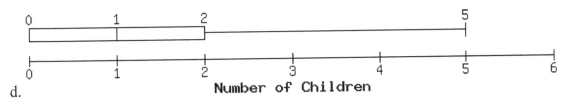

d.

11. Kendra makes $90,000. Kelsey makes $40,000. Kendra makes $50,000 more.

Chapter 10:   Probability

1. a. $\dfrac{6}{13}$ b,. $\dfrac{2}{13}$ 3. $\dfrac{150}{335} = 44.8\%$

5. $\dfrac{1}{6}$ 7. $\dfrac{26}{65}$

9. $\dfrac{3}{6} = \dfrac{1}{2}$ 11. $\dfrac{4}{52} = \dfrac{1}{13}$

13. $1 - \dfrac{1}{12} = \dfrac{11}{12}$

15. $1 - \dfrac{25}{65} = \dfrac{40}{65}$

17. $\dfrac{1}{6} \cdot \dfrac{1}{6} = \dfrac{1}{36}$

19. $\dfrac{1}{6} \cdot \dfrac{3}{6} = \dfrac{3}{36} = \dfrac{1}{12}$

21. $\dfrac{17}{49} \cdot \dfrac{16}{48} = \dfrac{17}{49} \cdot \dfrac{1}{3} = \dfrac{17}{147}$

23. a. $\dfrac{4}{52} \cdot \dfrac{4}{52} = \dfrac{16}{2704} = \dfrac{1}{169}$

    b. $\dfrac{4}{52} \cdot \dfrac{48}{52} = \dfrac{192}{2704} = \dfrac{12}{169}$

    c. $\dfrac{48}{52} \cdot \dfrac{48}{52} = \dfrac{2304}{2704} = \dfrac{144}{169}$

    d. $\dfrac{13}{52} \cdot \dfrac{13}{52} = \dfrac{169}{2704} = \dfrac{1}{16}$

    e. $\dfrac{48}{52} \cdot \dfrac{39}{52} = \dfrac{1872}{2704} = \dfrac{117}{169}$

25. $\dfrac{4}{52} \cdot \dfrac{4}{51} = \dfrac{16}{2652}$

27. a. $\dfrac{11}{25} \cdot \dfrac{14}{24} = \dfrac{154}{600}$

    b. $\dfrac{14}{25} \cdot \dfrac{11}{24} = \dfrac{154}{600}$

    c. $\dfrac{11}{25} \cdot \dfrac{10}{24} = \dfrac{110}{600}$

    d. $\dfrac{14}{25} \cdot \dfrac{13}{24} = \dfrac{182}{600}$

    e. no males = two females. Same as part d.

29. $P(\text{F and A}) = \dfrac{10}{65}$

31. $P(\text{red or odd}) = \dfrac{6}{14} + \dfrac{7}{14} - \dfrac{3}{14} = \dfrac{10}{14}$. Or 6 red and 4 odd-numbered blue marbles is 10 out of 14.

33. $P(F \text{ or } B) = \dfrac{26}{65} + \dfrac{22}{65} - \dfrac{4}{65} = \dfrac{44}{65}$. Or $P(F \text{ or } B) = \dfrac{18 + 4 + 10 + 12}{65} = \dfrac{44}{65}$

35. $P(\text{King of Hearts or Queen}) = \dfrac{1}{52} + \dfrac{4}{52} = \dfrac{5}{52}$

37. a. $P(\text{even} \mid \text{red}) = \dfrac{2}{5}$ b. $P(\text{even} \mid \text{red}) = \dfrac{2}{6}$

39. $P(\text{Heads on second} \mid \text{Tails on first}) = \dfrac{1}{2}$. They are independent events.

41. $P(\text{speak French} \mid \text{female}) = \dfrac{3}{14}$

43. Out of 4,000 people, 10 would have the disease. Out of those 10, 9 would test positive, while 1 would falsely test negative. Out of the 3990 uninfected people, 399 would falsely test positive, while 3591 would test negative.

   a. $P(\text{virus} \mid \text{positive}) = \dfrac{9}{9 + 399} = \dfrac{9}{408} = 2.2\%$

   b. $P(\text{no virus} \mid \text{negative}) = \dfrac{3591}{3591 + 1} = \dfrac{3591}{3592} = 99.97\%$

45. Out of 100,000 people, 300 would have the disease. Of those, 18 would falsely test negative, while 282 would test positive. Of the 99,700 without the disease, 3,988 would falsely test positive and the other 95,712 would test negative.

   $P(\text{disease} \mid \text{positive}) = \dfrac{282}{282 + 3988} = \dfrac{282}{4270} = 6.6\%$

47. Out of 100,000 women, 800 would have breast cancer. Out of those, 80 would falsely test negative, while 720 would test positive. Of the 99,200 without cancer, 6,944 would falsely test positive.

   $P(\text{cancer} \mid \text{positive}) = \dfrac{720}{720 + 6944} = \dfrac{720}{7664} = 9.4\%$

49. $2 \cdot 3 \cdot 8 \cdot 2 = 96$ outfits

51. a. $4 \cdot 4 \cdot 4 = 64$ b. $4 \cdot 3 \cdot 2 = 24$

53. $26 \cdot 26 \cdot 26 \cdot 10 \cdot 10 \cdot 10 = 17,576,000$

55. $_4P_4$ or $4 \cdot 3 \cdot 2 \cdot 1 = 24$ possible orders

57. Order matters. $_7P_4 = 840$ possible teams

59. Order matters. $_{12}P_5 = 95{,}040$ possible themes

61. Order does not matter. $_{12}C_4 = 495$

63. $_{50}C_6 = 15{,}890{,}700$

65. $_{27}C_{11} \cdot 16 = 208{,}606{,}320$

67. There is only 1 way to arrange 5 CD's in alphabetical order. The probability that the CD's are in alphabetical order is one divided by the total number of ways to arrange 5 CD's. Since alphabetical order is only one of all the possible orderings you can either use permutations, or simply use 5!. P(alphabetical) $= 1/5! = 1/(5 \text{ P } 5) = \dfrac{1}{120}$.

69. There are $_{48}C_6$ total tickets. To match 5 of the 6, a player would need to choose 5 of those 6, $_6C_5$, and one of the 42 non-winning numbers, $_{42}C_1$. $\dfrac{6 \cdot 42}{12271512} = \dfrac{252}{12271512}$

71. All possible hands is $_{52}C_5$. Hands will all hearts is $_{13}C_5$. $\dfrac{1287}{2598960}$.

73. $\$3\left(\dfrac{3}{37}\right) + \$2\left(\dfrac{6}{37}\right) + (-\$1)\left(\dfrac{28}{37}\right) = -\$\dfrac{7}{37} = -\$0.19$

75. There are $_{23}C_6 = 100{,}947$ possible tickets.

Expected value $= \$29{,}999\left(\dfrac{1}{100947}\right) + (-\$1)\left(\dfrac{100946}{100947}\right) = -\$0.70$

77. $\$48(0.993) + (-\$302)(0.007) = \$45.55$

Chapter 11: Normal Distribution

1. 49.9 or 49.85
2. 13.5
3. $15 \sim 81 \sim 97.35 \sim 84$
4. a) 0.9222 b) 0.9941 c) 0.9778
5. 0.54029
6. -0.55602
7. -0.2

8. 2.25
9. 7
10. 0.2451
11. 0.01426
12. 0.74064
13. $4.96
14. $2.76

Index

Projects

These projects are for Chapter 1: Problem Solving, collectively, Chapter 5: Measurement and Chapter 6: Geometry, Chapter 7: Finance, Chapter 10: Probability, and, collectively, Chapter 8: Statistics, Collecting Data, Chapter 9: Statistics, Describing Data, and Chapter 11: Normal Distribution. There are also additional projects in the body of the chapters for selected chapters. These four projects are meant for students to complete outside of the classroom, using technology, notes, or any resources outside of the textbook. After successfully completing a project, a student will be able to

- Apply research skills to obtain a result within reason and the scope of the problem
- Apply technology, if applicable, e.g., statistical software, calculator, financial software, word processor, etc., to obtain meaningful results
- Demonstrate critical thinking/problem solving skills

Chapter 1: Problem Solving

We use our problem-solving skills every day whether it's figuring out how much gas you can get with your last $20 or simply seeing if the dinner for $40 for a family of four is really a better deal than a la cart. Create a scenario in which you use problem-solving skills in everyday life.

Step 1. Define the scenario clearly.

Step 2. State the question or problem where you will use problem-solving skills.

Step 3. How will you solve the problem? Be sure at every step you are clear in how you are solving your problem.

Step 4. State your conclusion or results.

Step 5. Write a paragraph summarizing why you picked the scenario and what you learned from this project.

Chapter 5 and 6: Measurement and Geometry

Step 1. Think of your ultimate dream house and the floor plan for the house. Your dream house should have at least

- 2 floors
- living room
- a garage
- 3 bedrooms
- 3 baths
- a laundry room

- a pantry
- a kitchen
- 2 closets
- 2 more additions you would want on your dream house

Step 2. On plain white paper or graph paper, draw the floor plans to your dream house. ***Your floor plans should be drawn to scale.***[1] You may use a computer for your floor plans and this should be printed on plain white paper. In your floor plans, you should include

1. the dimensions of each room (length and width)
2. labels of each room
3. two floor plans- one page for each floor
4. labels and dimensions of the additions you chose for your house

Note, rooms include bedrooms, bathrooms, pantry, kitchen(s), closets, laundry room and garage(s).

Step 3. List each room and addition to the house with its dimensions and area for that particular room/addition, i.e., Master Bedroom: 20 ft x 15 ft = 300 square feet.

Step 4. What shapes and formulas for those shapes did you use to calculate the area of each room/addition? (Include the actual formula for each shape.)

Step 5. Take all the areas for each room and addition of the house and calculate the total square footage of your dream house. Only include rooms and additions.

Step 6. Reflection: The median number of rooms in a home in Orange County, CA is six. (ref: http://www.city-data.com/county/) This means half of the homes have less than 6 rooms and the other half has more than 6 rooms. Where were you? Did you have more or less than 6 rooms? Now knowing this information, would you have done a different floor plan? Why or why not?

---

[1] *When a map (or any drawing) is "drawn to scale" it means a specific amount of distance on the map equals a specific amount of distance in real life, and that this scale is consistent across all of the map. The ratio and proportions of all the features will carry over from the depictions on the map to what they are in reality or in real life. For example, if an island in a bay is half as long as the bay on the map, then it is half as long as the bay in real life. When drawing to scale in your floor plans, make sure if 1 unit = 1 ft in your floor plans, then that measurement should be consistent throughout the entire house's floor plans. This is why graph paper may be the best choice for this project.*

Chapter 7: Finance

Step 1. Choose one of the three following scenarios:

- Purchase a car
- Purchase a house
- Finance college

Step 2. Find a reasonable price for your purchase or future cost of college (*attach an example from the newspaper, internet, or other reputable source- this can be a cut-out from the newspaper, print out from the internet, etc.*).

Step 3. Find the current interest rate (*attach an example from a loan officer at a bank, internet, or other reputable source- this can be a cut-out from the newspaper, flyer from the bank, or a print out from the internet, etc.*).

Sources on the Internet:
- www.latimes.com
- www.bankrates.com
- www.ocreg.com
- www.collegeboard.com

Step 4. Assume there is no down payment: Calculate your monthly payments over the life of the loan/purchase.

Step 5. Calculate your total payments over the life of the loan/purchase, assuming there is no down payment.

Step 6. How much interest are you paying over the life of the loan/purchase?

Step 7. Assume you are making a 10% down payment or if you are looking at college loans, assume you are receiving 10% of the loan from your parents, i.e., you are borrowing less: Calculate your monthly payments over the life of the loan/purchase.

Step 8. Calculate your total payments over the life of the loan/purchase **assuming you are making a 10% down payment**.

Step 9. How much interest are you paying over the life of the loan/purchase?

Step 10. What is the absolute difference in interest from making no down payment to making a 10% down payment?

Step 11. Would you suggest making a down payment or not? Why or why not? *Use your results above to justify your answer.*

Step 12. Write a paragraph summarizing the reason(s) you picked the scenario and anything you learned from this project.

Chapter 8, 9, and 11: Statistics and Normal Distribution

Step 1. Choose an ***observational quantitative*** study. (Some examples: What is the average height of SCC students?, How many times do SCC students check their email daily?, How many times do SCC students check Facebook in one day?)

Step 2. Ask at least 30 subjects your question and record your data. (*This is called raw data and you will use this for calculations, so it is important that your data is accurate or your results could be skewed.*) Be sure to attach your raw data to your project. ******Do not rewrite your raw data! Raw data should be handwritten or a print out from the internet if you choose to use social media to survey your subjects.***

Step 3. Create a frequency distribution. (*You may need to use class intervals if you have a wide range of values and/or if you have no mode.*)

Step 4. Create a histogram and a line chart for your data. (*It is recommended for students to use Excel, or any other statistical software for this step.*)

Step 5. Calculate the mean, range, and standard deviation. Interpret your results in context of your study without using the words "mean" and "standard deviation". (*You may use the word "average" for the mean in your interpretations.*) **Hint:** *The average height of SCC students is 62 inches give or take 4 inches with a range of 21 inches. Note, the mean is 62 inches, the standard deviation is 4 inches, and the range is 21 inches. (It is recommended for students to use Excel, or any other statistical software for this step.)*

Step 6. Find the five-number summary. (*It is recommended for students to use Excel, or any other statistical software for this step.*)

Step 7. Interpret the median in context of your study without using the word "median" in the interpretation. **Hint:** *The median is the half-way mark of the data. Hence, half of the students are taller than 60 inches and the other half are shorter than 60 inches.*

Step 8. Draw a box plot. (*Be sure to follow the steps in your notes or, you may use technology to obtain your results. It is recommended for students to use Excel, or any other statistical software for this step.*)

Step 9. Looking at your histogram, would you say your data has a normal distribution? Why or why not?

Step 10. ***Despite your results from Step 9***, let's assume your data is normally distributed. Apply the Empirical Rule (a.k.a. 68-95-99.7 Rule) to determine the values in which 68%, 95%, and 99.7% of your data lie. **Interpret your results in context of your study.**

Step 11. After collecting, describing and analyzing your data, write a few sentences describing your conclusions from all your results. What can you conclude from your results above *in context of your study*?

Chapter 10: Probability

Step 1. Purchase a bag of M&Ms (or a bag of candy that is multicolored).

Step 2. Create a frequency table where each category is color:

Color	Amount of candy
Brown	
Blue	
Green	
Orange	
Red	
Yellow	

Step 3. Select a color. Then calculate the probability of randomly selecting 2 candies of that same color. **Hint:** Find the *P(A and B)*, where this is one less candy in the bag when grabbing the second candy.

Step 4. Select a different color. Then calculate the probability of randomly selecting 1 candy of that color.

Step 5. Using the probability from **Step 4.**, what is the expected proportion of candies in the color selected in **Step 4.** if you decided to buy a bulk bag that contains a total of 3,378 candies.

Step 6. Looking at the results from **Step 5.**, calculate the number of calories, fat, and carbohydrates for the amount of candies obtained in **Step 5.** using the nutrition label on your bag of candy.

Step 7. Let's say you are a pre-diabetic[2] and are allocated a number of carbohydrates per day. You are allocated 60 grams of carbohydrates for snacks throughout one day. Looking at your results from **Step 6.**, over how many days would it take for you to finish eating the number of candies found in **Step 5.** if you stayed on your recommended daily carbohydrate allocation for snacks?

Step 8. Using information from **Step 4.-Step 7.**, would you reconsider eating your candies as a snack if you were a pre-diabetic? Why or why not? Would you recommend eating your candies as a snack even if you weren't a pre-diabetic? Explain. *Feel free to read a little about pre-diabetic diets here: http://www.healthline.com/health/diabetes/prediabetes-diet#healthy-eating2*

---

[2] Reference: https://g.co/kgs/p33VHk